Chlorinated Insecticides

Volume I

Technology and Application

Chlorinated
Insecticides

Volume I
Technology and Application

Author:

G. T. Brooks
The University of Sussex
Brighton, Sussex
England

published by:

CRC PRESS, INC.
18901 Cranwood Parkway · Cleveland, Ohio 44128

This book represents information obtained from authentic and highly regarded sources. Reprinted material is quoted with permission, and sources are indicated. A wide variety of references is listed. Every reasonable effort has been made to give reliable data and information, but the author and the publisher cannot assume responsibility for the validity of all materials or for the consequences of their use.

International Standard Book Number 0-87819-043-0 Complete Set
International Standard Book Number 0-87819-044-9 Volume 1
Library of Congress Card Number 73-90535

Printed in the United States

To Ann, my wife and former colleague, and my many friends in Insecticide Toxicology whose work is referred to in these pages.

PESTICIDE CHEMISTRY SERIES – PREFACE

The literature on pesticides is voluminous, but scattered among dozens of journals and texts written or edited by experts. Until now, with the publication of Chlorinated Insecticides by G. T. Brooks, there has been no attempt to produce a single, comprehensive series on the chemistry of pesticides. CRC Press should be commended for having undertaken this Herculean task.

When asked by the publisher to serve as editor of the Pesticide Chemistry Series, I discussed the idea with some of my colleagues at the International Pesticide Congress in Tel Aviv in 1971. At that time, Dr. Brooks enthusiastically agreed to become the author of the treatise on chlorinated insecticides, which became such a comprehensive work that it is being published in two volumes.

As editor of this series, my goal has been choosing experts in their respective fields who would be willing to write single-authored books, thus assuring uniformity of style and thought for the individual text as well as the entire series. I would like to express my deep appreciation to each author for having undertaken the large task of writing in such a comprehensive manner without the aid of contributors or an editorial board.

Chlorinated Insecticides is the first contribution to this series; subsequent volumes under preparation will appear in the near future under titles such as Organophosphate Pesticides; Herbicides; and Fungicides. Looking into the future, we plan to include books on the chemistry of juvenile hormones and pheromones. Suggestions on other titles and possible authors are invited from the reader.

Gunter Zweig
Editor
Pesticide Chemistry Series

THE AUTHOR

Gerald T. Brooks is a Principal Scientific Officer, formerly in the Biochemistry Department of the Agricultural Research Council's Pest Infestation Laboratory (now the Pest Infestation Control Laboratory of the Ministry of Agriculture, Fisheries, and Food) at Slough, England, and currently in the Agricultural Research Council Unit of Invertebrate Chemistry and Physiology at the University of Sussex, Brighton, England. He holds B.Sc. and Ph.D. degrees of the University of London, and is a Fellow of The Royal Institute of Chemistry. His field of work is the chemistry, biochemistry, and toxicology of insecticides and hormones.

CHLORINATED INSECTICIDES
TECHNOLOGY AND APPLICATIONS

TABLE OF CONTENTS

FOREWORD

This definitive and most detailed work ever to have been published, devoted exclusively to the chemistry and biology of chlorinated insecticides, comes from the authoritative pen of a leading investigator in this field. It should serve for many years to come as the prime reference source both for instructional and research purposes, for it not only delineates that which is known about this valuable class of compounds, but also emphasizes those important biological and ecological areas in which our knowledge is still much too sketchy and in which further investigation may be expected to uncover facts of great interest and wide applicability.

In a way, it is ironic that this excellent treatise should appear at a time that seems to mark the twilight of the various uses of the chlorinated insecticides. It is not that they have been superseded by some other class or classes of chemicals that have proved themselves cheaper or more effective. Quite the contrary; the cost effectiveness of the chlorinated insecticides has remained unequaled since their introduction some three decades ago. Neither is it because the amounts of insecticide required are, from an agricultural or public health standpoint, so very great. Indeed, the very hallmark of the chlorinated insecticides, which characterized them originally as wonder chemicals, was the unbelievably small amounts needed over a given area to effect complete control of insect pests. It is the very persistence of trace quantities of this class of chemicals that has been seized upon by some environmentalists as ammunition in their war to eliminate totally, for whatever use, all chlorinated insecticides, which, but a comparatively few years ago, were lauded as the saviors of a hungry and disease-ridden humanity. Participants in these activities should observe the impact of the current world grain shortage on food prices and consider the vital role of well-proven control chemicals for cereal pests in balancing an increasingly precarious food supply in our favor.

By and large, the chlorinated insecticides have not been shown to have produced any symptoms of chronic toxicity in the general public. However (along with some of their metabolic products), they appear to have endangered the survival of certain species of predatory birds whose diets are restricted to animals which store the chlorinated insecticides of their derivatives in their tissues. Such findings, along with some oncological studies of more or less questionable applicability to humans, have prompted official departments of public health in some countries to proscribe the chlorinated insecticides for all uses. In this situation, it behooves investigators to keep the shelves of their stockrooms well filled with supplies of these compounds for possible experimental use, against the day when some of them may no longer be available on the market.

The extension of programs currently required for the biological screening of new insecticide candidates has made this sort of research so expensive that even the wealthiest manufacturers are now seriously considering abandoning the search for new insecticides as economically unrewarding. Further advances in the field of pest control in agriculture and public health will then depend, to an even greater extent, upon the interest of governmental agencies and upon university laboratories. These, however, must consider the shrinking purchasing power of their budgets and the concomitant inevitable expansion of their involvement in other interests. The upshot is that the outlook at the moment for effective insect pest control is bleak. Thus, once again, shallow, if well-intentioned clamor threatens to do great harm to national economies and to impressive public health advances.

Yet, despite the gloomy picture of the moment, a more balanced attitude toward the chlorinated insecticides may eventually prevail. Meanwhile, it should be of benefit to the chemist and biologist to keep up-to-date on all that has been done (and omitted) in this field. For, once the mode of biological activity of these compounds is more fully elucidated, it would seem quite likely that new types of chlorinated insecticides and, indeed, related structural types lacking chlorine may be devised, which will combine price effectiveness with environmental acceptability.

Julius Hyman
Berkeley, California

PREFACE

It is now nearly 23 years since the publication of the second and revised edition of the book by West and Campbell entitled *DDT: The Synthetic Insecticide*. Since then, although numerous texts have included sections on various aspects of chlorinated insecticide technology and biology, there has been no publication that has attempted to treat the subject as a whole, in spite of all that has been achieved by the use of these compounds and all that has been said about them in the controversial atmosphere of recent years. This two-volume book makes some attempt to fill the gap, although it does not deal with the chlorinated hydrocarbons used as fumigants.

The task of writing about chlorinated insecticides is a truly formidable one, since the amount of literature on the chemistry of these compounds and their effects, real or imaginary, on living organisms has grown phenomenally in recent years. While there is a clear need to present the complicated science of these chemicals in an understandable fashion, attempts to do so usually result in misleading and dangerous oversimplifications and generalizations. Thus, a statement such as "this chlorinated insecticide fed to rats in the diet produces liver damage" is totally meaningless without details of the level and duration of the feeding. Likewise, a text without adequate references is of little value because the reader is often unable to properly distinguish between what is actually in the scientific literature and what is the opinion of the author. This area above all is one in which those who are interested should, if they wish, have access to the original work in order to make their own judgments. Accordingly, the levels of treatments have been included whenever possible, and the text is as fully referenced as is feasible.

The subject falls conveniently into one volume dealing with chemistry and applications (technological aspects) and a second (biological and environmental aspects) which considers their interactions with living systems. However, rigid division is neither easy nor entirely desirable, and there is inevitably some degree of overlapping between those parts of the subject that can be treated from several points of view. Thus, there is a relationship between "applications" in Volume I and "development of resistance in the field," in Volume II, since resistance represents a failure of applications. Accordingly, it is emphasized that the division is merely one of convenience and the two parts are to be regarded as a single entity dealing with the *science* of chlorinated insecticides.

While the chemistry of DDT analogues is reasonably straightforward, that of the cyclodiene insecticides is unfortunately more complicated and a different order of presentation has been adopted for them. Thus, sections on the stereochemistry of the Diels-Alder reaction and on the nomenclature of cyclodienes have been included in Volume I in an attempt to clarify some of the more difficult points, this and other basic information being presented before the history and use aspects.

The development of insect resistance to organochlorine insecticides is an interesting and important part of their history and accordingly has been discussed in some detail in Volume II. All remaining aspects of their environmental behaviour are conveniently treated in Volume II under "action," dealing with topics ranging from biodynamics, through biochemical interactions, to mechanism of action. A final short section in that volume deals with some general matters, including current arrangements for regulating pesticide residues in food.

Since the combined volumes are intended to be of use to those whose interests are historical or in applications, as well as to chemists, toxicologists, and resistance geneticists, some parts are inevitably more readable than others. If readers from any one of these areas are so much as tempted to examine one or more of the others, then the book will have served a useful purpose.

The provision of copies of their books by Dr. T. F. West and Professor A. W. A. Brown is gratefully acknowledged, as also various items of information furnished by Mr. J. A. Smith, Dr. R. F. Glasser, and colleagues of the Shell International Chemical Company, by Dr. P. B. Polen of the Velsicol Chemical Corporation, and by Dr. A. Calderbank of Jealott's Hill Research Station, Imperial Chemical Industries, Plant Protection Limited. It is a pleasure to thank Dr. J. Robinson of Shell Research Limited, Tunstall laboratory and Professor J. R. Busvine of the London School of Hygiene, who kindly read various parts of the manuscript, and Dr. C. T. Bedford, also of the Tunstall laboratory, for valuable discussions relating to cyclodiene insecticide nomenclature.

Views or opinions expressed herein should not be taken to represent those of any official source, unless references are given.

G. T. Brooks
Brighton, England

Chapter 1

INTRODUCTION

There is a natural, but unfounded tendency to regard human life as being different from other forms in its degree of subjection to natural laws. It is, of course, true that because of his varied skills, man is probably better able than many other species to survive in a hostile environment, but beyond this, it needs to be pointed out rather forcibly that human life, like other forms, is a complex equilibrium between processes of chemical synthesis and degradation that is supported by the continual intake of gaseous, liquid, and solid chemicals of all kinds from the external environment. There is no doubt that some of these chemicals from natural sources (in the food) have deleterious effects on life in the long term and this is undoubtedly one reason why man's allotted lifespan is 70 years rather than some much higher figure.

To the scientist who, by his training, has become accustomed to thinking in chemical terms and whose work brings him into daily contact with chemicals, the realization that members of the general public apparently have scarcely appreciated the extent of their dependence on chemistry for actual existence as well as for comfortable survival comes as a considerable shock. Viewed against the inevitability of human involvement with chemicals, the popular objection to one particular group of toxic substances from among the immense number to which humans and other living organisms have undoubtedly been exposed over a long period of time seems almost irrelevant. Nevertheless, the controversy surrounding the chlorinated insecticides has served to stimulate renewed interest in the relationship between man and his environment. Furthermore, the ability to detect minute quantities of these compounds has given warning of the way in which stable, man-made materials can become distributed far beyond the spheres in which their use was originally intended. Since the effects of such distributions cannot readily be predicted, it is clearly prudent to use materials having these properties with as much restraint as possible, but bearing in mind the benefits that accrue from their use in relation to the actual hazards so far established.

The origin of chemical insect control seems to be lost in antiquity. According to one source,[1] pyrethrum was introduced into Europe from the Far East by Marco Polo, and another authority states that the use of pyrethrum as an insecticide by certain Caucasian tribes was observed in the early 19th century.[2] Extracts of roots of plants of the *Derris* species have long been used by primitive peoples as fish poisons, and it was suggested in 1848 that such extracts might be insecticidal. Preparations from the plant sabadilla contain veratrine alkaloids, and for centuries have been used as insecticides by South American natives, while the application of tobacco extracts (nicotine alkaloids) as insecticides dates from the mid-18th century.[1] The general toxicity of alkaloids from the South American plant *Ryania speciosa* has also been known for some time, but the insecticidal properties were first reported in 1945. Various oils such as petroleum, kerosene, creosote, and turpentine came into use in the 18th Century, but their use was limited due to phytotoxicity and they were succeeded by highly refined oils formulated as emulsions of low phytotoxicity.

Several elements and their salts have had longstanding use as insecticides. Plinius is said to have recommended arsenic for this purpose in 70 A.D., and arsenic sulfide was used by the Chinese in the 16th Century.[3] Paris Green (a complex of copper arsenite and copper acetate), calcium arsenate, and lead arsenate came into use in the 19th Century and a use was found for various salts of copper, zinc, thallium, chromium, lead, mercury and selenium, and cryolite (sodium fluoaluminate). Some of these compounds are highly persistent and give rise to considerable residue hazards. Hydrogen cyanide was used against scale insects by the year 1886 and dinitrophenols as insecticides as early as 1892. Methyl bromide and carbon disulfide have been used for many years and the use of naphthalene and *p*-dichlorobenzene dates from the beginning of the 20th Century.[1]

A few additional potential insecticides were recognized in the 1930's, but there seem to have been few developments of importance before the appearance of DDT. World food supply comes largely from cereal crops that require for their protection chemicals having a significant degree of persistence if economy in rates and frequency of application are to be achieved. Therefore, it seems a remarkable stroke of good fortune that within little more than 10 years after 1939, a whole series of chlorinated insecticides was discovered, and that several of them had the requisite properties for the efficient and economic protection of cereals, as well as collectively exhibiting an astonishing range of activity towards insect pests of other crops and insect vectors of disease.

Whether discoveries of this magnitude and value will occur again is open to speculation. In the decade prior to 1967, the total pesticide industry research budget probably increased up to three times over that in 1957, but fewer insecticides were introduced than during the preceding decade. With the ever-increasing restrictions placed upon new compounds on environmental and other grounds, the chances of achieving major sales and profits become much more difficult. This last factor can hardly be ignored, since research funds must be derived from the sale of previously successful compounds. Figures published in 1969 indicate that the average total investment in a compound that finally reaches the market (excluding the cost of developing manufacturing processes, marketing, advertising, etc.) rose from $1,196,000 in 1956 to nearly $3 million in 1964, and $4 million in 1969. The overall survival rate of compounds entering the complex evaluation procedure fell correspondingly from 1 in 1,800 to 1 in 3,600 and finally to 1 in 5,040.[4] Other figures indicate even longer odds of 7,000–10,000:1.[5] The time now required to bring a new compound into commercial use is estimated at 5 to 7 years, depending upon the number of difficulties encountered in the process.

It is sobering to read that in 1967 the ten best selling insecticides had been introduced, on average, 17 years before and none of them within the previous 10 years.[6] In view of the hazards, it is not surprising that manufacturers of insecticide chemicals, faced with the possibility that insect resistance to a successful new compound may develop within a short time, are becoming increasingly reluctant to invest in this area. Yet chemical pest control continues to play a vital role in the modern agricultural practices that nowadays produce crop yields unheard of 30 years ago. High yields are now quite vital for survival and there will be a requirement for chemical agents in one form or another for the foreseeable future.

While the object of progress is doubtless that modern, comfortable man should have leisure to contemplate his situation and to devise even better and safer ways to maintain it, great care must be taken to ensure that short-sighted action does not nullify the benefits already won, especially if there is the added risk of destroying the machinery that created them in the first place and might conceivably repeat the process. Thus, it is to be hoped that man will shortly come to terms with his chemical origins so that some middle way can be found through the controversy that currently surrounds the use of chemicals of all kinds.

The use of chlorinated insecticides should be viewed against this background. Whatever their shortcomings, and it is unlikely that any device created by man will be perfect, the story of their discovery and development is one of outstanding achievement which deserves due recognition.

Chapter 2

CHLORINATED INSECTICIDES OF THE DDT GROUP

A. HISTORY AND DEVELOPMENT

DDT is the common name for the technical mixture of compounds in which the major component is 1,1,1-trichloro-2,2-bis(p-chlorophenyl)ethane (also called 1,1-bis(p-chlorophenyl)2,2,2-trichloroethane, p,p'-DDT, 4,4'-DDT, or simply dichlorodiphenyltrichloroethane; Figure 1, structure 1). The name DDT will be used in the following account when distinction between the isomers is unnecessary. The first DDT formulations were introduced by J. R. Geigy in 1942 under the trade marks Gesarol® and Gesapon® (in continental Europe and the U.S.A.), Guesarol® and Guesapon® (in England), and Neocid®, and were protected by Swiss Patent 226180 and British Patent 547871.

1. Origins and Applications in World War II

The history of the development of DDT is particularly interesting because of the wartime circumstances surrounding the discovery of the insecticidal properties of the p,p'-isomer by Dr. P. Müller in the Basle laboratories of Geigy in the autumn of 1939.[7] The synthesis had actually been reported already by Zeidler in 1874, some 65 years earlier.[8] The research leading to the discovery of DDT as a biocide began in 1932 with investigations by P. Lauger, designed to produce a new moth-proofing agent.[9] At that time, it was noted that the Geigy triphenylmethane dye Eriochrome Cyanine R (Figure 1, structure 2) and the oxindole derivative Isacen (Figure 1, structure 3) of Hoffman La Roche had structural features in common with the colorless triphenylmethane moth-proofing agent Eulan New (Figure 1, structure 4) developed by I. G. Farbenindustrie in Germany.

These structural similarities led to the synthesis of some "mixed" structural types (Figure 1, structures 5 and 6) which were obtained by the reaction of isatin-5-sulfonic acid with 2- and 4-chlorophenols, respectively, and had moderate activity as stomach poisons for larvae of the clothes moth (*Tineola biselliella*); the N-4-chlorobenzyl derivative of the ketone (Figure 1, structure 6) proved to be even more effective. A number of sulfur-containing variants of Eulan New were prepared from benzaldehyde, but the requirement for light stability combined with good larvicidal activity was rarely met, compounds such as the mixed thioacetal of benzaldehyde-o-sulfonic acid (Figure 1, structure 7) being exceptional. The reader will notice that running through these syntheses is a trend of thought toward DDT, but a major requirement in this particular series of compounds was water solubility, and the final product, Mitin FF (Figure 1, structure 8) that resulted from tests of more than 6,000 chemicals, is hardly recognizable as a relative of DDT.

During the search for stomach poisons useful in mothproofing, the need for chemicals useful as contact insecticides in plant protection had not been forgotten. At this point, reference may be made to the intriguing discussions in the 1944 paper of Lauger, Martin, and Müller[7] in which the molecular structures of a number of naturally occurring toxicants are considered in terms of their content of (a) a moiety conferring lipid solubility and (b) a moiety conferring toxicity (that is, a toxophore). The division of bioactive molecules in this way has been challenged on many occasions and is doubtlessly an oversimplification. However, it must be realized that even today this is often the only approach available to the chemist involved in drug design; the authors of such ideas usually recognize them to be somewhat naive, but early reservations frequently disappear in the familiar aging process of "hardening of the hypotheses," especially if these lead to fruitful developments. Apart from the work on water-soluble moth proofing agents, the Geigy team was seeking effective lipophilic compounds resembling the benzene-soluble member of the Eulan series, Eulan BL (Figure 1, structure 9). It was noticed that this compound combines the polar, substituted sulfonamide moiety with 3,4-dichlorobenzene, which, besides conferring lipid solubility, is a good respiratory and contact poison. Accordingly, a chlorinated benzene nucleus appeared to be a desirable constituent of active molecules and a series of compounds containing two chlorobenzene nuclei separated by a polar group (Figure 1, structure 10; R = SO_2NHCO-, $-SO_3-$, $-OCH_2-$, $-O-$,

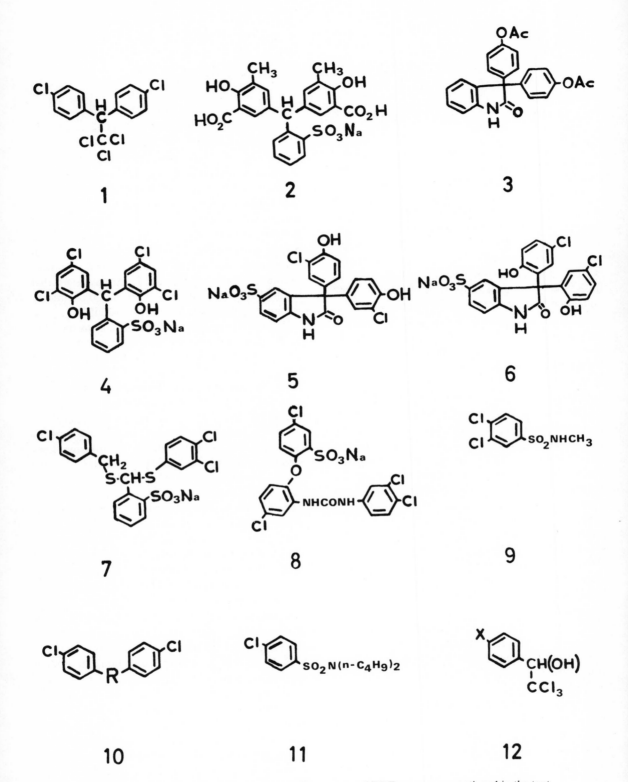

FIGURE 1. Structures of various moth-proofing agents and DDT precursors mentioned in the text.

-S-, -SO-, -SO$_2$-) were subsequently found to be active as stomach poisons. The sulfone (R = -SO$_2$-) was especially effective, and a number of outstanding acaricides in use today are derived from these basic structures. Thus, Eulan BL became the forerunner of a number of remarkable developments in insect control; its relative, N-di-n-butyl-4-chlorobenzenesulfonamide (Figure 1, structure 11) achieved fame some years later as WARF antiresistant (WARF stands for Wisconsin Alumni Research Foundation), a DDT synergist considered to act by inhibiting the enzyme effecting detoxicative dehydrochlorination of DDT in some resistant insects.

The precise sequence of events leading from the stomach poisons to the contact poison DDT is of great historical interest but not easy to determine for the question as to the amount of foresight involved in any discovery is not easy to answer solely from a consideration of published work. At any rate, the replacement of the electronegative -SO$_2$- group in 4,4'-dichlorodiphenylsulfone by the nonpolar group -CH$_2$- to give 4,4'-dichlorodiphenylmethane (Figure 1, structure 10; R = -CH$_2$-) gave weak activity as a stomach poison and not much contact activity, but, as detailed in the 1944 publication, this change might be offset in chemical terms by replacing -CH$_2$- with another lipophilic, but strongly electronegative group, namely -CHCCl$_3$, with the possible advantage that the -CCl$_3$ moiety is found in chloroform, a highly lipophilic inhalation narcotic. Thus it appeared possible to include the desirable toxic properties of chlorobenzene in a molecule having relatively low volatility but retaining the most significant structural feature of bioactive 4,4'-dichlorodiphenylsulfone, namely, the separation of the benzene nuclei by a strongly electronegative group, and all this with the added bonus of the chloroform structure. Viewed in this light, the DDT molecule must represent one of the most remarkably successful examples of all time of the fabrication of a bioactive molecule from simpler structures having individual biological effects. Beyond the certainty that DDT is a potent nerve poison, the precise relevance to its insecticidal action of the bioactivities of the component structures or of the general reasoning behind the synthesis is still unclear. However, this does not detract from the magnitude of the achievement, duly recognized by the award of the Nobel Prize to Müller in 1948.

Apart from the earlier Zeidler synthesis of DDT itself, the corresponding derivative from unsubstituted benzene had been prepared by von Baeyer, and its synthesis was also described by Chattaway and Muir in 1934.[10] Chattaway and Muir described, in a completely nonbiological context, the synthesis of carbinols (12, X = H, halogen or alkyl) as intermediates in the synthesis of DDT analogues, and also the preparation of a number of unsymmetrical DDT analogues (Figure 2, structure 13; R$_1$ = H, R$_2$ = CH$_3$, Cl, Br, I and R$_1$ = CH$_3$, R$_2$ = Cl, Br, I) together with the corresponding ethylene derivatives. Thus, the stimulus for synthesis in this area already existed, and a publication of Müller in 1946 details the preparation of a number of new DDT analogues, including some derived from ethylene and propylene, with an indication of their relative toxicities to houseflies and several other insects.[11] Compounds mentioned there include 1,1,1-trichloro-2,2-bis(p-methoxyphenyl)ethane (Figure 2, structure 13; R$_1$ = R$_2$ = OCH$_3$, also known as dianisyltrichloroethane, DMDT, or methoxychlor) and 1,1-dichloro-2,2-bis(p-chlorophenyl)ethane (DDD; Figure 2, structure 15), which were later to achieve commercial status.

Following his initial observation of the outstanding contact insecticidal activity of p,p'-DDT, Müller lost no time in establishing the range of application of the product, and patent application was made in Switzerland in March 1940. Since neutral Switzerland was at that time virtually cut off from the rest of Europe, communications were difficult. Two years passed before the news of the discovery reached the Allied powers, and an additional period of time passed before its significance was fully appreciated. In the meantime, Müller had prepared solution, emulsion, and dust formulations of DDT and had shown that the 5 or 10% active ingredient in dusts and spray powders, upon which the Gesarol preparations were based, gave successful control of a range of agricultural pests such as raspberry beetle, apple blossom weevil, apple sawfly, cabbage moth, cabbage flea beetle, carrot fly, and the Colorado potato beetle. The new Gesarol preparation proved particularly effective in controlling the infestation of Colorado beetle which threatened the Swiss potato crop in 1939. Formulations were also devised for the control of pests of stored products such as grain weevils.

In the public health area, laboratory tests by

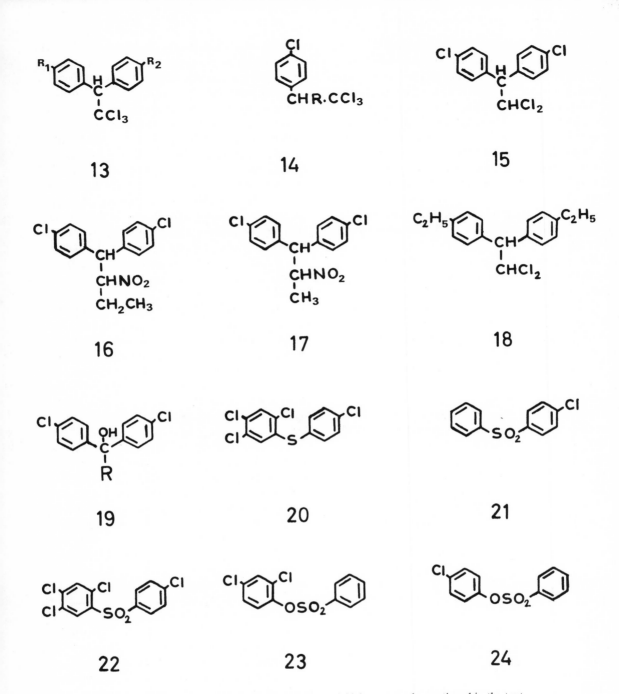

FIGURE 2. Structures of insecticidal or acaricidal compounds mentioned in the text.

Müller had established the efficiency of DDT as a control agent for houseflies, cockroaches, and mosquitos. Following laboratory experiments with body lice and fleas by Dr. R. Domenjoz during the winter of 1941–42, the dust formulation known as Neocid, containing 5% active ingredient, was successfully employed in schools and refugee camps. The possibilities for the control of diseases such as typhus were soon realized, and by 1942, Geigy, in collaboration with the Red Cross, was conducting field trials in the Balkans. The summer of 1942 saw the use of Neocid on Swiss airfields and it was used by the Swiss army from the beginning of 1943. In the meantime, extensive toxicological experiments in the pharmacological laboratories in Basle gave no evidence of hazard to man or warm-blooded animals arising from the use of Gesarol or Neocid preparations.

Thus, a great deal of information about DDT had already been obtained in Switzerland by the time the news of the discovery reached Britain and the U.S.A. In the U.S.A., news of the effective control of the Swiss Colorado beetle infestation was received by the American subsidiary of Geigy in September 1941 without special excitement since this pest could be successfully controlled by lead arsenate. A subsequent report received in June 1942 detailed some of the remarkable results achieved against agricultural pests by the Gesarol formulation, indicating that some 150 tons of the material had been used in Switzerland during 1942. In August of the same year, the great potential of Neocid as an agent for controlling body lice was brought to the attention of the American Military Attaché in Berne, so that the military significance of the finding was soon recognized. The active ingredient in samples of Gesarol had been extracted, analyzed, and synthesized in the Beltsville laboratory of the U.S. Department of Agriculture (U.S.D.A.) by the time the structure of DDT had been received from Switzerland, and other investigations at the Orlando laboratory of the U.S.D.A. firmly established the importance of the new product.

The entry of Japan into the war and the subsequent loss of the British and Netherlands East Indies had greatly reduced the availability of pyrethrum and derris, important control agents for malaria and typhus, and substitutes for these natural insecticides were urgently required by the time that DDT was brought to the attention of the British Government by the British Geigy Company in 1942. The situation was critical because of the need to deploy large numbers of troops in areas full of disease-carrying insects. Results subsequently obtained at the Rothamsted Experimental Station with samples of Gesarol confirmed the value of the product, which was further evaluated at the Plant Pathology Laboratory of the Ministry of Agriculture, and early in 1943 news was received of the remarkable success of Neocid against lice. At this stage, the investigations were taken up by the Insecticide Development Panel of Experts, set up by Professor Sir Ian Heilbron to find substitutes for pyrethrum and derris. Henceforth, the manufacture of DDT became a national priority of the highest order. By the very nature of discoveries of this sort, the excitement surrounding them is usually confined to a limited number of laboratory and development scientists. In the case of DDT, however, the need for rapid solutions to a host of manufacturing and formulation problems, and for extensive toxicological evaluations, soon resulted in the involvement of scientists and technologists on an unprecedented scale.[1 2]

Due to the wartime shortage of both manufacturing facilities and labor, many difficulties had to be overcome with regard to manufacture. A few pounds of pure DDT were first made in the Manchester laboratories of Geigy in January of 1943, the first pilot plant batch was forthcoming in April, and bulk production commenced at the Manchester plant in November of the same year. This was the first continuous large scale production plant for DDT outside of Switzerland, and its completion marked the climax of a remarkable coordination between scientists from several disciplines, technologists, industrialists, and government officials. Some idea of the excitement and enthusiasm surrounding the project is given in the book by West and Campbell,[1 2] which was first published in 1946, a time when these events were still recent news. In spite of attempts to maintain secrecy, the existence of a substance called DDT and some of its achievements became fairly well known to the American public during 1943. Large-scale production was in hand in that country also, and had reached the rate of 300,000 lb per month by the summer of 1944, with the prospect of 1.7 million lb per month by the following year.

In August 1944, the DDT story* was released by the British Government, and it is worth repeating most of this story as it was given to the press, since it shows why DDT became a household word along with penicillin and radar:

"The full story can now be told of what has been described as one of the greatest scientific discoveries of the last decade, a synthetic multi-purpose insecticide which has already stopped a typhus epidemic, threatens the existence of the malaria-carrying mosquito and household insect pests, and is capable of controlling many of the insects which now do untold damage to food crops. It is *p,p′*-dichlorodiphenyl-1,1,1-trichloroethane, DDT for short. DDT is lethal to the body louse which transmits typhus fever to man and is capable of killing mosquitoes, thus helping to control the spread of malaria. Dysentery, enteric, and cholera will be capable of better control than heretofore, as DDT is deadly towards the various species of flies, whilst it has already been used successfully to destroy bugs, fleas, cockroaches, beetles, cabbage worms, apple codling moth, and aphids. Its efficacy is almost unique, as on insects it acts both as a contact and as a stomach poison, although it is non-toxic to man and other warm-blooded animals in the concentration normally used. It also has the remarkable property of being effective for many weeks after application. For instance, when sprayed on walls, it kills any fly alighting thereon, in some cases for as long as three months afterwards; a bed sprayed with DDT is deadly to bed-bugs for 3 to 6 months; clothing impregnated with it is safe from lice for one month, even after several launderings; whilst a swamp properly treated may be freed from breeding mosquitos for a considerable period.

"The first full-scale use of DDT in a war sector was in Naples at the turn of the year. Here, in December 1943, typhus broke out in the overcrowded civilian population which in the main was poverty-stricken, dirty and louse-ridden. As soon as the Allied forces were in control, vigorous steps were taken to suppress the outbreak by mass disinfestation. This was first done by dusting with non-DDT-containing lousicides, but as soon as DDT became available it was used solely and with signal success. During January 1944, 1,300,000 civilians were dusted at two delousing stations (72,000 on the peak day) and within weeks, the outbreak in the city of Naples was completely under control, the weekly number of civilian cases reported falling sharply from 305 in the peak week ending 11 January to 155 the following week.

"DDT has already made medical history of tremendous significance, as never before has a typhus outbreak been arrested in mid-winter. It will, therefore, be the main protection for Allied troops in liberated Europe where typhus is endemic over a large area. With this end in view, British troops going to the continent now have a standard issue of DDT impregnated shirts which effectively protect the wearer against lice for the first two months, even after regular washing. For the troops, however, protection against malaria and dysentery is even more important, and in the operations which are being carried out in the Far East, DDT is going to find its most important war use. In this theatre of war large areas are made practically untenable by the enormous population of malaria-carrying mosquitoes. Added to oil, however, which has been used against mosquito larvae ever since the Panama Canal was built, DDT produces a larvicide of such potency that only a fraction of the oil previously employed will henceforth be required and this new preparation will remain toxic to the mosquito larvae for days afterwards.

"The majority of casualties in all wars are not directly due to enemy action. In Sicily the 7th and 8th Armies suffered more casualties from malaria than from battle. Many more people are afflicted by disease than are killed or wounded, and so far as our knowledge extends, more people die from epidemics following war than are killed by enemy action during it. Of the Serbians who fought in the last war 25% perished from typhus, and the Russian losses during 1914—18 are officially estimated at several millions, while in Poland there were some 400,000 cases of typhus with a 10% mortality rate.

"When, after the Armistice in 1918, a vast army of soldiers and refugees returned from malarial districts to Europe the disease spread in epidemic form over the whole of south-eastern Europe, affecting as much as 90% of the population in some regions. A distinguished scientist wrote recently that the post-war ravages of malaria and dysentery probably accounted for well over a million deaths. After the last war epidemics of typhus swept Poland, Rumania, Lithuania, the Near East, and Russia. It has been computed that during this period there were more than 25 million cases of typhus in the territories controlled by the Soviet Republic, with 2½ to 3 million deaths.

"Before an insecticide can be used on a large scale, however, particularly as a hygiene measure, a great deal has to be known not only about its power to kill insects and its methods of application and the strength in which it should be applied, but even more important, the degree of risk to health which may attend its use. The early laboratory tests carried out in England by groups of chemists, entomologists, and other scientists concentrated the work of several years into a slightly higher number of weeks, and the faith of those research workers was justified, as DDT was shown to be a unique compound, with properties superior to those of any insecticide yet made. It thus had obvious immense military possibilities. Pilot-scale production was immediately commenced and in collaboration with the British Geigy Colour Company, plans for larger scale production were made. Its full potentialities and methods of application were simultaneously worked out by teams of Government, university, and industrial scientists, in collaboration with experts from the three Services. Close liaison was established with American and Dominion scientists, who were already working on similar lines, and now many hundreds of

---

*West, T. F. and Campbell, C. A., Synthetic insecticide which stopped a typhus epidemic, in *DDT and Newer Persistent Insecticides,* 2nd ed., Chapman and Hall, London, 1950, 7.

workers are collaborating in developing all aspects of its use and application.

"Large scale experiments in this country, in Panama and West Africa, are now nearing completion. On the basis of these experiments work is proceeding on methods of air and land sprayings which, as soon as sufficient DDT is available, may be capable of producing a tremendous diminution in the dangers arising from insect vectors over huge areas before operations are carried out. Before the advent of DDT this would have been quite impossible. It is obvious that DDT opens new ways of protecting man from his hereditary enemies and it is going to help in giving the Forces of the United Nations a higher degree of security against the dangers and discomforts of insects than any army has ever had. In the United Kingdom and the U.S.A., big manufacturing projects are in train and a steady flow of this life-saving compound is now ensured. In the United Kingdom all production is under the direct control of the Ministry of Supply, which has already set up a number of factories for this purpose and which has greatly simplified the task of production by pooling the ideas and experience of all the separate manufacturers. All the output is at present reserved for Service uses.

"Whilst Service requirements must take prior place at the moment, the post-war possibilities of DDT are enormous, as it is known that the peace-time death rate from malaria in India is about 2 million annually and the annual economic loss to that Empire alone is over 80 million cases . . .

"Immediately when the European war is over DDT will be used to prevent the spread of epidemics among refugee populations and to re-establish healthy conditions in devastated areas."

Apart from the early use of DDT by the Swiss against several agricultural pests, the foregoing account relates entirely to the hygiene and public health uses of DDT and demonstrates the impact it had already made in this area in the remarkably short space of 2 years. Reviewing the prospects at that time for its general application in agricultural pest control, it is interesting to note that the problems which might arise from the widespread use of such a persistent, broad spectrum toxicant had already been foreseen. The Entomological Society of America, and the American Association of Economic Entomology, meeting in New York in December 1944, issued a statement recognizing the promise of DDT for the control of most potato pests, many vineyard and orchard pests, and the major pests of vegetables, seed crops and cotton. It was recognized, however, that DDT does not kill all insect pests and may in some cases kill their natural enemies selectively, so that the control of such insects might later become a major problem.[12] At that time there was also a lack of information about suitable formulations of DDT for agricultural use, since only small amounts of the compound had been available, and there was poor information about its compatability with other pesticide formulations and uses. There had already been some reports of phytotoxicity, and although the accumulated information regarding toxicity to animal life indicated a reasonable margin of safety, it was felt that heavy deposition of DDT on edible plants must be avoided. Toxicity had been noted to cold-blooded animals such as fish and frogs but there was no information about possible long-term effects on the composition of soils.

By 1944, the United States Department of Agriculture had concluded that before DDT could be recommended for use by farmers, more information should be obtained about phytotoxicity and the possible danger of the insecticidal formulations to livestock, wild animals, birds, fish, or honeybees and other beneficial insects. Bearing in mind the abundantly available evidence of long DDT-persistence on many surfaces, it was necessary to know whether DDT, ingested by man in small quantities along with food commodities, would persist in the tissues and produce eventual cumulative poisoning. Necessary information should be obtained on the following: formulations and permissible combinations with other pesticides; the best types of machinery for application; the cost of formulations, combinations, and application methods relative to that of other satisfactory pesticides; and the performance of DDT relative to insecticides already available for particular pests. Thus, a great deal of evidence exists that during the early period of DDT use, consideration had been given to most of the possible limitations of this chemical with the possible exception of insecticide resistance, which was to appear within a fairly short time. Making recommendations for a few new uses of DDT for the 1946 crop year,[12] the U.S.D.A. observed that no case of human poisoning resulting from the use of DDT in insect control had been brought to the attention of the Bureau of Entomology and Plant Quarantine (the agency overseeing much of the development work on this compound), and that DDT had a markedly smaller effect on higher animals than other insecticides such as arsenicals and nicotine. By this time, the use of water-dispersions of DDT or DDT powders on the skin had been found to cause no irritation or other adverse effects, but persons using oil formulated solutions or emulsions were advised to avoid unnecessary exposure to these preparations, since it was already known that DDT passes readily through the skin. There were also reports that DDT fed to animals might be stored in the fat and excreted in milk, so that the need for strict adherence to rules for the safe use of DDT was emphasized. Using the analytical methods available at that time, the Bureau's chemists had found no uptake of DDT by the tissues of treated potato plants. On the basis of the information then available, the Bureau made definite recommendations for the use of 10% DDT dusts in talc or pyrophyllite, emulsions or suspensions of 2.5% DDT dispersible powders in water, and solutions of 5% DDT in kerosene or fuel oil, against home and outhouse pests such as houseflies, mosquitoes, fleas, bedbugs, cockroaches, lice, ants, ticks, and stable flies. The oil solutions were for use against insects inside houses, while dusts, emulsions, or suspensions were to be used elsewhere. Dusts or water-dispersions were recommended for animals. Aerosols containing 3% DDT with purified pyrethrum extract were recommended for space application to control flying insects in houses, but not for applying residues to surfaces for the control of crawling insects.

For controlling various lepidopterous larvae and other pests of forest and shade trees, 0.1% emulsions of DDT made by diluting 1 lb of DDT in 1 qt of xylene plus emulsifier with 100 gal of water, were to be sprayed on the leaves until the spray began to run off. Cabbage caterpillars were to be controlled by applying 3% DDT in talc or pyrophyllite at a rate of 20 lb per acre, but not within 30 days before marketing the crop. Sugar beet grown for seed was to be treated at the time of full bloom with 5% DDT dust in pyrophyllite or talc at 30 lb per acre for the control of lygus bugs.

Since the main consideration was to avoid applications of DDT which would result in the contamination of foodstuffs, DDT was not recommended for use on stored grains or cereal products to be used for food. Provided that such contamination was avoided, however, sprays of not more than 5% DDT in deodorized kerosene, water suspensions, or emulsions could be applied to walls and woodwork of storage areas at the rate of up to one gallon of spray per 1,000 ft². Stored seed insects could be controlled by incorporating 0.5 oz of 3% DDT dust with each bushel of seed. For the control of weevils infesting hairy vetch

grown for seed, one application of 3% DDT dust in talc or pyrophyllite was recommended at 25 lb per acre when the first pods appeared. Livestock were not to be pastured after harvest in fields so treated unless the crop straw had first been buried by ploughing.

A dust of 5% DDT in not less than 75% sulphur was recommended at a rate of 12 to 15 lb per acre for the control of cotton pests such as the cotton flea hopper, and application could be repeated weekly until control was achieved. For bollworm control, the recommendation was for two or more applications of 5% DDT dust at 15 to 20 lb per acre, with a 5-day interval between applications.

By this time in 1946, DDT had shown great promise for the control of a wider variety of insects including the Colorado beetle, European corn borer, various species of leafhoppers, flea beetles, alfalfa weevils, pea aphids, and some serious fruit pests such as the Japanese beetle, codling moth, and oriental fruit moth. However, the Bureau of Entomology and Plant Quarantine considered at that particular time that, in spite of the extensive information already available, at least one more growing season would be needed for full evaluation not only of the residue hazard but also of possible phytotoxic effects arising from these applications. These recommendations show that the introduction of DDT into agricultural practice was made cautiously and with full regard for all the environmental information available about it at that time.

The discovery of DDT provided a tremendous worldwide stimulus to insecticide research and a great many structural analogues were made in a very short time. In Müller's 1946 paper,[11] the p,m'-dichlorodiphenyl analogue of DDT received the same rating as DDT as a housefly toxicant, while the o,p'- and o,o'-isomers were indicated to have decreasing toxicity in that order. Other investigations of the o,p'- and o,o'-compounds indicated a similar order for fourth-instar larvae of the mosquito *Anopheles quadrimaculatus,* and adults of this insect. The o,p'-isomer was much less effective than DDT against body lice.

The p,p'-difluoro analogue (Figure 2, structure 13; R_1, R_2 = F) of DDT was indicated by Müller to be more toxic than DDT to houseflies but other investigators found it to be less toxic to lice and bedbugs, and to fourth-instar larvae of *A. quadrimaculatus* when formulated either as a dust or oil spray. Although more expensive to manufacture than DDT, this analogue was claimed to emulsify more readily, and was developed by the Germans who also used the unsymmetrical p-phenyl, p-chlorophenyl analogue (Figure 2, structure 13; R_1 = Cl, R_2 = H), mentioned as a housefly toxicant by Müller. According to a recent report of Metcalf,[52] the latter compound is actually about 10 times less toxic than DDT to houseflies (*Musca domestica*), blowflies (*Phormia regina*), and larvae of the mosquito *Anopheles albimanus,* and somewhat less toxic than DDT to larvae of another species, *Culex fatigans.* The corresponding unsubstituted compound (Figure 2, structure 13; R_1, R_2 = H), was indicated by Müller to have moderate activity towards houseflies. More recent comparisons with DDT indicate it has an activity about 200-fold, more than 100-fold, 100-fold, and 1.5-fold lower, respectively, toward the same four insects just mentioned.[52] It is interesting that the difference in toxicity between DDT and these two analogues is much smaller for *Culex fatigans* larvae than for the other insects, illustrating the point that compounds inactive towards some species

may be quite effective toxicants against others. Chloromethyl-4-chlorophenylsulfone, developed in Germany as "Lauseto neu," was found to be more toxic than DDT toward lice and bedbugs, but less effective against flies. Its use was limited by low solubility (0.25%) in refined mineral oils, but the Germans used it mainly in the form of powder dispersions, or emulsions. A mixture called "Lauseto" was also used in Germany to impregnate clothing for protection against body lice. This material was prepared by reacting benzene, chlorobenzene, and chloral, so that besides DDT and the impurities usually present in technical DDT, it probably contained the 2,2-diphenyl analogue, as well as the expected unsymmetrical derivatives. The oily nature of the product made it easier to emulsify than DDT but it was about four times less toxic than pure DDT to lice (Table 1).[13]

A number of simple compounds containing only one benzene or chlorobenzene ring were mentioned in the literature in the 1940's. The result of attaching a trichloromethyl group directly to different nuclear positions of chloro-

TABLE 1

Toxicity of DDT and Some Related Compounds to Lice and Bedbugs

Compound tested	Median lethal concentration (%) in oil spray[a]	
	Lice	Bedbugs
1,1,1-Trichloro-2,2-bis(p-chlorophenyl)ethane (DDT).	0.3	0.53
o,p'-DDT (iso-DDT).	5.5	>20
1,1,1-Trichloro-2,2-bis(p-fluorophenyl)ethane (DFDT).	1.4	5.0
"Lauseto" mixture	1.2	—
1,1,1-Trichloro-2,2-bis(p-methoxyphenyl)ethane (methoxychlor).	0.9	0.55
1,1-Dichloro-2,2-bis(p-chlorophenyl)ethane (DDD).	0.9	1.2
1,1,1-Trichloro-2,2-bis(p-methylphenyl)ethane (methylchlor).	1.7	3.6
1,1,1-Trichloro-2,2-diphenylethane	7.5	>20
1,1-bis(p-chlorophenyl)ethane	8.5	>20
Chlorophenyl chloromethyl sulfone	0.1	0.2
"Gammexane"	0.016	0.051

(Data adapted from Busvine.[13,46])

[a]Samples were dissolved in refined white oil and sprayed on the insects in a Potter tower. Deposit kept constant at 0.36 mg/cm² and concentration varied.

benzene or dichlorobenzenes was explored, and of attaching two trichloromethyl groups to a benzene ring to give the corresponding hexachloro-*o*-, *m*- and *p*-xylenes. These compounds were inactive, but 2-acetoxy-2-*p*-chlorophenyl-1,1,1-trichloroethane (Figure 2, structure 14; R = OCOCH$_3$), described as "weakly active,"[7] was the subject of a German patent and the related chlorohydrocarbon (Figure 2, structure 14; R = H) is described elsewhere as "weakly active."[12] Another German product called "Lucex," made by chlorinating the side chain of ethyl chlorobenzene, was stated to be cheaper, but less effective than DDT.

None of these modifications to the DDT structure produced a better general toxicant, but, as mentioned previously, Müller[11] described the compound now known as methoxychlor (Figure 2, structure 13; R$_1$, R$_2$ = -OCH$_3$), and a number of DDT analogues in which the trichloromethyl group was replaced by dichloromethyl (-CH.Cl$_2$), one of these being DDD (Figure 2, structure 15). The latter is also referred to in the 1944 paper of Lauger, Martin, and Müller[7] as being moderately active and the subject of a patent application. Methoxychlor is about three times less toxic than DDT to houseflies, nearly equitoxic with DDT to *Phormia regina* and *Culex fatigans* larvae, and about ten times less toxic to larvae of *Anopheles albimanus*.[52] According to the comparative toxicity data of Busvine, obtained for several DDT analogues with lice and bedbugs (summarized in Table 1), methoxychlor is somewhat less toxic than DDT to lice and has about the same toxicity to bedbugs under the testing conditions employed. It is a nonsystemic contact and stomach insecticide which has a range of insecticidal activity coinciding roughly with that of DDT. Methoxychlor has some advantages over DDT from an environmental standpoint; the oral LD50 for rats of 6,000 mg/kg compares favorably with 113 mg/kg for DDT and, unlike DDT, it shows little tendency to store in the body fat of animals or to be excreted in the milk. Thus, one application of methoxychlor is as a replacement for DDT for fly control in dairy barns. Methoxychlor was first introduced about 1945 by the Geigy Company, and by E. I. du Pont de Nemours under the trade name, Marlate®.

DDD [1,1-dichloro-2,2-bis(*p*-chlorophenyl) ethane known also as TDE] was produced by I. G. Farbenindustrie at Leverkusen under the code number ME 1700 during the war and was intro-duced in 1945 by the Rohm and Haas Company under the trade mark Rhothane®. It is a non-systemic contact and stomach insecticide which does not have the broad-spectrum insecticidal activity of DDT, but has equal or greater potency against the larvae of some mosquitos and lepidoptera. The acute oral toxicity of DDD to the rat (3,400 mg/kg) is much less than that of DDT, but some species, such as the dog, have been found to undergo degeneration of the adrenal cortex when deliberately fed this compound at moderate levels (50 mg/kg daily). Rats, on the other hand, show no such effect when given high levels of this compound in the diet, which is just one example of the profound species differences often to be observed in drug studies. DDD is found in the fatty tissues of animals both as a result of its direct use in pest control and because it can arise metabolically from DDT. There is no evidence that the levels produced in this way have any effect on adrenal function.

DDT analogues containing nitro-groups in the aromatic nuclei were described in the 1874 paper of Zeidler,[8] and *m,m'*-dinitro-DDT is mentioned as being without insecticidal activity by Lauger, Martin, and Müller.[7] However, certain compounds containing a nitro-group in the aliphatic portion of the molecule are effective insecticides. The toxicant Dilan® is a mixture containing 53.3% of 1,1-bis(*p*-chlorophenyl)-2-nitrobutane (known as Bulan®; Figure 2, structure 16), 27.6% of 1,1-bis(*p*-chlorophenyl)-2-nitropropane (known as Prolan®; Figure 2, structure 17) and 20% of related compounds, these mostly being the corresponding *o,p'*-dichloro-analogues, as found with technical DDT. Dilan was introduced in 1948 by the Commercial Solvents Corporation and is a non-systemic insecticide with a similar range of action but generally lower insect toxicity and persistence than DDT. The acute oral LD50 for rats is stated to be from 475 to 8,073 mg/kg, while figures of 330 and 4,000 mg/kg, respectively, have been given for the individual major constituents Bulan and Prolan. Another report gives 1,320 and 620 mg/kg, respectively, for these individual components, but in any case, they seem to be rather less toxic than DDT.

The 1944 paper of the Geigy researchers[7] stated that the *p,p'*-dimethyl analogue of DDT (Figure 2, structure 13; R$_1$, R$_2$ = -CH$_3$) was toxic to clothes moth larvae, and Müller rated the corresponding analogue of DDD along with DDD itself as a housefly toxicant. The corresponding *p,p'*-diethyl analogue (Figure 2, structure 18) was introduced as experimental insecticide Q-137 or Perthane® by the Rohm and Haas Company in 1950. It is a nonsystemic compound, with generally less insecticidal activity than DDT or DDD, which finds some application for the control of pests of fruit and vegetables as well as against pests on livestock and domestic insects such as clothes moths and carpet beetles. Another advantageous feature is the low toxicity to mammals as indicated by an acute oral LD50 for rats of 8,170 mg/kg and for mice of 6,600 mg/kg.

In the early developments leading to the synthesis of DDT, the most effective stomach poisons were those compounds in which electronegative groups such as sulfoxide (-SO-) and sulfone (-SO$_2$-) moieties separated the *p*-chlorobenzene rings. The high contact insecticidal activity obtained by replacing these groups by the trichloroethane moiety, as in DDT, disappears entirely when the replacing group is derived instead from trichloroethanol; the product, called dicofol (Figure 2, structure 19, R = CCl$_3$), was introduced in 1955 by Rohm and Haas under the code number FW-293 and trade mark Kelthane®. This replacement of the tertiary aliphatic hydrogen atom in DDT by hydroxyl largely destroys the insecticidal activity, but confers on the dicofol molecule good acaricidal properties. Dicofol is used today for the control of mites on a wide range of crops; it is less toxic to rats than DDT. Two structurally related compounds, chlorobenzilate (Figure 2, structure 19; R = COOC$_2$H$_5$) and chloropropylate (Figure 2, structure 19, R = COO*i*-propyl), were introduced by Geigy in 1952 and 1964, respectively, with the code numbers G23992 and G24163 and trade marks Akar® and Folbex® (chlorobenzilate), and Rospin® (chloropropylate). The acaricidal properties of esters of *p,p'*-dichlorobenzilic acid were first described by Gasser in 1952.[14] Like difocol, chlorobenzilate has little insecticidal activity but is a nonsystemic acaricide recommended for use against phytophagous mites on fruit. Chloropropylate has similar properties and is used on fruit, nuts, cotton, tea, vegetables, sugar beet, and ornamental plants for mite control. Both compounds have low acute oral toxicity to rats, the LD50 for chloropropylate being more than 5,000 mg/kg.

At this point, it is convenient to make further mention of acaricidal compounds which were developed commercially on the basis of the structure-activity relationships of the stomach poisons described by the Geigy group in 1944. Based on the ether or thioether type of structure examples are Neotran® (Figure 1, structure 10; R = -OCH$_2$O-), a Dow Chemical Company product, tetrasul (Figure 2, structure 20), a product of Philips-Duphar, and chlorbenside (Figure 1, structure 10; R = -SCH$_2$-) by the Boots Pure Drug Company. Neotran® was first described as an acaricide in 1946, and is no longer of particular commercial interest. Tetrasul was introduced in 1957 as a nonsystemic acaricide, is highly toxic to eggs and all stages of phytophagous mites and is recommended for their control on fruit and vegetables. Chlorbenside, introduced in 1953 as Chlorparacide® or Chlorsulphacide® is a nonsystemic, persistent acaricide having little insecticidal action and low mammalian toxicity, which is used mainly for the control of tetranychid mites on fruit.[15]

Other acaricides were made by preparing unsymmetrical variants of the original bis-*p*-chlorophenylsulfone or sulfonate structures. Thus, Sulfenone® (Figure 2, structure 21), was introduced by the Stauffer Chemical Company, and tetradifon (trade mark Tedion®; Figure 2, structure 22) by Philips-Duphar. The latter compound is still recommended for use on fruits, vegetables, and ornamental crops against all stages of phytophagous mites except adults and has low mammalian toxicity. The stomach poison *p*-chlorophenyl *p*-chlorobenzenesulfonate described by the Geigy group in 1944,[7] later become known as chlorfenson or ovex and was developed as an acaricide by the Dow Chemical Company. Variants of this structure with unsymmetrical, aromatic rings are 2,4-dichlorophenyl benzenesulfonate (Figure 2, structure 23), marketed as an acaricide in 1947 by the Allied Chemical Corporation under the trade names Genite® and Genitol®, and fenson (Figure 2, structure 24) introduced for crop protection in Britain by the Murphy Chemical Company in 1952 under the trade mark Murvesco®. Fenson was originally reported to be less effective than ovex against eggs of the two-spotted mite, but subsequent tests showed it to be more effective than ovex against fruit tree red spider mite and red spider mite. The mammalian toxicities are low, as with most other structures of this type.

2. Development of Application Methods

a. Sprays, Dusts, Emulsions, etc.

Once a compound proves to be toxic to insects in laboratory tests conducted under carefully controlled and therefore ideal conditions, it is necessary to devise ways of applying it efficiently to pests in the field. Even in laboratory tests, striking differences in toxicity towards a particular insect are sometimes seen when the toxicant is applied in different carriers. Volatile solvents containing the toxicant evaporate after being applied to the insect cuticle, so that the toxicant is left in free form as a liquid spread on the cuticle or as a solid deposited on it. In this case, the toxicant is in direct contact with the epicuticular wax, which can therefore become quickly saturated with it and will act as a primary source of toxicant for the underlying phases of the cuticle. However, if the toxicant is applied in a nonvolatile inert carrier such as mineral oil, the oil spreads on the surface of the cuticle and the toxicant must partition between the oil and the epicuticular wax before it can gain access to the lower layers of the cuticle. The difference produced by these effects is illustrated by toxicity data obtained by Busvine (Table 2) for houseflies, using DDT topically applied in mineral oil or in acetone; the toxicity to a normal, DDT-susceptible strain of houseflies which does not metabolize DDT is four times greater (0.12 μg/fly) with the acetone application than with the oil application (0.44 μg/fly). The resistant strain metabolizes DDT to a less toxic compound (Vol. II, Chap. 2B.2a and Chap. 3B.3a,b), so that more externally applied toxicant is required to provide the lethal internal concentration. The much larger difference between the median lethal doses for the two strains of house-

flies with the acetone application is attributed to the poor efficiency of the DDT deposit left when acetone evaporates for maintaining the supply of toxicant to the insect cuticle when an active metabolic detoxication process is occurring, as in the resistant Sardinian strain. As shown by the ratio of the toxicities towards the two strains measured by the different methods, the acetone application suggests a much higher resistance level than the oil application, which is a rather important consideration in regard to the practical measurement of resistance levels (Vol. II, Chap. 2A.).

Thus, the preparation of suitable formulations for a new toxicant is a vital part of its development for practical use and will largely determine whether it can be satisfactorily exploited in a variety of pest control situations. The formulation possibilities are clearly more limited with an unstable toxicant than with a stable one, and it was soon recognized that the considerable stability and persistence of DDT would enable it to be formulated not only in solutions or dusts for spray applications, but also in more permanent ways. A very large number of different formulations of pesticides are currently being manufactured for use in industry, agriculture, and public health applications; more than 1,200 formulations based on DDT alone and about 1,500 based on other organochlorine compounds were once made in the United States.

For the less durable applications, DDT may be formulated (a) in powder form for spraying or dusting applications, (b) in solution in an organic solvent selected to meet the requirements of spraying, impregnation, or other method of

TABLE 2

Toxicity of DDT to Susceptible and DDT-resistant Strains of Housefly, *Musca Domestica*, as Measured by Topical Application of the Toxicant in Different Solvents

DDT applied in	Median lethal dose (μg per fly)		Ratio: Sardinian/normal
	Normal strain	Sardinian strain (resistant)	
Mineral oil	0.44	7.2	16
Acetone	0.12	36	300

(Data adapted from Busvine.[16])

application, (c) as an aqueous suspension prepared for spraying from a wettable powder, (d) as an emulsion, in which case the solvent and emulsifying agent have to be carefully chosen to give the formulation suitable stability, toxicant particle size, etc., or (e) as an aerosol. In addition to these formulations, there are more durable ones in which the toxicant is incorporated into paints, lacquers, polishes, cloths, papers, etc.[12] The development of some of these will be described later.

Many experiments have been conducted to determine the minimum concentration of DDT in a formulation that would be effective for a particular application. It became fairly standard procedure to use 5% DDT powders and in the United States, louse powders used by the army commonly contained up to 10% DDT. The potency of the new compound was so great that the complete eradication of many serious insect pests appeared possible. With this object in mind, there was probably a tendency in hygiene and other applications to use more DDT than was strictly necessary for successful pest control. Intense insecticide selection pressures imposed upon insects in this way are exactly those conditions which may eventually produce resistant strains. Some of the present difficulties in this regard might have been avoided if today's knowledge of the development of insect resistance could have been applied during the very early years of DDT usage.

Because of the extreme water-insolubility of DDT, liquid formulations are most conveniently made with an organic solvent that is compatible with the particular application. Thus, concentrated solutions can be made in benzene, cyclohexanone, or methylcyclohexanone and diluted, as required, with deodorized kerosene, white spirit, solvent naphtha, or other petroleum fractions. DDT has a limited solubility in saturated hydrocarbons (3 to 4% in white spirit and liquid paraffin) and is more soluble in cyclic hydrocarbons and olefins or in the cruder fractions such as solvent naphtha. These solvents are flammable so that chlorinated hydrocarbons would find favor for some applications in spite of their toxicity. Because of the absence of a rapid knock-down effect with DDT, it has been standard practice until very recently to provide this effect by incorporating pyrethrum in sprays containing DDT. A recipe recommended in Britain about 1945 for a suitable spray for application at

the rate of 10 ml/1,000 ft^3 contained 0.3 to 0.5% of DDT and 0.03 to 0.05% of pyrethrins.

For the preparation of DDT powder formulations, it is preferred to use a crystalline form which is fairly pure, because technical DDT tends to lump. DDT and its own weight or more of talc (a form of magnesium silicate), gypsum, Kieselguhr (a porous amorphous silica), or various graded clays such as China clay or pyrophyllite are mixed to give a powder concentrate which can then be added to more diluent by simple mixing as required. Talc is particularly suitable for medicinal dusting powders, whereas agricultural uses require heavier carriers such as the coarser clays, chalk, basic slag, etc. to minimize wind dispersal. The choice of a suitable diluent is important from the point of view of chemical and physical properties, because' some diluents may adversely influence the chemical stability of the incorporated toxicant whilst in storage or in the environment after application (Vol. I, Chap. 3C.3b).

When a solution of DDT in a miscible solvent is added to water, DDT precipitates in a flocculent form even in the presence of a good dispersing agent. For this reason, wettable powders are formulated for conversion into aqueous suspensions. The toxicant is mixed with a diluent such as China clay or another hydrophilic carrier material, together with wetting agents, e.g., alkyl naphthalene sulfonic acids and other additives, to insure that the aqueous suspension remains stable for some time. The original Geigy preparation known as "Guesarol® Spray Powder" consisted of DDT carried on a very fine China clay, with wetting, suspension and sticking agents of the Belloid type coformulated to make the powder more easily wettable, to provide high stability of aqueous suspension, and to give even distribution over the treated surface (for example, foliage or walls), together with good adhesive properties to prevent rapid washing off by rain or surface condensation.

Very dilute solutions of the alkali salts of various alkyl naphthalene sulfonates, mainly the isopropyl and butyl derivatives (Belloid agents), produce a more than 50% reduction in the surface tension of water, so that wetting of the powder and spreading of the suspension over the sprayed surface are greatly facilitated. Enhanced stability of the suspensions may be achieved by the incorporation of substances such as gelatin or a glue which, when adsorbed on the particles, helps to maintain a hydrophilic surface layer. Alternatively, settling may be delayed by the incorporation of colloidal electrolytes. These compounds are adsorbed on the suspended particles, conferring like charges upon them so that they repel one another instead of flocculating. The early wettable powder formulations of DDT contained only five, ten or at most 25% of active ingredient, had poor qualitites of suspendibility and low biological efficiency. These formulations deteriorated rapidly on storage, so that in the 1950's it was common experience for hundreds of tons of DDT wettable powder to become unusable when stored in tropical areas between spraying seasons. For this reason, much effort was expended in developing the high quality wettable powders containing up to 75% of DDT that are in use today.[17] In addition to an increase in effectiveness, there is reduction in bulk and transportation costs and such powders provide the most economical use of DDT in malaria control. Other formulations such as solutions and emulsions are rather less convenient. A number of physical factors greatly affect the toxic efficiency of deposits from wettable powders. Thus, an inverse relationship has been shown to exist between the size of DDT particles and their effectiveness against mosquitoes and tsetse flies exposed for short periods. Particles less than 20 μ in size are picked up more readily and are less easily lost again by the insect than are larger particles. Furthermore, the amount of insecticide acquired during a given contact time depends on the morphology and behavior of the species. The relationship between particle size and effectiveness holds for DDT analogues such as methoxychlor and DDD, but is less marked with aldrin, dieldrin, and γ-HCH because these compounds are intrinsically more toxic than DDT, and mortality can be produced by the pickup of only a few particles. Apart from the question of contact with the insect, small particle size obviously results in good suspension qualities, uniform cover of the surface and minimum blockage of spray nozzles.[18]

Unfortunately, deposits of solid particles are readily absorbed by a number of surfaces which have to be treated in the field. For example, when mud blocks made from African soil were treated with DDT or γ-HCH at a rate of 25 mg/ft^2 at a particle size of 10 to 20 μ, the DDT particles disappeared in less than one week and those of γ-HCH disappeared within 3 to 4 hr at 78°F. When sorption was complete and the surface deposits were no longer visible, mosquitoes were unaffected by contact with blocks treated with DDT (or dieldrin) but there was a fumigant effect from blocks treated with γ-HCH (or aldrin) which gave kills for several weeks. The toxicants could be recovered from inside the blocks, and, even following high dosages, they remained mostly in the top 1 mm. The rate of sorption increases as the temperature increases and the particle size of the insecticide decreases. Small particles in the size range 0 to 10 μ are most toxic initially, but their effect disappears more rapidly than with larger particles.[18]

Formulations of the wettable powder type are generally useful in horticulture, for fly control on farm outbuildings, spraying primitive dwellings, and in situations in which the powder coating effect can be tolerated. It is clear that the loss due to absorption which occurs when oil solutions and emulsions are applied to vegetation and to the mud and plaster of dwellings is less evident with dispersible powders. On the other hand, the dispersible powders produce deposits on vegetation that are fairly readily removed by tropical rain and are less stable towards tropical sunlight than residues from emulsions. A 1952 report by van Tiel[19] describes trials with two new DDT formulations "Supona-D" and "Supona-DB," which were "thin, creamy pastes, easy to pour and dilute with water." They contained 50% DDT, and 39% DDT plus 13% technical HCH, respectively, the active material being suspended in a complex colloidal oil-water system. The oil used had only poor solvent properties for the toxicants, which were mainly suspended as minute solid particles. Dilution of this concentrate gave suspensions of particle size less than 25 μ, which were considerably more stable than the suspensions of the dispersible powders available at the time. Comparative trials against tsetse flies with "Supona-D" and a comparable 30% mineral oil emulsion of DDT, both diluted to contain 5% of active ingredient, indicated a superior residual effect of "Supona" on vegetation. As a residual spray against mosquitoes, "Supona-D" residues were more effective than a deposit at the same rate from a DDT dispersible powder.[19]

The type of formulation just discussed appears intermediate between a wettable powder and an emulsion. Emulsions are mixtures of two immiscible liquids compounded in such a way that one, the internal or dispersed phase, is suspended as fine globules in the other, called the external or continuous phase. An emulsifying agent may be added to stabilize the system by preventing the dispersed globules from coalescing and so breaking the emulsion. Since organochlorine insecticides are readily soluble in organic solvents, the most widely applicable formulations consist of concentrated solutions in suitable organic solvents dispersed in water as the continuous phase, to give "oil in water" emulsions. The emulsifying agents used to bring about dispersion may be colloids such as agar, glue, gelatin and casein, or colloidal electrolytes such as soaps and sulfonated oils. Finely divided powders such as lampblack, gelatinous alumina, and various colloidal clays are also effective emulsifying agents. An emulsion of DDT in water may be prepared by adding a solution of the compound in a suitable organic solvent to water already containing the emulsifying agent, with vigorous stirring to assist dispersal. Alternatively, the emulsifying agent may be created *in situ* by dissolving a fatty acid with the DDT and adding the organic solution to water containing sodium hydroxide or ammonia. Vigorous stirring produces the fatty acid salts and the anions are adsorbed in an orientated monolayer at the interface between the phases, so that the suspended globules are stabilized.

As an example, the Guesapon® emulsion of DDT was prepared by adding a solution of 5 parts of DDT in 20 parts of solvent naphtha containing 7 parts of linseed oil fatty acid to 65 parts of water containing three parts of ammonia (specific gravity 0.94). The resulting emulsion can be diluted with water to give a 0.1% DDT content suitable for soil applications, etc.[12]

By 1945, a great deal of ingenuity had been exercised in devising suitable means for the preparation of stable emulsions from DDT and related compounds, and several publications of that time list the most suitable commercial emulsifying agents available and also the properties of a number of DDT emulsifiable concentrates. As mentioned earlier, sulfonated emulsifiers such as sodium isopropyl naphthalenesulfonate, sulfonated methyl oleate, and sulfonated castor oil (as the sodium, potassium, or ammonium salts) were commonly used and the application of detergent agents of the Span (sorbitan esters of higher fatty acids) and Tween (polyoxyalkylene derivatives of the corresponding Span) types has been described.

As a convenient basis for the preparation of dilute emulsions, various types of concentrated emulsion, or miscible oil may be prepared. For example, an aqueous solution of sulfite waste liquor is mixed with a solution of DDT in oil or organic solvent (preferably having a density near unity) and the coarse dispersion thereby obtained is mechanically homogenized to produce a stable emulsion which can be suitably diluted at the site of operations. Miscible oils consist of homogeneous mixtures of toxicant, solvent and emulsifier, which can be diluted with water to give stable emulsions. Organic solvents used may be hydrocarbons or halogenated hydrocarbons, various petroleum products, coal tar oils, etc., and these are combined with emulsifiers such as ethers of polyethylene and polypropylene glycols, calcium sulfonates, Spans, and Tweens. Calcium alkyl benzenesulfonates are more useful than the corresponding sodium salts, since unlike the latter, they are retained by organic solvents and therefore have less tendency to crystallize from the emulsifiable concentrates. As an example, a solution of 15 parts by weight of DDT in a warm mixture of 15 parts of methylhexaline and 30 parts of toluene is treated with 36 parts of sulfonated castor oil and four parts of ammonia with good stirring. A clear miscible oil is formed which can be diluted with 50 times its volume of water to give an emulsion containing 0.3% of DDT.

Concentrated emulsions may tend either to "break" or "cream" on standing. Breaking involves separation of the phases and is not readily reversible. Creaming is a separation due to the difference in density between the continuous phase and the disperse phase but the disperse phase can usually be redistributed by vigorous agitation. It is also possible for emulsions to change their form by "inverting." For example, an oil-in-water emulsion may become a water in oil emulsion due to change in storage conditions resulting in an alteration of the physical interactions in the emulsion. Another cause of phase separation is the use of hard water to dilute emulsions containing an emulsifying agent that can form an insoluble calcium salt. The result is a greasy mass which can no longer be used.

In a list of DDT emulsion concentrates given in 1946, DDT (25%) was formulated with Triton

X-100 (15%) and 60% of various solvents.[12] The best of these concentrates were those in which mixtures of aromatic petroleum fractions with cyclohexane or isophorone (in 3:1 or 4:1 ratio), or standard coal tar fractions mixed with cyclohexanone in these ratios, were used as solvents. These concentrates did not deposit DDT after standing for three months at 5°F, and 1% emulsions prepared from them by dilution with distilled water, hard water containing magnesium, calcium and bicarbonate ions, or sea water, did not break within 24 hr. Since the constituents of the formulation are carried into the environment along with DDT, the use of dispersing agents from natural sources, rather than synthetic detergents, is obviously advantageous. A Spanish patent of 1969 describes the treatment of pressed olive skins by acid hydrolysis, exposure to steam, neutralization, and decantation to give mucilage, gum, pectin, starch, and dextrin which are stable emulsifiers for DDT and other insecticides.

Another important question is that of compatibility of the formulations with other insecticides that may be added by chance or deliberately. A Dutch patent of 1970 describes the effect of a stabilizing mixture for aqueous emulsions: a mixture of tetradifon (15%), dioxan (10%), xylene (70%), and an emulsifier (5%) consisting of a mixture of calcium dodecylbenzenesulfonate and a Tween was diluted with water to give a 0.5% emulsion of tetradifon. When 100 ml of this emulsion was mixed with 100 ml of a 0.5% suspension of 80% carbaryl (1-naphthyl-N-methyl-carbamate) dust, flocculation occurred with the formation of a sticky mass. Such occurrences may be prevented by the addition to the emulsion of a stabilizer containing a hydrated animal protein such as blood albumin, gelatin or casein. Thus, 55 parts by weight of dried sodium caseinate is mixed with 10 parts of sodium bicarbonate and the mixture added to 35 parts of a 70% solution of calcium dodecylbenzenesulfonate in butanol, with stirring to give a clear solution. If 0.2 g of this stabilizer solution is added to the tetradifon emulsion before addition of the carbaryl suspension, the final mixture remains completely homogeneous, although it contains both an emulsified phase and suspended phase.[20]

Aerosols produced by the atomization of solutions of insecticides in volatile solvents are recommended for the control of flying insects in closed premises. The solutions are contained in metal cylinders fitted with atomizing devices and the pressure within is maintained by carbon dioxide or low boiling solvents such as Freon's® or methyl chloride. A typical aerosol might contain 3 to 5% of the principal insecticide used in the formulation, a similar amount of a second insecticide (for example, pyrethrum with DDT), 85% of Freon 12 (dichlorodifluoromethane) as propellant and the balance made up of other, less volatile solvents such as cyclohexanone and a lubricating oil. When this solution is sprayed into the air through the atomizer, the solvent evaporates, leaving the pesticide finely dispersed in the air. In some cases, mixtures of methyl chloride and acetone replace the Freon. In these limited situations and also in the open ones involving aircraft spraying in the field, physical conditions, especially particle size distributions, are of great importance in determining the toxic efficiency of the application. Since insects in flight are moving relative to suspended droplets, they encounter more of insecticide droplets than they would if resting, and the beating wings are efficient collectors of these droplets. In the treatment of rooms with space sprays, the droplets should not be so large that they fall too rapidly, nor so small that they are swept around the flying insects by moving air rather than impacting efficiently. Compounds such as pyrethrins, which stimulate violent flight, are additionally useful, since they promote greater contact with the toxicant. When spraying insects that are protected by foliage, the droplet sizes must not be so large that they do not pass the first barrier of leaves; a number of experiments indicate that sizes in the range 10 to 40 μ are likely to be best for this purpose, with an optimum of 10 to 20 μ.[18]

b. More Durable Formulations

Following the early laboratory tests with houseflies by Müller, development to the field trial stage was continued by Wiesmann,[21] who showed that walls treated with the Guesarol spray (a 1% suspension of powder containing 5% of DDT) retained their insecticidal activity for 2 months. Furthermore, other investigations showed that garments impregnated with DDT from solution were still insecticidal after 6 to 8 weeks' wearing and 6 to 8 washings, while beds sprayed with a solution containing 10% of DDT and 5% of cyclohexanone in kerosene still killed bedbugs several months later.

However, there are many circumstances in which the highly active surface deposit left by the Guesarol type spray is not acceptable, for example in houses, and the unique stability of DDT appeared to afford unusual opportunities for its incorporation into more permanent surface layers such as paint. The question whether such surfaces would have insecticidal activity was investigated in a series of experiments described by West and Campbell.[1,2]

In these experiments, zinc mesh fly cages were fitted with plywood sheets which covered the floor, the bottom half of each side, and two thirds of the lid of each cage. The boards in each of one set of cages were then painted, respectively, with dry distemper (a water-based paint), oil-bound water paint, white lead oil paint, and synthetic varnish enamel, each paint being white in color. A corresponding set of cages was treated with the same paint types into each of which 5% of DDT had been ground. When flies were reared in the cages containing DDT and the control cages, good kills were obtained with both the dry distemper and the oil-bound water paint containing DDT. In the case of the distemper, the control cages also showed fly mortalities, which were attributed to the effect of loose distemper on the flies, since inert powders are known to kill some insects by cuticle abrasions leading to dehydration or by other effects. The cages treated with the oil paint and the synthetic enamel gave no kills since the pigment particles containing DDT are presumably covered with an adsorbed oil or resinous layer, thus preventing the insect from coming into contact with the toxicant. With the oil-bound water paint, the insecticidal effect hardly diminished after 1 year and the test was extended to small rooms, one treated with a paint containing 0.5% of DDT and one with normal paint as the control. Even at this lower concentration of DDT in the paint, a high kill was obtained in 24 hr and the painted surface was still quite active 9 months later.

Oil-bound water paints are based on a lipophilic phase such as linseed oil dispersed as an emulsion in water as the continuous phase. Pigments and extenders are ground with this emulsion to give a concentrated paint which is marketed as a gel for dilution with water to the required consistency before application. The water evaporates after application and the surface becomes increasingly durable as the oil hardens through oxidation and polymerization. It is assumed that with this type of paint, the DDT incorporated by grinding is at the oil-water interface in the emulsion and remains at the surface of the sprayed layer after drying, rather than being covered with oil as in the case of the oil paint. The same insecticidal effect was obtained when DDT was dissolved in linseed oil before emulsification so that the order of addition is not critical and the effect is clearly not much changed by any aging processes that occur in the paint layer with time. Experiments with both oil-in-water and water-in-oil emulsion paints containing DDT, with houseflies, German cockroaches (*Blattella germanica*), and Oriental cockroaches (*Blatta orientalis*), indicated that the phase reversal had little effect on the toxicity of the final paint layer. Surfaces treated with varnishes incorporating 1% or 5% DDT were ineffective against cockroaches.

Emulsion paints of the type described above and containing preferably not less than 5% DDT were also effective against bedbugs confined on plywood strips treated with the coatings. It was shown that DDT incorporated with a flat oil paint (Walpamur Muromatte, drying with a matt surface because of the high pigment content) or into wax polishes containing only 2% of the toxicant, gave effective surfaces against houseflies and bedbugs, but not against cockroaches. Samples of rubber containing 1% and 5% of DDT also killed houseflies and bedbugs and showed some toxicity towards cockroaches although the effects were much delayed with the last insect. In the meantime, methods for incorporating DDT into oil paints and varnishes had been improved, so that eventually surface finishes with these materials could be obtained which showed good toxicity to houseflies and were moderately effective against bedbugs. To kill cockroaches efficiently, a Walpamur varnish had to contain up to 10% of DDT but higher contents of toxicant were usually required to produce surfaces effective against other insects. As usual, best results were obtained with houseflies.

The above results showed that the incorporation of DDT into potentially perishable materials would protect them from damage by insects, and different surface treatments were tested. A surface coated with a film of coumarone polymeric resin formulated with an equal proportion of DDT killed houseflies very effectively. The DDT was found to be present on the surface treated in this way as microcrystals, which may have provided a greater surface for contact with resting insects, since, as indicated in the discussion on wettable powders, particle size is of considerable importance. When the crystals were rubbed off the surface, microcrystalline material reappeared. Although Campbell and West indicated that the coumarone resin asserted its effect through an influence on crystal size, it appears in much subsequent work that the resin was added to formulations as a binder of DDT to the surface treated, rather than as a modifier of crystal size.[12]

Van Tiel[22] conducted a detailed examination of this phenomenon and found that coumarone resin at a concentration as low as 0.005% in a 1% solution of DDT in kerosene produced a notable reduction in the size of deposited DDT crystals. Since there was no increase in the speed of knock-down or in final mortality when DDT and the resin together were topically applied to houseflies, the effect was not attributable to a synergistic effect of the resin on DDT action. It was also found that the effect extended to several other resins, all of which reduced the size of deposited DDT crystals; resins that did not produce this effect had no effect on toxicity. Thus, the effect certainly seems to be associated with an influence on the crystal size of DDT deposits and this is supported by the increase in toxicity seen when crystals are deposited from xylene, which evaporates more rapidly and gives smaller crystals rather than kerosene (Table 3). There is a marked species variation in the effect of coumarone resin

on DDT toxicity. It is pronounced with *Drosophila melanogaster,* and more noticeable with mosquitoes (*Aedes aegypti*) than with houseflies. These differences may be related to the size of the insect, since no effect was observed, for example, on DDT toxicity to the American cockroach (*Periplaneta americana*).

The effect might be somewhat equivalent to topical applications of solid toxicant in a volatile solvent. Tarsal contact with microcrystalline material results in a more rapid transfer of toxicant into the epicuticular wax than that which occurs from coarse material. In this connection it may be significant that the vapor pressure at a convex surface is greater than that at a flat surface, and the vapor pressure of small particles of solid is greater than that of larger crystals of the same material. Effectively, the thermodynamic activity of a finely divided substance is higher than that of the coarse material, and this will result in the former dissolving more rapidly in the nonpolar epicuticular wax of a contacting insect. This sort of effect, together with that on vapor pressure, may contribute in the present case, especially with small insects in close proximity to the deposit.

In contrast to *Periplaneta americana* and the common cockroach (*Blatta orientalis*), German cockroaches (*Blattella germanica*) are rather susceptible to the DDT-resin combination, and so the DDT content of the formulation could be reduced for this insect. Bedbugs were slowly killed by a resin with a DDT content as low as 5%. When putty containing this concentration of DDT was

TABLE 3

Percent Knock-down and 24-hr Mortality of Houseflies (*Musca domestica*) Following a 30-Sec Exposure to DDT Residues Deposited from a Solution in Xylene, or from Solution in Kerosene With and Without Added Coumarone Resin

DDT conc^n (%)	Coumarone resin conc^n (%)	Solvent	Hr after exposure			
			0.5	1	2	24
2.5	—	Xylene	13	61	84	100
2.5	—	Kerosene	0	4	29	100
2.5	0.02	Kerosene	0	19	46	100
2.5	0.1	Kerosene	51	74	96	100
2.5	0.5	Kerosene	41	57	95	100
	Control		0	0	0	0

(Data adapted from Van Tiel.[22])

spread on boards, the resulting surface was effective against houseflies and grain weevils (*Calandra granaria*), and also showed toxicity towards German cockroaches and bedbugs. A linoleum surface was insecticidally active towards houseflies when the DDT concentration was 2% and toxicity toward other insects was observed as the DDT content was increased. Thus, a linoleum containing a 2 to 4% concentration showed toxicity toward grain weevils, and with 4 to 6%, effects were obtained against bedbugs; cockroaches were only affected after a long period of exposure to the surface containing 6% of DDT.[12]

On the basis of these results, it seemed likely that DDT might be used to render other permanent surfaces toxic to insects. By treating building materials, for example, it might be possible to prevent crawling insects from entering dwellings. Accordingly, German cockroaches or bedbugs were exposed to treated breeze blocks in pinhole zinc mesh cages. The DDT formulation (called Guesapon® EXM) used to treat these breeze blocks was a mixture of the toxicant with coumarone resin, xylene, oleic acid, and emulsifying agents dispersed in water containing sodium hydroxide. The ratio of DDT to coumarone resin was 1:1 in this preparation, which was diluted eight times for spray treating the breeze blocks to give a deposit of 0.12 mg of DDT/cm². In other experiments the blocks were dipped directly in the diluted emulsion. Blocks given the spray treatment killed *Blattella germanica* or bedbugs in 2 to 3 days, but the immersion treatment appeared to give a deposit which was somewhat less effective.

By 1945, many investigations had been undertaken in regard to the incorporation of DDT into semipermanent surfaces for long-lasting control of flying and crawling insects. One of the other most significant wartime applications of DDT was undoubtedly the control of insects of hygienic importance infesting clothing, especially lice and fleas.

It will be remembered that the concept of DDT as a contact insecticide evolved from the use of stomach poisons in mothproofing, and involved the transition from water-soluble compounds to structures such as Eulan BL which were soluble in organic solvents. In the course of the early investigations with *p,p'*-dichlorodiphenyl derivatives of various types, it was observed that wool fibers took up some of these compounds from organic solutions, and the attraction between fabric and toxicant appeared to be greater than expected from the lipophilic nature of the compounds. Indeed, the binding capacity turned out to be strong enough to resist removal during washing. This property was obviously a considerable advantage for mothproofing investigations since it permitted many compounds to be tested by following a simple impregnation procedure from organic solution. It was soon found that woolen fabrics could be impregnated by exposure to solutions of DDT in white spirit to give a temporary mothproofing effect.

Obviously, a compound which combines the above properties with powerful and persistent contact insecticidal action is ideal for controlling insects infesting clothing, and no time was lost in developing these properties for the protection of the armed forces personnel. Large-scale impregnation was carried out on garments by a dry cleaning type process. Shirts or fabric used in shirt manufacture acquired about 1% of the fabric weight of DDT when immersed in a 5% solution of the toxicant in white spirit. The treatment was still effective in controlling lice after up to eight washings and a DDT content as low as 0.1% was reputed to be effective for this purpose. Washing tests were carried out with a standardized soap solution at 40°C for 30 min and it was demonstrated that both wool and cotton, impregnated initially from white spirit to give a DDT content of 1.63 and 1.28%, respectively, lost about 90% of their initial DDT content after four washings. For retreatment of the garments, it was more convenient to use DDT emulsions, which could be readily prepared in the field by diluting concentrates with water, but the binding achieved by this method was not as strong as that using solutions.[12]

The wartime successes obtained with DDT-impregnated textiles suggested many other applications and mothproofing with DDT during the normal dry cleaning process quickly became popular after the war. Woolen goods are specially susceptible to serious damage by moths and, since DDT is soluble in the types of oil used in the manufacture of woolens and worsteds, it is not difficult to insure protection by incorporating about 5% of DDT into these oils during the process. The DDT treatment is satisfactorily stable to most of the wool dyeing processes.

The developments so far mentioned refer mainly to the control of insects of hygienic importance and those actually attacking textile fibers. Besides the various house moths, which include the well-known clothes moth (*Tineola bisellella*), there are the larder and hide beetles (*Dermestes lardarius* and *Dermestes maculatus*, respectively), and the related carpet beetle dermestids, *Attagenus* and *Anthrenus* (including the well-known woolly bear (*Anthrenus vorax*), as well as silverfish (*Lepisma saccharina*), firebrats (*Thermobia domestica*), and various termites. However, the effectiveness of DDT impregnated fibers in insect control also has significance for the protection of bagged or stored products such as grain, against insect infestations. Several families of beetles and weevils infesting stored grains and cereals cause a great deal of damage to these products and therefore are of considerable economic importance (Vol. II, Chap. 2C.3c). Since the most convenient mode of storage is usually in sacks, the impregnation of these containers with an insecticide such as DDT can prevent the invasion of insects from the exterior and may also be effective for controlling insects already within. An account by Hayhurst in 1945 dealt with laboratory tests of small sacks made of twill or Hessian and containing wheat (when the test insect was the rice weevil, *Calandra oryzae*) or flour (for the confused flour beetle *Tribolium confusum*, or the spider beetle, *Ptinus tectus*). These were placed in larger jars with the insects either inside or outside the sacks.[23] In one series of tests, the sacks were impregnated with DDT before being filled while in a parallel series the sacks were untreated. Treatment of either type of sack with a DDT emulsion gave the highest DDT content (1.6%). Impregnation from white spirit gave a content of about 0.5% and impregnation with a DDT-coumarone resin combination from white spirit gave 0.63% and 0.75% contents for the Hessian and twill fabrics, respectively. Each of these treatments gave good control of *C. oryzae* and *T. confusum*, whether the insects were inside or outside the sacks, most insects being dead within 48 hr. *P. tectus* was more resistant, however, but prolonged exposure (3 months or more) caused mortality.

As in the case of textiles, paper may be impregnated with pesticide either for protection against direct attack on the fibers by insects, or when used as a container, to protect the contents from attack by insects which may bore through the wrapping to reach the interior. DDT was found to be effective for such purposes and impregnation was effected by either dipping or spraying with a solution in an organic solvent or by incorporating a preparation such as the Guesarol Spray powder into the actual paper manufacturing process. Insecticidal papers containing up to 2% of DDT were made by this last method. DDT was also applied to the surface of the paper by incorporating it into the glue size for application in the normal way.

3. Applications in Pest Control

The scale and intensity of research on DDT during World War II and in the immediate postwar years were such that all of the major uses of the compound had probably been established by the middle to late 1940's. The following pages mention only a selection of the applications of DDT in pest control. The full list is much longer and comprehensive accounts with many references to the original work are to be found for DDT and related compounds in the books by West and Campbell[12] and Brown.[24]

a. Toxicity to Plants and Beneficial Insects

One of the foremost considerations in the development of a new insecticide for agricultural use is the question of its possible effects on plants and beneficial insects. These effects may be either acute or chronic; the acute effects strongly influence the development of the compound since they soon become apparent as phytotoxicity or in the disappearance of non-target insects. On the other hand, chronic effects may not become apparent until the toxicant has been in practical use for some time and are very much a part of the controversy that has surrounded the chlorinated insecticides in recent years. The most spectacular of the chronic effects (although in some cases it manifests itself quite quickly) is undoubtedly insect resistance. This phenomenon is considered in detail in Volume II but some mention of the acute effects is made here, as they relate to the development of DDT applications.

The possibility of acute effects arising from the use of DDT was recognized at an early date, as is evident in a statement issued by the American Association of Economic Entomology in December 1944. Problems which might arise from the pronounced residual effect of DDT were recognized in early statements made about the compound by the Swiss investigators. By the late 1940's, it seems to have been well established that high-grade commercial DDT was quite harmless to the foliage of a wide range of plants and their immature stages when DDT was applied in suspended form or as an oil-free emulsion at concentrations effective for insect control. Indeed, it seems that many observed cases of phytotoxicity were attributable to the detergents used in wettable powder formulations or to impurities present in the technical grade of the toxicant.

Ten percent DDT dusts proved to be harmless to many kinds of crops, although the toxicant cannot be used with safety on cucurbits, some varieties of which are more susceptible to injury than others. Much depends on the purity of the DDT used, since dusts containing DDT of setting point 104°C produced no phytotoxicity in any variety (in contrast to results with formulations containing DDT of setting point 90°C),[25] while the addition of sulfone impurities normally contained in technical DDT to the higher quality formulations restored phytotoxicity.[26] Nevertheless, it seems likely that some stunting and yellowing effects seen with cucumber and marrow are produced by p,p'-DDT itself. DDT applied in greenhouses in aerosol form at 5 mg/1,000 ft^3 caused some etiolation and chlorosis in cucumber but did not affect other species; the effect on cucumbers seems to be related to the formulation, since the damage was produced with cyclohexanone but not with methylnaphthalene as solvent.

A perusal of the published work on the above effects illustrates the problems associated with the development of adequate formulations for various purposes and environmental conditions.[24] Thus, research at the U.S. Department of Agriculture Bureau of Plant Quarantine indicated that injury was caused to the foliage of tomato, velvetbean, soybean, peanut, cowpea and upland cotton, and to 11 species of herbaceous flowers treated with a DDT spray (0.5 lb DDT and 3 lb raw linseed oil/100 gal of water). However, a lengthy dry spell following these applications may have aggravated the observed phytotoxic effects.

Other reports from the same source state that no foliage injury was observed on velvetbeans, peanuts, cotton, or corn in field plots that received repeated treatments of 2.5% DDT dust, diluted or concentrated spray. DDT, applied as dust or wettable powder, is more liable to be washed off the foliage than as formulations containing oils. In the absence of oil there is little tendency for absorption from the formulation into the tissues of the crop, unless these tissues themselves are of a particularly oily nature. Phytotoxic effects might therefore be expected to be lower with dusts or wettable powders on these grounds. There seems to be much more tendency for the translocation of DDT applied in oil solutions because the toxicant penetrates the tissues following such applications. Thus, DDT is translocated into the terminal foliage of beans following its application in lanolin pellets or by means of wicks treated with DDT suspensions. In some cases, there seems to be definite evidence of the synergism of phytotoxic effects resulting from a joint action between DDT and the formulation vehicle. For example, cocoa pods blackened rapidly and became subject to fungal attack when treated with a 5% solution of DDT in kerosene, an effect which was not observed when either carrier or insecticide was applied alone. Oil solutions applied from the air during mosquito control programs have destroyed squash plants and severely damaged papaya. DDT emulsions containing 0.3% of white oil have caused damage to pear and apple foliage. DDT penetrates citrus foliage rapidly when applied in oil solutions but there seems to be no deleterious effect on the trees.[12,24]

DDT is well known to be highly persistent in soil and still capable of killing wireworms 5 years after an application at 20 lb/acre. The question therefore arises whether possible deleterious effects are produced on plant growth following soil applications, either as a result of direct contact with DDT or from its effects on soil microorganisms associated with normal development. Early references cited by Brown indicate no effect of DDT on soil pH or on soil bacteria, there being no effects on bacterial nitrification or ammonification at up to 5,000 ppm in soil; even 20,000 ppm of DDT was said to have no effect on legume nodulation.[27] However, trials conducted by the U.S. Bureau of Plant Quarantine showed that soil treatments with DDT at 25 lb/acre against larvae of the Japanese beetle (*Popillia japonica*) retarded the growth of onions, spinach, tomatoes, lima beans, bush beans, and soybeans, while some retardation of development was seen with potatoes, muskmelons, beets and carrots at higher rates of application. At the lower rate of application, only scabious, alyssium, lobelia, and gaillardia were affected out of some 270 species of ornamentals treated, the effect again being a retardation of growth. Crop yields of beans and rye were reduced by 30 to 60% following soil treatment at 200 lb/acre. These results were obtained with technical DDT and neither lima beans, bush beans or tomatoes were injured at 200 lb/acre when refined DDT was applied. Indeed, some experiments showed normal germination and development of a number of species exposed to 20,000 ppm of pure DDT in the soil. Early reports of the effects of DDT on plant development indicated a growth stimulating effect of the technical product when applied at subphytotoxic levels, the results being similar to those seen with sublethal doses of 2,4-dichlorophenoxyacetic acid. Analogues of DDT such as DFDT, methoxychlor and DDD are generally harmless to plants, except cucurbits, when applied at the rates recommended for insect control. All chlorinated hydrocarbons are potentially injurious to cucurbits, especially under moist conditions, but methoxychlor appears to cause least trouble in this respect, some of the effects produced being only transient.[24]

DDT undoubtedly influences the ecology of soil insects, as has been clearly shown by Edwards,[28] and the effect is related to the persistent nature of the compound. Thus, a single soil application of DDT caused a progressive fall in the average population of predatory mites during the following 5 years in spite of seasonal increases in numbers. After four years the number of mites found was only 5% of those found in untreated soil; there was a corresponding inverse relationship of the numbers of springtails (*Folsomia candida*) which increased progressively to a level four times greater than that found in the untreated soil. The effect of DDT on predator populations has become evident in the upsurge of mite populations in orchards where DDT has been widely used.

Aquatic insects are particularly susceptible to the toxic effects of DDT, but, fortunately, the larvae of culicine and anopheline mosquitos and of other noxious diptera (*Simulium* spp.) are even more susceptible. Among the beneficial insects of immediate concern in control programs, the honeybee (*Apis mellifera*) must always rank highly because of its role as a pollinator. DDT is quite toxic to bees, either by contact or as a stomach poison; the oral LD50 is about 5μg per bee. In laboratory experiments with bees, concentrations of DDT as low as 0.01% in nectar (queen-cage candy) can kill a whole colony in 48 hr, but large quantities are tolerated when given in the form of a pollen paste with sugar syrup or soybean flour.[12] In spite of its potential toxicity to bees,

DDT seems to present remarkably little hazard especially if precautions are taken to avoid using DDT during flowering periods. DDT is certainly less dangerous than the highly toxic arsenicals which were widely used before the advent of DDT. Some reports indicate that even when crops in full bloom have been treated with DDT, the destruction of bees has been limited due to an apparent repellant effect of the application. For example, application of 3% or 5% dusts to alfalfa in bloom killed some bees, but the working population in the fields was mainly reduced due to the temporary repellent effect and soon returned to normal. Temporary loss in productivity was balanced by increased nectar availability that later resulted from destruction of pests on the crop.[29] The literature contains a number of similar reports of this apparent lack of effect of DDT on bee populations, especially when contrasted with the effects of previously used arsenicals; others suggest, however, that introduction of DDT for use on legumes in the middle 1940's caused a reduction in the wild bee population in the United States. These results emphasize the need for careful control of toxicant application rates and due regard to species differences in sensitivity.

b. Agricultural Applications

The spectrum of insecticidal activity shown by DDT is remarkably broad and it is far easier to discuss the pests which are not controlled satisfactorily than it is to discuss the whole range of DDT's practical application. In the section of their book on the application of DDT against agricultural pests, West and Campbell mention tests of DDT against some 240 species, and there are many more applications against pests affecting man and animals.[12]

During 1943–44 the U.S. Department of Agriculture examined the effects of DDT on 170 insect species. Various DDT preparations were found to be more effective than existing control agents against about 30 pests attacking agricultural crops, trees, man, or animals; equally effective against 18 species; and without effect against 14 others. The 30 species against which DDT had proved to be effective included codling moth, gipsy moth, cotton bollworm, and various other lepidoptera; several kinds of sucking bugs; Japanese, elm bark, and white-fringed beetles, and cattle hornflies, as well as various insects of public health significance. In the "no improvement" category were other lepidoptera, including some attacking corn, grasshoppers, cockroaches, and a number of stored product pests. Among insects unaffected by DDT treatments are the boll weevil, scale insects, aphids, and mites. These findings roughly illustrate the pattern of DDT efficacy that soon became established.

Writing in 1951, Brown[24] concluded that in agricultural applications, DDT is most effective against plant bugs (Mirids, Tingids), tree hoppers (Membracids), leaf hoppers (Cicadellids), thrips (Thysanopterans), butterflies and moths (Pierids, Lasiocampids, Geometrids), Hymenoptera (Tenthredinids), beetles (Bruchids, Buprestids, Cerambycids). Many of the most serious agricultural pests are lepidopterous larvae that are generally well controlled except for a few refractory species such as tobacco hornworm (*Protoparce secta*), salt-marsh caterpillar (*Estigmene acraea*), cotton leafworm (*Alabama argillacea*), orange tortrix (*Argyrotaenia citrana*) and other, *Argyrotaenia* spp., and sugar cane borer (*Diatraea saccharalis*). Differing susceptibilities are found among the various species of beetles, and for some of these, such as certain asparagus beetles (*Crioceris duodecimpunctata*) and scolytids (ambrosia beetles), HCH, chlordane, and toxaphene are more effective than DDT.

Following the groups of pest insects which are best controlled by DDT are others for which DDT can be used as control agent but with less effect than certain alternative control agents. These groups include some weevils (Curculionids), cutworms (Agrotids), frog-hoppers (Cercopids), Tenthredinids and Agromyzid leaf miners, and fruit flies (Trypetids). Finally there are various insects which are either unaffected by DDT or at best, rather poorly controlled. These include grasshoppers (Orthopterans), pentatomid bugs (larger Hemiptera), Cicadas (larger Homoptera), jumping plant lice (Psyllids), white flies (Aleyrodids), aphids, scale insects and mealy bugs (Coccids), adult Hymenoptera, wireworms (Elaterids), root maggots (Dipterous larvae), Mexican bean beetle (*Epilachna varivestis*), white grubs (Scarabaeid larvae), powder post beetle larvae (Lyctids), boll weevil, plum curculio, and mites. The relative immunity of the last group of insects appears to relate either to their morphology (heavy sclerotization), as in the case of grasshoppers, bugs, cicadas, the Mexican bean beetle, boll weevil and plum curculio, to a rapid life cycle in the case

of aphids, plant lice, scale insects, and mites, or to the soil habitat of some of the larval forms of Diptera and beetles mentioned. For insects with heavy sclerotization, the rapid contact action of DDT appears to fail, while the satisfactory control of wireworms, root maggots, and other soil inhabitants requires a fumigant action which will enable the toxicant to penetrate into the cracks and crevices of the soil. DDT has low volatility and no fumigant action (in contrast to other organochlorines such as chlordane, aldrin, or γ-HCH) and is therefore a poor soil insecticide. The fumigant properties required in a soil insecticide may equally well apply if the toxicant is to control insects on foliage with many crevices or convolutions into which insects can retreat. Thus, DDT is effective towards the aphid, *Myzus persicae*, on foliage of tomato or peach, but not on cabbage with its deep recesses, or on the convoluted leaves of plants such as celery or spinach. DDD has a somewhat more spreading and penetrative effect than DDT, and is therefore more effective than the latter against certain tortricid larvae and other species frequenting sheltered places.

A characteristic of DDT which was soon noticed is its selective toxicity towards certain pest predators. It frequently happens that the pest is imperfectly controlled while the predators are mostly destroyed, resulting in a later damaging increase in the pest population. When for example, DDT is used in orchards for the control of the codling-moth, mite predators are destroyed and mite infestations frequently follow because the mite eggs are not affected by DDT. The nymphs and adults may be killed by the treatment, but the life cycle is very short and the unharmed eggs quickly give rise to more individuals which proliferate rapidly in the absence of natural biological control. For this reason, the use of DDT in orchards has to be accompanied by applications of mite ovicides to restore the population balance.[24]

A number of the insects mentioned above which proved to be susceptible to DDT had been particularly difficult to control before its introduction, infestations of various plant bugs and hoppers having been suppressed by treatments, depending on the species, with calcium arsenate, calcium arsenate plus sulfur, sulfur alone, sabadilla, thiocyanates, pyrethrum, or nicotine. The tarnished plant bug (*Lygus pabulinus*) for example, was formerly controlled rather inefficiently by applications of nicotine, pyrethrum, or sulfur to infested potato plants. Slow kill, but effective final control was obtained with 1% DDT dusts. Aerosols containing 5% or 10% DDT, when applied at about 11 lb/acre gave a rapid knockdown and a high 24-hr mortality.[12] Dusts containing up to 5% of DDT have generally given good control of Lygus and other bugs although the effect is usually delayed. Sulfur may be incorporated in the formulation to increase the speed of action. Sprays containing 0.1% DDT controlled the Mirid (*Plesiocoris nigricollis*) on apples in the United Kingdom and became the regular treatment for control of the related apple capsid bug and for apple blossom weevil. The apple red bug (*Lygidea mendax*) proved to be efficiently controlled by DDT as a substitute for nicotine. Thiocyanates and nitro-compounds had previously proved to be ineffective against this pest. Flea hoppers on cotton were originally controlled by dusts containing calcium arsenate with sulfur, but frequent application of DDT dusts proved more effective and gave better cotton yields than any of the previous control measures. DDT dusts, sprays, or aerosols were found to be valuable replacements for Paris green baits, rotenone formulations or fumigation (using naphthalene or hydrogen cyanide) for the control of greenhouse thrips (*Heliothrips haemorrhoidalis*). Many species of thrips are controlled by 5% DDT aerosols (at 100 mg/ft^3), but dust or spray applications are more effective in some cases and lindane (γ-HCH) or chlordane have frequently been co-formulated with DDT to give better control.[24]

Lepidopterous larvae were most readily controlled by the application of stomach poisons to the foliage before the advent of the modern insecticides. Larvae of the various cabbage caterpillars (e.g., *Pieris rapae*) were killed by applications of calcium arsenate, lead arsenate, or Paris green. There are obvious residue hazards to the consumer arising from these treatments, which, therefore, had to be applied early in the season. On the other hand, dusts containing rotenone, nicotine, or pyrethrum can be used at all times without such hazard. Rotenone is the best of these because of its persistent contact and stomach toxicity, but dusts of up to 3% DDT, or 0.05% sprays, soon proved to give much more efficient control. DDT is as effective against *P. brassicae* as against *P. rapae*, and complete control of the alfalfa caterpillar (*Colias eurytheme*), was given by a 1% DDT dust applied at the rate of 25 lb per

acre. DDD is also effective against some of these insects and gave complete control of *C. eurytheme* when applied as a 5% dust at the same rate. Methoxychlor is toxic to cabbage caterpillars but is significantly less efficient than other available treatments. Hawk moth (Sphingids) larvae infesting tobacco were traditionally controlled, before the introduction of DDT, with dusts containing Paris green, cryolite, or arsenates. There are species differences in susceptibility in this family, *Protoparce sexta* being much less readily controllable by DDT dusts and aerosols than *P. quinquemaculata*.[30] DDD (10% dust) is more effective than the traditional cryolite against either of these species and toxaphene can also be used. The arsenates of lead or calcium were also used to control the gipsy moth (*Lymantria dispar*) by aerial spraying of these compounds at up to 40 lb per acre over forest land. DDT oil sprays at 1 lb per acre or even at 0.5 lb per acre for young larvae were later found to be adequate for efficient control.[31] The aerial spraying of forests with DDT proved to be a very effective way of controlling insects infesting trees, although it clearly made a major contribution to contamination of these environments with the compound. As indicated previously, the salt-marsh caterpillar (*Estigmene acraea*), infesting cotton, proved to be refractory to DDT and to chlorinated insecticides generally. However, toxaphene in the form of a 20% dust gave efficient control and the mixed nitroalkane analogues of DDT ("Dilan") turned out to be even better than toxaphene against this insect.

The various highly destructive lepidopterous larvae referred to as "armyworms," owing to their habit of moving from one food source to another in large numbers, were controlled by sweetened bran baits poisoned with 1% of sodium arsenite, Paris green, or arsenious oxide before the introduction of organochlorine insecticides. Organochlorine insecticides provide control that is generally superior to that given by these inorganic agents, although the efficiency varies with the species and the type of formulation applied. For control of the fall armyworm (*Laphygma frugiperda*), chlordane proved to be the best bait poison, but both HCH and DDT gave better performances than Paris green. DDT, HCH, and chlordane were about equitoxic as sprays to *Cirphis unipuncta* (armyworm), *Prodenia eridania* (southern armyworm), and *L. frugiperda,* but only DDD was equal to DDT in efficiency when formulated as a dust.[32] Some lepidopterous larvae (cutworms) spend most of their time underground, emerging only at night to feed on the crops. One traditional control practice was to distribute poisoned baits in the evening, before the cutworms emerged to feed. When DDT became available, it was found that dusts containing up to 10% of DDT applied to the crops gave satisfactory control of insects such as the black cutworm (*Agrotis ypsilon*) and several other species, although both DDT and methoxychlor had little contact effect on *Agrotis orthogonia* (the pale western cutworm). Chlordane and especially lindane were much more effective against this species. DDT provided an effective replacement for Paris green in bran baits for control of cutworms on tobacco.[24]

Another notorious cutworm is *Heliothis armigera,* which causes widespread damage to crops in North America and has various common names, depending on the particular crop attacked (cotton bollworm, corn earworm, tomato worm, etc.). Rotenone has no effect on this insect which, at one time, was controlled by the inorganic arsenates and cryolite. DDT proved to be more effective than any of these, the actual efficiency of control depending on the crop infested and therefore the accessibility of the pest. On tomatoes, for example, 2% dusts or 0.1 to 0.2% suspension sprays gave good control, but the larvae become difficult to control on corn when they enter the ears. Dusts containing up to 5% DDT have proved effective in this situation, but DDD, having somewhat more fumigant action, and especially lindane, are superior, except for the taint caused by the latter.[33] Toxaphene and methoxychlor by themselves, as well as Dilan, appear to be less effective towards corn earworm than DDT. However, a 2:1 mixture of toxaphene and DDT combines some of the individual properties of the two toxicants and has been used at the rate of 2 to 7 lb per acre against this pest on corn or cotton, and for a variety of other applications (see section on toxaphene). The inaccessibility of pests of this type in some infestation situations required extremely laborious control measures before the introduction of the organochlorine insecticides. For example, the tobacco budworm was controlled by placing poisoned cornmeal (containing lead arsenate) in each bud. DDT 10% dusts applied at the rate of 9 lb per acre subsequently gave control of both the tobacco budworm and the tobacco flea beetle. The economic effect of this

improvement in technique is not difficult to imagine.

The degree of control obtained with tortricid and similar larvae is much influenced by the habits of the various species, so that the mechanics of insecticide application become very important. For example, DDT is much better than the calcium arsenate which was originally used for control of the spruce budworm, but much depends on catching the larvae at a stage in which they are exposed. Orange tortrix (*Argyrotaenia citrana*), which attacks raspberries in some areas of North America, is better controlled by DDD than by DDT, and the predator toxicity of the latter causes mite outbreaks in this situation. DDT also kills parasites attacking the red-banded leaf roller (*Argyrotaenia velutinana*), and outbreaks of this tortricid which followed the early use of DDT in American orchards were attributed to this effect, aggravated by the poor practical control of the rather inaccessible pest by this toxicant. DDD has better ability to penetrate the crevices inhabited by these larvae and gave good control as a 0.003% suspension or 0.05% emulsion. One of the best known of all fruit pests is the codling moth (*Carpocapsa pomonella*), which, in larval form, is mainly responsible for wormy apples. Much effort has been spent in devising effective ways to kill the young larvae as they leave the egg and travel to the fruit which they attack. Lead arsenate (applied as a 0.2% suspension) after petal fall is a well-established remedy, and this treatment has to be given at rather frequent intervals to kill successive generations of the pest, so that a considerable quantity of arsenic residues may accumulate on the fruit. After the introduction of DDT, many trials were conducted to compare the efficacy of the standard 0.1% DDT treatment with one of 0.3% lead arsenate, and DDT proved to be superior in the majority of cases. Neither DDD nor methoxychlor gives effective control of codling moths. Chlordane and toxaphene are ineffective and the use of HCH requires care due to the possibility of tainting the fruit.[34,35] Dinitro-*o*-cresol has been included with DDT in oil emulsions used to control this insect, and the combination of the two compounds gives good control of tortricid larvae during the winter.

Among lepidopterous larvae of the Pyralid family, the European corn borer (*Pyrausta nubilalis*) is a notorious corn pest in North America and on the European continent. For its control, DDT (as a 3% dust) proved to be superior to the nicotine or rotenone treatments used earlier. Most other organochlorines were inferior, although aldrin and DDD gave good control when applied as sprays from aircraft.[24] As mentioned previously, the sugar-cane borer (*Diatraea saccharalis*) proved to be refractory to DDT treatments but was well controlled by dusts containing 1% of lindane or 10% of toxaphene. Cryolite was the traditional control agent for this insect. However, like DDT, cyrolite tends to cause aphid infestations, a problem not found with toxaphene. However, all organochlorine compounds tend to destroy natural predators of *D. saccharalis* and so, favor resurgence of the pest in cases in which control is incomplete. DDT and methoxychlor have been used to control the pink bollworm (*Pectinophora gossypiella*), which resists control by many chemicals. Cryolite, the best of the inorganic control agents for this purpose, gives no more than 50% control, whereas up to 90% control can be achieved by the use of 10% DDT dusts. The disadvantages attending the low volatility of DDT are evident in its failure to control psychid moths, commonly known as bagworms. The larvae of these species spin "bags" which protect them from toxicants lacking penetrating power. However, toxic action does not appear to be related mainly to fumigant effect since lindane is less effective than DDT. In this case, effective control can be obtained with toxaphene, and parathion was found to give complete control.

Among the Hymenopterans, larvae of the various sawflies are responsible for a share of the damage done by insects to fruit, forests, and shade trees. Lead arsenate, rotenone, pyrethrum, or nicotine have been traditional control agents for this group. DDT proved to be highly effective against sawflies infesting forest trees, and treatments of 2 to 3 lb per acre have been used for this purpose. The adults of some species, such as wheat-stem sawfly (*Cephus cinctus*), are tolerant to DDT, which makes control by it difficult. The immature stages are passed in galleries within the plant so that they are protected from the toxicant. The dipterous larvae which attack plant roots, such as cabbage maggot (*Hylemyia brassicae*) and onion maggot (*Hylemyia antiqua*) have been effectively controlled by the use of seed dressings incorporating mercurous chloride or mercuric chloride. Seed treatments with DDT give protection against *H. antiqua,* but only moderate or variable success

seems to be experienced with *H. brassicae,* for which treatments with HCH or cyclodienes are more effective.[36] A heavy application of technical HCH is necessary (for example, 28 lb per acre) because this material contains only 10 to 14% of the active γ-isomer (lindane), but a corresponding reduction of the total chemical applied is, of course, possible if the γ-isomer itself is used. The possibility of crop tainting also arises and the technical material is too phytotoxic to be used as a seed treatment, so that this is an application for which aldrin and dieldrin soon proved valuable. DDT proved to be superior to nicotine sulfate as a control agent for various leaf mining insects such as French bean fly (*Agromyza phaseoli*), but for some species, HCH, chlordane or toxaphene are better. The crane fly or daddy-longlegs is a well-known member of the Tipulid family whose larvae (leatherjackets) infest turf. In Britain, a common form of control was to bring the larvae to the surface by treating it with emulsions of *o*-dichlorobenzene. However, DDT dusts or emulsions proved to give effective control, and lindane provided control at rather lower concentrations.

Many pest beetles are well controlled by DDT, but the Mexican bean beetle (*Epilachna varivestis*) is particularly recalcitrant and even survives aerosols and concentrated (1%) sprays. Even 20% DDT dusts give incomplete kill and are inferior to derris (rotenone). Methoxychlor has been reported to be better than DDT in dusts, but not as aerosols, for which pyrethrum or rotenone seem to be more effective ingredients. The mixture of nitroalkane analogues, Dilan, proved to be particularly effective against this beetle. DDT controls Japanese beetle larvae (*Popillia japonica*) at 25 lb per acre. Although it is less effective than lindane or the cyclodienes, it is about 100 times better than lead arsenate in terms of effective application rates.[12] Wireworms (*Limonius, Agriotes, Ludius* spp.) require lindane or the cyclodienes for effective control. Parathion is an excellent soil insecticide for some of these applications. This compound's lack of residual toxicity was formerly considered a disadvantage but is now a favorable characteristic, although there is a distinct operator hazard with this potent organophosphorus insecticide.

Adult Japanese beetles may be controlled on crops by DDT sprays (0.04%), dusts (5%), or aerosols (5%). Beside the Japanese beetle, there are many other beetle species (both adults and larvae) for which economical control became possible with the introduction of the organochlorines. The persistent nature of these chemicals is highly valuable, and indeed, essential for many such applications. Before the emergence of DDT, the Colorado potato beetle had been controlled since the mid-19th century by foliage applications of 0.4% suspensions of Paris green mixed with lime and subsequently with various inorganic materials such as calcium or lead arsenates. In Switzerland in 1939, the spectacular application of DDT for its control was the first example of the powerful practical impact of DDT on pest control technology.[12] A 4% DDT dust is as effective as one containing 30% of calcium arsenate. DDT gives satisfactory control when applied as dusts, sprays, or aerosols to various leaf-beetles and flea-beetles. Lindane is usually more effective on immediate application but lacks residual activity while toxaphene and chlordane appear to be as effective as DDT at equivalent concentrations for many applications. In some cases, sprays of combined DDT and calcium arsenate have been used to enhance control obtained with DDT, the arsenate retaining favor for use against other species on tobacco when the DDT-tolerant hornworm (*Protoparce sexta*) was present.

Larvae of the various cucumber beetles (*Diabrotica* spp.) are called corn rootworms and are difficult to control because they feed underground. Moderate control is obtained by treating the soil surface with DDT dusts. Methoxychlor and toxaphene are about as effective as DDT when applied in this form. Lindane is more effective than these against both larvae and adults but is likely to taint the crop and is phytotoxic. The incidence of Dutch elm disease, which is caused by various elm bark beetles, can be greatly reduced by spraying large elms annually with 0.1 to 0.2% DDT emulsions at the rate of 4 lb of DDT per tree, and various other bark beetles have been controlled by aerial spraying with DDT.[24,37] Methoxychlor has also been recommended for the control of elm bark beetles. HCH emulsions or solutions in fuel oil have been used to protect sawn timber from infestations and seem to be superior to DDT formulations for this purpose.

DDT has been used to control a number of species of weevils but is generally inferior to the cyclodienes or lindane for this purpose. DDT or methoxychlor alone, or in combination with other insecticides such as malathion, has been used for the control of alfalfa weevil. The white fringed beetle (*Pantomorus leucoloma*), which infests a variety of vegetation in parts of North America, can also be controlled by methoxychlor (sprayed at 2 lb per acre) as well as by DDT. Methoxychlor controls the larvae, when applied to the soil at 10 lb per acre, and adults, when applied at frequent intervals as an emulsion at 1 lb per acre. The cotton boll weevil presents a particularly difficult control problem because the larvae inhabit the interior of the bolls and are thereby protected from contact with the insecticide. The traditional control method involved treatment with up to 10 lb per acre of calcium arsenate dust (not diluted with any carrier), and resulted in extensive poisoning of the soil in some areas.

DDT is not a very efficient control agent for the boll weevil, although 10% dusts are generally more effective than calcium arsenate, as are combinations of DDT and lindane in dusts. In contrast to DDT and toxaphene, chlordane and lindane appear able to reach the larvae within the bolls, undoubtedly due to their greater volatility.[24] Lindane is effective against most cotton insects except *Heliothis* spp. but lacks persistence while toxaphene controls with lesser effect all the significant pests. These deficiencies have been offset by using the DDT-lindane combination and also a 2:1 combination of toxaphene and DDT which assumed prominence following the appearance of toxaphene resistance in the boll weevil in the southern United States about 1955. In addition to the extensive use of DDT-toxaphene, large quantities of DDT-parathion and DDT-methyl parathion combinations are used mainly for the control of pests on cotton and oil seed crops in Mexico, Central America, and Brazil. In the last two countries, more modest quantities (in the order of 1000 tons annually) of DDT-HCH, DDT-toxaphene-methyl parathion, DDT-HCH-methyl parathion and DDT-HCH-parathion have also been used on cotton, while a few hundred tons of DDT-endrin, DDT-endosulfan and DDT-dimethoate are used annually on cotton crops in various African countries. Following their introduction for insect control in 1948, aldrin and dieldrin also proved to be effective against the boll weevil.

At an early stage of its development, it was discovered that DDT tends to be stored in body fat (Vol. II, Chap. 3B.1c) and is excreted in cow's milk. In contrast, methoxychlor shows little tendency to store in adipose tissue and the plateau levels achieved in tissues following high dietary exposure decline rapidly when exposure ceases. For this reason, the compound provides a favorable replacement for DDT for fly control in dairy barns. It is less toxic than DDT toward many insects and nearly four times as expensive (66¢/lb), so that its use has been somewhat restricted. By 1961, only 81 agricultural uses had been registered in contrast to 334 uses for DDT. It is estimated that during the 25 years of its use up to 1970, about 100 million lb have been used in the United States. Interest in it and related alkoxy derivatives has increased in relation to current searches for more biodegradable and therefore less persistent insecticides.[38]

Various applications of methoxychlor have already been mentioned in the preceding account of the agricultural applications of DDT. It is not very effective against soil insects and, at one time, was used primarily on fruit and vegetables. More recently, methoxychlor has been widely recommended for use on alfalfa and clover for the control of pests such as alfalfa weevil larvae, clover leaf weevil, alfalfa caterpillar, meadow spittle bug, potato leaf hopper, flea beetles, and fall armyworm. The Japanese beetle and Colorado potato beetle are also controlled by methoxychlor. For this purpose, it is frequently applied at 1 to 1.5 lb per acre in combination with malathion emulsifiable concentrate (1 to 1.25 lb active ingredient per acre). Another application of methoxychlor is for the elimination of infestations of various stored product insects in empty grain storage bins. A spray made from an emulsifiable concentrate such as "Marlate" 2-MR is applied to give a cover of 0.4 lb of technical methoxychlor/1000 ft². In the veterinary and public health area, the use of methoxychlor in dairies and other farm buildings has already been mentioned. For the control of hornflies and lice on cattle (except dairy animals) and lice on hogs, Marlate® spray is diluted with oil and used to coat cattle "back rubbing" devices from which self-application is effected.[39] A moderate increase in methoxychlor usage is likely in the future.

A number of applications of DDD and DDT have been mentioned already. Although DDD lacks the generally high toxicity of DDT it is equally toxic or more toxic as a mosquito larvicide, or for the control of caterpillars such as corn earworms, or tortricids on fruit. Apart from the mosquito control application, these are situations in which the pest is protected from direct contact with the toxicant and the effect may be due to the somewhat higher volatility of DDD as compared with DDT. The fact that it is three- to fourfold less toxic to honeybees by oral administration is also an obvious advantage. It is non-phytotoxic except to cucurbits and is used at rates from 1 to 12 lb/acre of active ingredient in agricultural applications.[40] Although its mammalian toxicity is generally much lower than that of DDT, its current fortunes seem to parallel those of the latter. For example, in Britain, the two chemicals are viewed similarly and use restrictions have been made on DDD, but the compound is still retained for some purposes on soft fruit and pre-blossom on fruit trees.

Perthane is another DDT analogue of a much lower mammalian toxicity and generally lower insecticidal activity which, nevertheless, has some use against specific insects such as leaf hoppers, various caterpillars on vegetables, and Psyllids on fruit. Perthane is not phytotoxic and is of intermediate persistence in soil, normal application rates being from 1 to 16 lb per acre on vegetable crops.[40]

Dilan, the mixture of nitroalkane analogues of DDT, mentioned before, is of interest since it cannot be dehydrochlorinated and is expected to be toxic against insects whose natural tolerance or acquired resistance to DDT is associated with this detoxication mechanism in vivo. The Mexican bean beetle (*Epilachna varivestis*) is an example of an insect having natural tolerance to DDT and several analogues that is apparently related to their metabolic dehydrochlorination. Accordingly, these compounds are poor toxicants, while Dilan is highly effective. It is also effective against the salt-marsh caterpillar (woolly bear) on cotton and against certain thrips and aphids which are unaffected by DDT.[40,41] The structurally related compounds dicofol and DMC, especially the former, are noted for their different type of action from DDT. Both difocol and DMC are nonsystemic acaricides with little insecticidal activity. Dicofol is widely used for the control of mites on a wide range of crops at 0.5 to 4.0 lb per acre.[40] In the present context, most interest is attributed to their synergistic action with DDT, which appears to be associated with the inhibition of DDT-dehydrochlorinase. This interest is mainly derived from an experimental, rather than a practical point of view, however, since insects quickly develop further resistance to DMC-DDT mixtures (DMC having been most used for this purpose).[41]

c. Public Health, Veterinary and Other Applications

The applications of DDT in the area of public health have attracted most attention and many of these are discussed in the sections on the early development of this compound as a practical pesticide and in Volume II in connection with the development of insect resistance to organochlorines. Although its toxicity as measured, for example, by topical application to insects in the laboratory, is often lower than that of other insecticides, it has useful properties other than toxicity which, when combined under conditions of practical use, make it an ideal control agent. The fact that such combinations are only rarely found in one chemical is not widely appreciated and it may take years of practical application before such facts emerge. Thus, its lack of odor becomes an important factor in practice and the lack of skin irritancy makes it ideal for the control of ectoparasites. Unfortunately, the persistence which made DDT so economical in use results in a tendency to accumulate in areas of application and has undoubtedly contributed to the maintenance of resistance in some insects once selection has occurred. However, even insect resistance to DDT is frequently a type that does not necessarily make further control with it impossible, which is not always the case with other insecticides.

The inadequate control given by DDT against mites has already been mentioned, and here HCH, or the γ-isomer, lindane, is a valuable and more potent substitute, except that its famous odor precludes its use on man. On the whole, dipterous larvae are inadequately controlled by DDT, as seen in the discussion on soil applications. This means that as far as the organochlorine group is concerned, HCH or the cyclodienes are the most useful control agents for housefly larvae. Ants and cockroaches represent another deficiency in the versatility of DDT, and for these, hexachlorocyclopentadiene derivatives such as mirex (for ants), chlordane, and the chlorinated terpenes such as toxaphene find valuable application.

The history of DDT development traces its origin to the early water-soluble mothproofing agents, but DDT is less effective than these in practical use due to eventual removal from impregnated cloth during cleaning processes, although its insecticidal action towards the pests involved is excellent. For mothproofing, clothing is usually impregnated to about 0.1% of its weight with DDT and at this level it is fairly resistant to removal and survives a number of washings. This is not a sufficiently permanent attachment to favor its use on carpets or furniture upholstery, for which members of the Eulan group or Mitin FF are more effective. DDT is effective against pests in stored grain and has been used by admixture with grains intended for seeding. In the past, wheat intended for food uses has been protected by the addition of DDT, but this practice is no longer recommended. Since it is effective at lower concentrations, γ-HCH is better for this purpose from the point of view of the residues produced, but here, as in many other applications for HCH, there is the risk of tainting the commodity. Another means of protecting stored produce is by impregnating the containing sacks or packages with DDT either before or after packing.[23] DDD, which is less persistent than DDT, has been found to be a useful alternative for protection against spider beetle (*Ptinus tectus*). One effective way of providing protection against stored product insects which does not require treating the commodity or its wrapping is to spray the structure itself with DDT. All the treatments outlined give control of insects such as granary weevils (*Sitophilus granarius*), rice weevil (*S. oryzae*), rust red flour beetle (*Tribolium castaneum*), confused flour beetle (*Tribolium confusum*), Indian meal moth (*Plodia interpunctella*), and others. Certain larvae, such as those of the Mediterranean flour moth (*Ephestia Kuehniella*), are rather resistant to DDT, and those of the golden mealworm (*Tenebrio molitor*) are resistant to HCH. Others are highly refractory to treatment with a number of insecticides.[24]

In spite of its generally poor performance against mites, DDT does appear to be effective against a few species and it can be used to control various ticks.[24] Thus, effective and persistent control of chicken mite (*Dermanyssus gallinae*) has been obtained by treating henhouses with 1% suspensions, and various species of cattle ticks (*Boophilus* spp.) have been controlled by 0.5% dips given at monthly intervals or by 1% sprays. In the case of cattle ticks, considerable reduction of the immature stages was also formerly achieved by area sprays at 2 lb per acre to effect control before the pest reached its host. DDT suspension sprays (2%) or emulsions (5%) have been found to control the Gulf Coast tick (*Amblyomma maculatum*) on cattle or sheep, and the lone star tick (*A. americanum*) on dogs. Area sprays at 2 lb per acre also control this species in pastures for a time. Temporary control of the brown dog tick (*Rhipicephalus sanguineus*) has been achieved by treating dogs with 2% emulsions of DDT or with 10% dusts, but the treatment of both animal and habitat has to be very thorough in order to prevent rapid reinfestation. The *Ornithodorus* species (argasid ticks), vectors of relapsing fever, are naturally resistant to DDT but can be controlled by HCH.[42-44]

When first used for cockroach treatment, DDT was found to be somewhat better than traditional measures involving the use of sodium fluoride or pyrethrum, but even a 10% DDT dust kills rather slowly. Surface treatments of DDT are relatively ineffective against cockroaches, which pick up a lethal dose from them rather slowly. With lindane or cyclodienes, pickup is much faster and these compounds are efficient control agents. Pyrethrum has the advantage of giving rapid and spectacular knock-down effects, especially when formulated with one of the methylenedioxyphenyl synergists. Lindane possesses the knock-down property to a smaller extent, while DDT gives no such effect but has a longer residual action.[24,45] There are evident species differences in the susceptibility of cockroaches, since the American cockroach (*Periplaneta americana*) is more susceptible to DDT than the German cockroach (*Blattella germanica*). Ants have been controlled with varying efficiency, by treatment of trails with inorganic toxicants such as sodium fluoride or fluosilicate, by use of poisoned baits, or by fumigant treatments using volatile organics such as carbon disulfide or ethylene dichloride. DDT effects some degree of control when used at high doses, but generally is poorly effective in comparison with insecticides derived from hexachlorocyclopentadiene. Chlordane appears to be the best general ant toxicant and the compound mirex, produced by the dimerization of hexachlorocyclopentadiene, has also been extensively used in recent years.

As indicated earlier, DDT is remarkably effective for the control of various ectoparasites infesting man and animals.[12,24] Sucking lice of the order *Siphunculata* (sometimes called *Anoplura*) are exclusively found on mammals. So far, about 225 species have been described of which the well-known body louse and head louse infest man, about twelve are found on domestic animals, and the remainder on a wide range of other species. Biting lice or bird lice of the order *Mallophaga* are found mainly on birds, but a number of species infest livestock. The biting lice on livestock can be readily controlled by DDT, which replaced earlier treatments involving the use of dips containing nicotine, creosote, or inorganic agents such as arsenicals and sulfur. Dusts containing rotenone or sodium fluoride were also used at one time. Many species of the genus *Bovicola* (also called *Damalinia*) are controlled by dipping livestock in emulsions containing 0.2 to 0.3% DDT or by treatment with dusts containing 5 to 10% of this toxicant, although dusts, sprays, or dips using lindane as the active ingredient frequently give effective control with much lower concentrations of chemical. Various biting lice on birds have been eliminated by the use of dusts containing 4 to 10% of DDT.

Cleanliness is clearly the best protection against the human body louse (*Pediculus humanus corporis*), but treatment of clothing with anti-louse powders is a possible solution in the event that such measures are not possible. During World War I, a louse powder consisting of a mixture of naphthalene, iodoform, and creosote (9:2:2) was used to kill the lice and their eggs. The habits of lice were subsequently exploited by the use of pleated belts offering a suitable creviced habitat as "bait" and which could be poisoned with lauryl thiocyanate. Various powders were devised at the beginning of World War II; diphenylamine was used by the Russians and was said to be as effective as a British product (AL63) which combined derris (rotenone) with naphthalene and other tar extracts. An American powder (MYL) combined synergized pyrethrum for control of adult lice with an anti-oxidant (since pyrethrum is subject to oxidative deterioration) and an ovicide (2,4-dinitroanisole). In view of the importance of this particular area of insect control, the introduction of DDT was sensational. Ten percent DDT powder produced complete kill of lice within 24 hr and protected for 3 weeks. With such a dust, rapid and complete disinfestation of large numbers of people, such as refugees and prisoners of war, was achieved by simply blowing it into the orifices of the clothing worn. Impregnation of clothing was a later effective alternative, although the chemical is gradually removed by washing.[12,24,46,47]

Subsequent to the introduction of DDT, more than 7,000 chemicals were tested as lousicides in the United States, and out of these, only four compounds proved to be more toxic than DDT in field practice. These were the organochlorines lindane, toxaphene and chlordane, and 2-pivalyl-1,3-indandione (valone). Toxaphene proved to be more toxic than DDT but is less desirable for skin treatments, as is chlordane. Lindane, like chlordane, gives greater toxicity (tenfold) than DDT and rapid knock-down, but neither compound is fast on garments (toxaphene is said to be more like DDT in this respect), and lindane preparations usually have an unpleasant odor. Thus, although some other compounds give faster knock-down, DDT has continued to be the most effective lousicide. Since it is not ovicidal, a number of ovicides compatible with human applications had to be developed. For this purpose, 2,4-dinitro-anisole was used in the American MYL powder, and several other nonchlorinated compounds also proved effective; diallyl succinate and adipate, 2-benzylpyridine and 2,4-dinitrophenyl propionate are examples, but a member of the sulfone series incorporating chlorobenzene proved to be the most effective ovicide at the time. This compound was chloromethyl p-chlorophenyl sulfone, developed in Germany under the name *Lauseto-neu*. Liquid formulations are required for use against the head louse (*Pediculus humanus capitis*), 0.5 to 1.0% of rotenone in suitable oils or creams, or pyrethrum formulations having been frequently used for this purpose before the advent of the modern insecticides. Subsequently, 1% emulsions or solutions of DDT in suitable carriers have been used. Lindane has some advantages over DDT for eradication of this insect, but both compounds continue to be used effectively apart from a recent indication of lindane resistance in one area of the United Kingdom. The extensive use of DDT dust for body louse control has led to resistance to this toxicant which has somewhat reduced its value as a lousicide (Vol. II, Chap. 2C.2d).

In such cases, control has been obtained by using pyrethrum, HCH, DDT-HCH mixtures or the carbamate insecticide carbaryl.[48]

DDT, applied as spray, dust, or even in aerosol form proved to be a highly effective and long-lasting control agent for bedbugs (*Cimex lectularius*).[49] Methoxychlor is about equally effective, while *Lauseto-neu* and lindane are better. Reduviid bugs such as *Triatoma infestans*, which are carriers of Chagas disease, are naturally tolerant to DDT although the adults tend to be more susceptible than the immature stages, and these insects are normally controlled by applications of lindane or dieldrin.[24,44] The various fleas infesting man and animals are readily eliminated in normal circumstances by the application of dusts containing up to 10% of DDT to premises, persons, or animals. Thus, the oriental rat flea (*Xenopsylla cheopis*), vector of bubonic plague, can be controlled by treating rat runs with 8 to 10% DDT dusts, and the combination of such treatment with clothing applications resulted in full protection of Allied Services personnel against the plague during a 1945 outbreak in Casablanca.[12,24]

One of the most important applications of DDT for disease vector control is against the malaria-carrying mosquito, and it is in this area that its impact has been most evident and profound. Before the introduction of DDT, control of the larvae relied upon applications of larvicidal oils to the surface of breeding pools at the rate of about 25 gal per acre. Naturally, this form of control can have undesirable effects on vegetation and non-target forms of aquatic life. Pyrethrum treatments have been used when the areas of water involved are small, and for larvae of anophelines, 1 lb per acre treatments with Paris green formulations (dusts or suspensions) are effective. DDT is as toxic as pyrethrum to mosquito larvae and about 100-fold more toxic than Paris green. According to costs in 1947, an effective DDT application against *Anopheles quadrimaculatus* cost 30 times less than an oil treatment, 26 times less than a pyrethrum treatment, and 3.5 times less than a treatment with Paris green or calcium arsenite. It is understandable that the new organochlorine insecticide completely changed the outlook for malaria control and, in fact, offered the prospect of total eradication of this disease for the first time.[12,24,50]

Immense possibilities for rapid and widespread DDT treatments in inaccessible places were provided by the technique of spraying from aircraft. Thus, nearly complete control of larvae was achieved in areas of the South Pacific by aerial spraying of DDT as a 5% solution in kerosene or fuel oil at the rate of 0.25 lb per acre (active

·ingredient). The important factor in larval control is complete coverage of the water surface, so that parameters such as spray droplet size are not very critical for this type of application. Oil solutions are ideal since the oil forms a complete film on the surface which the larvae must penetrate in order to breathe. Therefore, formulations producing aqueous suspensions of solids are understandably less effective than oil solutions or emulsions. Methoxychlor and DDD are also useful mosquito larvicides, being nearly as effective as DDT for some applications. *Culex* mosquitoes are generally more difficult to control with insecticides than *Anopheles*. For larvae of some of the former species, DDD is a better practical toxicant than DDT. Aerial sprays of DDT at 0.5 lb per acre provide excellent control of adult *Anopheles* or *Aedes* spp. Oil solutions are inferior for providing toxic deposits on vegetation since the toxicant appears to be absorbed into foliage from these formulations. Emulsions, and especially aqueous suspensions, are better for this purpose, but oil solutions giving suitable droplet size when sprayed are required to control flying insects. A good method for controlling adult insects, either flying or at rest, is by generating an aerosol fog of the toxicant. No problems arise with this method in open country where there are few obstacles, but areas offering protection to insects, such as woodlands, may present difficulties if wind conditions are such that there is insufficient time of contact with the toxicant passing through.

For the treatment of dwellings against mosquitoes, DDT as a 5% solution is sprayed on walls to give a deposit of about 100 mg/ft^2 providing a residual effect for about 6 months.[44] More immediate relief from the insect is obtained by the use of space sprays from 0.5% DDT solutions containing pyrethrum to give rapid knock-down. Evaluation of a number of DDT analogues for residual efficiency when applied to walls indicates that methoxychlor and 1,1,1-trichloro-2(bis-*p*-methylphenyl)ethane come after DDT in this respect. DDD is less effective than the others and DFDT least effective of all.[51] The 4,4'-dimethyl analogue (known as methyl-DDT, or more recently as methylchlor) is quite toxic to a number of insects, especially with a synergist, and has been reexamined recently as a result of the current interest in biodegradable DDT analogues.[52] DDT is usually superior to lindane as a residual toxicant on surfaces, but on some media

the toxicants are quickly absorbed and those such as lindane which can still give a vapor effect at the surface in these circumstances are then found to be more effective.

The housefly (*Musca domestica*) was the first victim of the practical application of DDT and also the first insect to become resistant to it. Pyrethrum, well known to householders due to its spectacularly rapid knock-down effect on this insect, had been widely used previously, but the material is expensive and the knock-down can be reversible if the concentration reaching the insect is insufficient to kill. The use of insecticide potentiators or synergists gives remarkable enhancement of pyrethrum activity and is intimately associated with the history of pyrethrum use. In some ways, DDT-pyrethrum represents an ideal combination; DDT knock-down takes much longer but the residual effect is far better, so that the combination provides both rapid knock-down and lasting efficiency. A formulation of 0.1% DDT with 0.03% of pyrethrum in odorless kerosene has these properties, and also has economic advantages, since without the DDT, more than double the concentration of the rather expensive pyrethrum is required. As a residual spray on surfaces, DDT has more staying power than other organochlorines such as toxaphene, chlordane, or lindane, and the same applies to its analogues, methoxychlor and methylchlor, which are more effective in laboratory tests than DDD or DFDT but less so than DDT, as in the case of mosquitoes.[51] However, comparative tests on analogues and other compounds in the laboratory do not necessarily reflect the practical situation; methoxychlor, DDD and DDT give equally good results against houseflies in spray applications in cowsheds.[24] Following the appearance of DDT-resistance in houseflies (Vol. II, Chap. 2C.2a) recourse was had to HCH sprays containing 0.25% of lindane until characteristic cyclodiene resistance appeared.

The blackfly (*Simulium* spp.) vectors of onchocerciasis are readily controlled at the larval stage by minute concentrations of DDT in water, 0.1 ppm being sufficient for complete destruction of the larvae after an exposure period greater than about 15 min.[24] Thus, rivers which pass through areas treated for mosquito control are likely to lose their blackflies as well, and the effect of an application in one area can extend for many miles downstream. DDT is not toxic to the pupae at these concentration levels, whereas lindane at 0.2 ppm is lethal to both larvae and pupae. Methoxychlor and DDD are also effective against blackflies, and methoxychlor is increasingly used as a replacement for DDT against these insects because of its biodegradability and lower toxicity to fish.[44] DDT fogs produced by aerosol generators have been used to control adult blackfly. Biting midges of the genus *Culicoides* are frequently difficult to control at both larval and adult stages because of the difficulty in adequately covering their swamp habitats with insecticide. Somewhat heavy applications of 10% DDT dusts have been successful however, and the use in buildings of fly screens treated with DDT solution has proved effective in reducing the number of insects reaching the interior.[44] *Phlebotomus* sandflies, vectors of sandfly fever, are controlled effectively and for long periods when dwellings and the outdoor haunts of these insects are sprayed with DDT (5%) in kerosene to give a surface deposit of 100 mg/ft². Aerosols containing 10% DDT have also been used to control adult sandflies in some circumstances.

The regular treatment of cattle with DDT, usually in the form of suspension sprays (0.1 to 1.0% DDT), which present less toxic hazard than skin-penetrating oil solutions, has effected pronounced improvements in milk yields in cases in which the herds were troubled by hornflies (*Haematobia irritans*) and stableflies (*Stomoxys calcitrans*).[53] Reports of the efficiency of DDT analogues in such applications vary; according to some methoxychlor acts faster than other toxicants against the hornfly and is as toxic as DDT under practical conditions, while deposits of DDT and methoxychlor are reputed to be equally long lasting and toxic towards *S.calcitrans*. Tsetse flies (*Glossina* spp.) are rather susceptible to organochlorine insecticides, although DDT is seven to nine times less toxic than dieldrin (LD50 0.008 μg/fly for *G.morsitans*) to these insects. Control with DDT has been effected in various ways, including hanging DDT-treated cloth screens in the bush to trap moving flies, spraying from the air or on the ground, and generating aerosol fogs. Cattle have been sprayed with DDT solutions to kill tsetse flies feeding on them. No organochlorine resistance has so far appeared in tsetse flies and DDT is still the most widely used control agent, followed by dieldrin.[44] The relative inactivity of DDT toward dipterous larvae is a disadvantage, but treatment of larval infestations means that the toxicant is present *in situ* when the adult flies appear and flies visiting such areas to deposit eggs may be killed. The same principle applies in the use of sheep dips containing 0.5% of DDT for the prevention of attack by the sheep blowfly (*Lucilia sericata*) and other species. Such treatment does not prevent attack completely, but the number of "strikes" is reduced and the period of protection increased in comparison with the results of arsenic-sulfur dips which, unlike DDT, are larvicidal. The high tolerance of the larvae of *Lucilia* is shown by their survival after crawling in solid DDT or following immersion in kerosene solutions or aqueous suspensions of the chemical. Treatment for protection against *Lucilia* also kills the sheep ked (*Melophagus ovinus*) and affords protection against further infestation for several months. A DDT concentration of at least 1% in the dip is required to give adequate control of the sheep tick (*Ixodes ricinus*), which is the vector of several cattle diseases.[12,24]

Mention has been made of the natural tolerance to DDT shown by a number of pests, but the account has mainly considered the effective applications without the attendant complications. The matter of species differences in sensitivity to toxicants and the somewhat related subject of insect resistance are topics of great fundamental importance and these are considered in Volume II. In effect, resistence is accelerated evolution produced by chemicals, and is a remarkable example of their effect on the environment.

d. Current Status of DDT

At the height of DDT production, about 1964, over 400,000 metric tons (since the figures are estimates, the small difference between the metric and the British ton is ignored hereafter) was used annually in worldwide agriculture, forestry, public health, and other pest control programs. In the United States in 1962, some 28,000 tons of DDT and 35,000 tons of cyclodienes (plus toxaphene) were included in an estimated 160,000 tons of pesticides used to treat 90 million acres (about 5% of the continental area).[2] In England and Wales, at about the same time, some 4.5 million acres were being treated annually with an estimated 100 tons of HCH, 180 tons of cyclodienes, and 260 tons of the DDT group compounds.[54] Since the area quoted for the U.S. includes all pesticides, it represents a maximum possible area treated with organochlorines and it is clear that the usage of these compounds in the U.S. has been at least fivefold more intensive than in England and Wales. In both countries, the DDT group compounds appear to have accounted for about half the organochlorine usage in the early 1960's. Clearly, the use pattern of organochlorines will vary widely between countries, depending on the particular insect control problems encountered. The above figures are given as an illustration of such patterns for two countries with different land mass and pesticide requirements. Matsumura[58] has recently discussed pesticide usage in the United States.

Since 1964, global production of DDT has decreased progressively. In 1971, it was estimated to be 200,000 to 250,000 tons, of which 40,000 to 50,000 tons is used for the control of human disease vectors.[55] A major proportion of the chemical used for this purpose is employed in the global malaria eradication program, which at its peak required 60,000 tons of technical DDT annually, and still requires about 35,000 tons per year.[55] An estimate for 1967 indicates that the global use of all organochlorine insecticides and acaricides was 331,000 tons,[56] and the agricultural applications of DDT totaled 34,000 tons, from which it appears that the agricultural and vector control uses are now of a similar order. The decrease has undoubtedly resulted from the widespread discussion of environmental problems that followed the appearance in 1962 of the book *Silent Spring* by Rachel Carson,[56a] and in a more practical sense by the advent of widespread DDT-resistance, which created a need for replacement toxicants that has been partly met by the development of new organophosphorus and carbamate insecticides.

Therefore, it seems that the problem of DDT persistence is solving itself by a process of natural evolution. In these circumstances, the World Health Organization is anxious to ensure that the action taken in 1969 by the Scandinavian countries, the U.S., and Canada to accelerate the process in their own environment by placing severe restrictions or total bans on the use of DDT does not damage the global vector control program for which this compound is essential. It is of particular concern that the countries currently manufacturing this and other essential control agents should continue to make them available to the developing nations for malaria control or eradication. The situation is complicated politically and requires a good deal of common sense and delicacy to achieve a rational solution. Thus, international trade can be disrupted if nations impose restrictions on the importation of each other's agricultural produce due to differences between their governments regarding the permissible residue levels of chlorinated (or other) pesticides in the commodities involved. For this reason, it is quite important that the possible hazards from persistent compounds in the environment should be investigated and widely understood (a problem of general education), and especially that those who gather and distribute such information should use it with the utmost moderation and regard for the consequences of their actions.

Between 50 and 75% of the world's pesticides are produced in the United States, and it was estimated in 1969 that the production of synthetic organic pesticides in the U.S. was increasing at the rate of 15% annually, in contrast to a growth of only 37% during the 5-year period 1963–67. In 1967, synthetic organic insecticides accounted for 38% of the total pesticide sales, and exports of insecticides worth 150 million dollars represented an increase of some 14% over the figures for 1966. This increase was largely due to exports of technical organochlorine and organophosphorus compounds, and reflects the increasing usage in other countries. In 1964, DDT and toxaphene together accounted for 46% of the pesticide volume used in the United States with toxaphene slightly predominant. A survey in that year revealed that 66% of all insecticides were used on farms to combat pests of apples, corn, and cotton (uses on cotton accounted for 70% of the DDT, 86% of the endrin, and 69% of the toxaphene). By 1966, pesticide usage had increased by 10%, with DDT, aldrin, and toxaphene accounting for more than half of all compounds used. A slight decrease in the use of organochlorine compounds between 1964 and 1966 was compensated by increased use of organophosphorus compounds, and in 1967, organochlorines constituted some 50% of U.S. production, about half of this being DDT. The production of DDT itself fell 27% to 103 million lb between 1966 and 1967, and there was a concurrent 10% reduction in exports, more than half of which consisted of the 75% wettable powder used in malaria control. In the peak production year, about 1963, 179 million lb was manufactured annually so that the production figure of 95 million lb (about 70 million lb exported) indicated for 1970 represents a near

50% reduction. Domestic use decreased by about 50% between 1958 and 1966, so that 70% or more of production in recent years has been exported.[57,58]

By 1969, a number of American states were considering restrictions on the use of organochlorine insecticides,[59] a 12-month ban on the use of DDT being in force already in Arizona. In Illinois, a proposal to ban DDT carried some distance before its final rejection by the Senate Agricultural Committee. The Coho salmon episode occurred in 1969 in which a large quantity of the fish (from Lake Michigan) were seized by the FDA after representative samples were shown to contain up to 19 ppm of DDT products. The result was a ban on the use of DDT in the State of Michigan. A Federal Commission (headed by Emil M. Mrak) was appointed in April 1969 by the U.S. Department of Health, Education and Welfare to weigh the benefits and risks attending the use of pesticides, especially DDT, and it reported in December 1969.[57] In the meantime, a National Research Council Committee concluded (May 1969) that some use of persistent pesticides is essential, that the use of pesticides generally is unlikely to decrease, and that there is no evidence that current dietary and environmental levels are inimical to humans, while accepting that some species of wildlife might be threatened. DDT was banned in California from the beginning of 1970 for crop dusting, farm livestock treatments, or home and garden use, while New York State proposed to legislate for the elimination of DDT by 1971. In July 1969, the U.S. Department of Agriculture's Federal Pest Control Program involving major organochlorines was temporarily suspended; it was resumed in August and the use of chlordane was adopted wherever possible as a replacement for dieldrin and heptachlor; the use of dieldrin was henceforth to be restricted. The Mrak Commission's report[1] of December 1969 also concluded that there would still be a need, albeit declining, for persistent pesticide use. It recommended the elimination, within 2 years, of all U.S. DDT and DDD uses except those "essential to the preservation of human health or welfare" and "approved unanimously" by the interested Departments. The uses of DDT in residential areas were thereupon banned (November 1969), with a proposal for similar action in regard to the other persistent organochlorines in 1970. However, substitute control agents are usually more expensive and frequently less effective than the organochlorines and there is no doubt that DDT will be extremely difficult to replace for some uses on agricultural crops. The last battle is yet to be fought in this area, with the insects possibly the only victors.* Events in some other countries paralleled those in the United States. The use of DDT was heavily restricted in Canada, Denmark, and Italy from November 1969, especially with regard to outdoor uses involving large-scale applications. In Sweden, the domestic uses of DDT and lindane were banned in January 1970, with a 2-year suspension on agricultural use of DDT and the uses of aldrin and dieldrin were banned completely. The use of DDT in agriculture and forestry ceased in Norway late in 1970.[59]

In Great Britain, the status of pesticides came under the surveillance of a number of working parties in the early 1950's. These later gave rise to the Advisory Committee on Pesticides and other Toxic Chemicals, which reports to the British Department of Education and Science, and the Agricultural Research Council's Research Committee on Toxic Chemicals.[54] Examining the benefits and risks associated with the use of DDT, DDD, HCH, toxaphene, and the cyclodienes in 1964, the Advisory Committee concluded that there was insufficient evidence to justify a complete ban on any of these chemicals, but the use of aldrin and dieldrin was restricted as an aftermath of the bird poisoning incidents involving seed dressings in the early 1960's. At this time, an annual use of about 600 tons of organochlorines (Table 4) was equally divided between HCH plus cyclodienes and the DDT group (mainly DDT itself, the use of DDD and dicofol being small). The DDT uses, mainly on fruit and vegetables, in England and Wales were more concentrated than those of the other organochlorines. In Scotland, DDT accounted for about 80% of the organochlorine usage and was used mainly on barley, oats, and fruit, especially raspberries. While the use of cyclodienes fell drastically between 1964 and 1968, the decline in DDT use was much slower (perhaps about 16%), and the difficulty of finding economical and efficient replacements for DDT is

---

*For summaries of recent hearings on DDT in the United States, the reader should see Jukes, T. H., DDT stands trial again, *Bioscience,* 22, 670, 1972;[58a] and Washington Correspondent of Nature, DDT condemned, *Nature,* 237, 422, 1972.[58b]

TABLE 4

Estimated Annual Usage (Tons) in the British Isles of Organochlorine and Other Insecticides in the 1960's

Compound	England and Wales (1962–64)	Scotland (1964–66)	Great Britain (1966–68)
DDT group	262	57	220
HCH	101	11	
Aldrin	135	3	
Dieldrin	28	3	80
Endrin	1		
Heptachlor	5		
Endosulfan	6		
Organophosphates			200 (1960–64)
			240 (1966–69)
Carbamates			10
Others			50

(Data adapted from Strickland,[501] Agricultural Research Council,[502] J.M.A. Sly, personal communication.)

evident from the Advisory Committee's report. Where the use of organophosphorus alternatives is possible, the timing of applications is frequently more critical and the treatment more expensive. Satisfactory alternatives (except other organochlorines) for cutworms and leaf-eating beetles are not easy to find.

The Advisory Committee's 1970 review of persistent organochlorine insecticides recommended that the use of DDT on grassland, brassica seed crops, peas (except for weevil control), soft fruit, and post blossom on top fruit should cease, DDD being treated like DDT for the purpose of these strictures.[60] Uses of DDT in food storage practice in the form of lacquers and paints is discontinued, as also the use in smoke generators or thermal vaporizers in situations in which food, people, or animals may be exposed. It was recommended that the use of DDT in houses and gardens should cease, since adequate alternatives are available for these purposes, and so should its use in commercial dry cleaning and moth-proofing of woolen goods in the home. A 1964 survey showed that DDT, chlordane, dieldrin, and lindane were widely used for hygiene and public health purposes in Britain and a curb on such uses has now been recommended, especially since the treatment of refuse tips for fly control tends to produce resistant strains. In spite of the difficulties in finding adequate replacements for some uses, a downward trend in DDT usage is evident (Table

4), and further reductions may be expected as a result of the recommendations made in 1970. The inevitable consequence of these measures is expressed in a recent survey, which indicates that pest control in Britain is now more expensive and often less effective than in the years during which organochlorines were in common use.[711]

The crucial role of organochlorine insecticides, including DDT, elsewhere in the world is well illustrated in the appendix on usage (relating to 1966) given in the report of the Mrak Commission.[57] Of the insecticides used on 12 major crop types listed there, organochlorines provided from 38 to 92% of the total usage, depending on the crop, the lowest proportion being on tea (19%). Their vital contribution towards production of vegetables (46% of usage), rice (57%), and other cereals (85%) is self evident, while a 38% contribution to the total insecticide usage on cotton (60,000 tons in 1966) represents a well-recognized unrivalled cost-performance effectiveness. With regard to pesticide usage the situation is perhaps best documented for India, where a 5.5-fold increase in area treated (to 85 million hectares; 54% of the total crop area) was predicted to occur between 1965 and 1969. A concurrent four fold increase in DDT usage for plant protection to 2,400 tons was expected to be more than offset by a reduction in public health uses during the same period.

Since the story of DDT as an insecticide began

with its spectacular success in the public health area, it is both fitting and important to conclude this discussion of applications with some account of its current status in vector control. The situation was recently described by the World Health Organization,[55] and some of the conclusions are summarized here.

DDT has been the main agent for control of vectors of an impressive list of diseases, including malaria, Chagas' disease, plague, typhus, yellow fever, dengue/haemorrhagic fever, encephalitis, filariasis, African trypanosomiasis, onchocerciases, and leishmaniasis. Except for vectors of malaria and African trypanosomiasis, suitable alternatives are now becoming available so that DDT use has passed its peak, as indicated by the decline in annual tonnage used for vector control. The spread of resistance in the anopheline mosquito vectors of malaria control has accelerated the move to alternatives in some places, but the change affects only 1% of the total area involved in eradication programs, for which DDT is still paramount. The culicine mosquito disease vectors have generally been more difficult to control with DDT and the emergence of wide-spread resistance among both the disease carrying and pest mosquitoes of this type has led to the replacement of DDT by other control measures. DDT continues in favor for tsetse fly control because of the low cost and the lack of resistance in these insects; about 60 tons are used annually. Control of blackfly larvae (*Simulium* spp.) with DDT continues by frequent applications to waterways each year, but the more biodegradable methoxychlor is increasingly favored for this purpose. The widespread use of DDT dust for the control of rat fleas (vectors of plague and murine typhus) and body lice (vectors of epidemic relapsing fever and epidemic typhus) has produced resistance in them necessitating its replacement by organophosphorus insecticides against fleas, and by lindane, carbaryl, or malathion for lice control.

In the developing countries, communicable diseases, especially gastrointestinal infections and the vector-borne diseases, make a most significant contribution to total ill health. These diseases are closely related to the local environment, but a significant improvement of environmental quality would be so costly as to be out of the question for most of these countries. Hence, immunization in some cases, but mainly insecticides to control the disease vectors, are the only immediately available

and financially feasible alternatives for improvement of the general situation. Before the introduction of DDT, malaria, (which is responsible for a remarkable amount of mortality and malaise) could only be effectively controlled in limited areas of dense population where continuous mosquito control and appropriate sanitary engineering could be supported on economic grounds. The advent of insecticides such as DDT with their lasting (residual) effects changed this picture completely and, for a number of reasons, DDT has proved to be the superior agent for mosquito control.

Between 1959 and 1970, the risk of contracting malaria was lifted from more than 1000 million people — a success of enormous magnitude (Table 5). However, this success, which represents about three quarters of the total task, has been obtained in the temperate and semitropical areas at the periphery of the problem; the hard core of malaria lies in the tropics and is proving much more difficult to handle. In South America, Central America, and Asian countries, national malaria eradication programs operate, but much remains to be done in these countries and in Africa, where the approach is through malaria control rather than eradication at present. Malaria has existed in 124 countries and territories in the tropics, threatening a population of 1,724 million. In 19 countries (37 million people) the disease has been eradicated, and another 48 national eradication programs currently operate, covering a population of 1,230 million. Extensive control operations are being carried out in 37 (412 million people) of the other 57 countries, while in the remaining 20 (16 in Africa; population 45 million) the problem is being tackled on the basis of control rather than eradication.

The direct effect of DDT on the health of a population in terms of reduction of malaria morbidity for a number of countries in which eradication is complete or proceeding successfully is shown in Table 6. Various pilot projects in some most difficult situations in Africa have also proved that the spraying of DDT can reduce the incidence of malaria and counteract the effects of the disease. The impact of control operations on the general health of populations is well illustrated by some examples. Following the introduction of extensive DDT spraying operations, the number of malaria cases in Ceylon fell from 2.8 million in 1946 to 110 in 1961; deaths from the disease fell

TABLE 5

Status of the Malaria Control Program in 1959 and 1970

	1959		1970	
	Population in millions	Percentage of total population	Population in millions	Percentage of total population
Maintenance phase (disease reported as eradicated)	279	21.5	710	39.4
(of which lived in tropical areas)	(22)	–	(361)	–
Consolidation phase (freed from endemic malaria)	55	4.2	296	16.4
Attack phase (protected through spraying operations)	505	38.9	329	18.3
Total	839	64.6	1335	74.1
Not yet protected by eradication operations (including population in areas in the preparatory phase of malaria eradication programmes and that where extensive malaria control operations are being undertaken)	459	35.4	467	25.9
Total population[a]	1298	100	1802	100

(Data courtesy of The World Health Organization.[55])

[a]Excluding China (mainland), the Democratic People's Republic of Korea, and the Democratic Republic of Vietnam.

from 12,587 to zero in the same period. In parallel, the general death rate fell from 20.3 to 8.6 per thousand and the infant mortality rate from 141 to 57 per thousand. Similar operations in Mauritius resulted in a fall in the number of deaths from malaria from 1589 in 1948 to three in 1955. Meanwhile, the general death rate and infant mortality rate were nearly halved. In Venezuela, the number of people treated for malaria fell from 817,115 to 800 between 1943 and 1958.

The economic gains and social benefits resulting from malaria control have amply justified investment in control or eradication programs, although the means of assessing the results are still very imperfect. By preventing the incapacity due to the disease, the high socioeconomic cost of prolonged treatment and medical care is reduced and the working capacity of the population is increased. Thus, increased rice population has resulted in Venezuela, Thailand, and the Phillipines and it has become possible to commence agricultural production in large areas of India, Nepal, and Taiwan, for example, where this was formerly impossible. In other areas, as in Kunduz in Afghanistan and

Mindanao in the Phillipines, it has been possible to improve agriculture from the existing subsistence level, with an attendant increase in land value. It is reported that malaria eradication in Syria has completely changed the economic outlook there, since a continuation of the incidence of this disease at the levels prevailing in the early 1950's would have totally negated that country's development effort due to loss of vigor and efficiency among the populace, as well as bringing possible higher death rates.

Malaria control requires residual insecticides with lasting efficiency applied at relatively high levels, unlike most other vector control operations, for which low dosages of less persistent compounds are satisfactory. The World Health Organization has tested more than 1,400 new insecticides as vector control agents since 1960, but only two of them, the organophosphorus compound malathion and the carbamate propoxur can be recommended as being safe and efficient in large scale use at present. However, the organophosphorus insecticides fenitrothion (dimethyl 3-methyl-4-nitrophenyl phosphorothioate),

TABLE 6

Changes in Malaria Morbidity in Countries Before and After Malaria Was Controlled or Eradicated

Africa		
Mauritius	1948	46,395 cases
	1969	17 cases[a]
The Americas		
Cuba	1962	3,519 cases
	1969	3 cases
Dominica	1950	1,825 cases
	1969	Nil
Dominican Republic	1950	17,310 cases
	1968	21 cases
Grenada and Carriacou	1951	3,233 cases
	1969	Nil
Jamaica	1954	4,417 cases
	1969	Nil
Trinidad and Tobago	1950	5,098 cases
	1969	5 cases
Venezuela	1943	817,115 cases
	1958	800 cases
Southeast Asia		
India	1935	over 100,000,000 cases
	1969	286,962 cases
Europe		
Bulgaria	1946	144,631 cases
	1969	10 cases[a]
Italy	1945	411,602 cases
	1968	37 cases
Romania	1948	338,198 cases
	1969	4 cases[a]
Spain	1950	19,644 cases
	1969	28 cases[a]
Turkey	1950	1,188,969 cases
	1969	2,173 cases
Yugoslavia	1937	169,545 cases
	1969	15 cases[a]
Western Pacific		
China (Taiwan)	1945	over 1,000,000 cases
	1969	9 cases

(Data courtesy of The World Health Organization.[55])

[a]Imported or induced cases.

phenthoate (0,0-dimethyl-S-a-ethoxycarbonyl) benzyl phosphorodithioate), and iodofenphos (dimethyl 2,5-dichloro-4-iodophenyl phosphorothioate), and the carbamate Landrin® (a mixture of 3,4,5-trimethylphenyl N-methyl carbamate and 2,3,5-trimethylphenyl N-methylcarbamate) may also prove to be useful biodegradable substitutes for DDT, although their evaluation is not yet complete. The widespread resistance to lindane among mosquitoes prevents its use as a DDT replacement. Several organophosphorus compounds are currently recommended as larvicides for culicines; diazinon and malathion control rat fleas, while carbaryl and some organophosphorus compounds are effective against blackfly larvae.

The need for more elaborate protection for control operators has to be faced when anticholinesterases are used as control agents. Malathion is one of the best organophosphorus compounds from the point of view of hazard and has been safely used in field operations, but other compounds are more hazardous and some have been known to produce acute, though usually nonlethal, symptoms in spraymen. The adoption of either malathion or propoxur as a replacement for DDT would also involve extensive changes in program organization and a higher cost of control. It has been estimated that the cost due to the insecticide alone would be five times greater for malathion and 20 times greater for propoxur, while operational factors, including more stringent safety precautions, would involve threefold and eightfold increases, respectively. The present cost of spraying operations with DDT is estimated at 60 million dollars and would rise to approximately 184 million dollars with malathion use and to 510 million dollars with propoxur as replacement. In any case, the present supply of alternatives would not meet the demand.

It is quite clear that DDT continues to be essential for the maintenance of current benefits and for their extension to populations currently unprotected against the ravages of malaria. Recent events in Ceylon show the results to be expected if prompt action is not taken to eliminate fresh foci of malaria transmission; in that country the remarkably improved situation mentioned earlier has reverted, and more than 2.5 million cases of malaria were reported during the two years 1968 and 1969. A recurrence of malaria transmission in India also shows that the temporary lack of DDT for "fire-fighting" operations can seriously threaten years of painstaking and costly progress. There is the distinct danger that the resurgence of malaria transmission in areas in which it is prevented by spraying (the attack phase of control programs) might result in its rapid spread into peripheral areas already freed from the disease. If this occurred on a large scale, it would result in fresh exposure of the vast populations of the formerly malarious areas to the ravages of endemic and epidemic malaria.

Here, it is appropriate to mention one important aspect of DDT usage in antimalaria programs that may not be widely appreciated. In the majority of treatments, DDT is applied as a residual spray inside buildings, and the amount of chemical escaping to the exterior is much less than that added to the environment during standard agricultural applications. House spraying may take place twice a year, while in the same period there may be multiple heavy agricultural applications to a particular crop (for example, cotton) which make a much larger contribution to environmental contamination than that due to vector control programs. Undoubtedly, the agricultural treatments also contribute to the spread of resistance among disease vectors, due to the insecticide selection pressure in their outdoor environment.

B. SYNTHESIS, PHYSICAL AND CHEMICAL PROPERTIES

1. General Synthetic Routes

In the following pages, aromatic substitution in DDT analogues is represented interchangeably by numerals or letters; thus, p,p'-DDT or 4,4'-DDT. The symbol Ph represents the variously substituted phenyl group, $C_6H_{6-n}X_n$, (for example, PhBr is equivalent to bromobenzene and 4Br-Ph to 4-bromophenyl-) and Ar is used for the often present 4-chlorophenyl moiety, in order to simplify the presentation of equations. The name "DDT" is frequently used to represent 4,4'-DDT unless distinction of the substituent positions is essential. DDT is synthesized by the Baeyer condensation of chlorobenzene with chloral (or an equivalent amount of chloral hydrate) in the presence of concentrated sulfuric acid, as indicated in Figure 3.[8] The first compound prepared by Zeidler appears to have been the

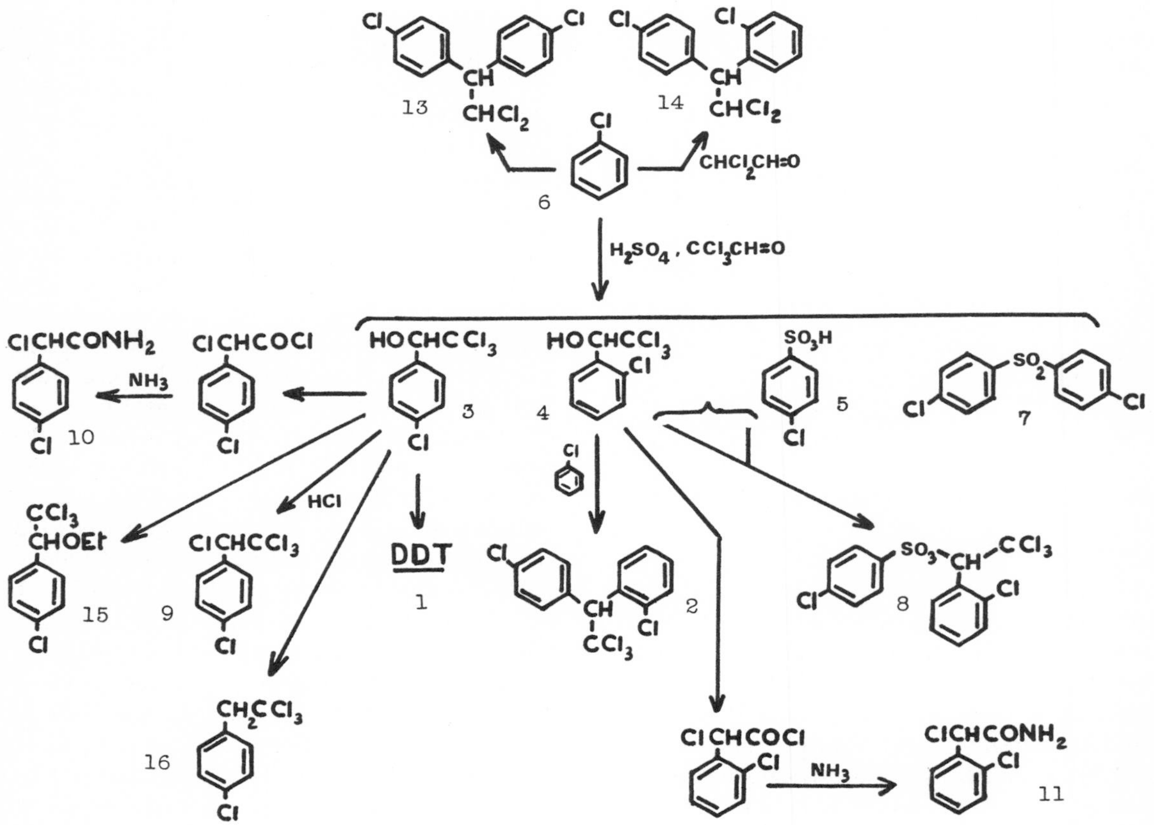

FIGURE 3. Products formed during the manufacture of DDT from chlorobenzene and chloral.

4,4'-dibromo-analogue of DDT, obtained from bromobenzene and chloral, and an excess of chloral was used in the reaction, giving conditions that are likely to lead to the production of an intermediate carbinol (the mono-condensation product with bromobenzene):

$$PhBr + O=CHCCl_3 \xrightarrow{H_2SO_4} 4Br\text{-}PhCH(OH)CCl_3$$

(4Br-Ph: 4-bromophenyl-)

$$4Br\text{-}PhCH(OH)CCl_3 + PhBr \xrightarrow{H_2SO_4}$$

$$(4Br\text{-}Ph)_2CHCCl_3 + H_2O$$

The product was further characterized by conversion to a dinitro-compound, and Zeidler mentions the preparation of DDT in the same manner from chlorobenzene and its similar conversion to a dinitro-compound.[8] In the normal synthesis, in which the formation of the intermediate carbinol in undesirable, 2 mol of chlorobenzene are mixed with 1 mol of chloral or chloral hydrate and the vigorously stirred mixture is treated with an excess of concentrated sulfuric acid; the reaction is exothermic and is cooled to maintain it at room temperature, since higher temperatures favor the formation of 4-chlorobenzenesulfonic acid (Figure 3, structure 5). The product frequently separates as a solid and is isolated by pouring the whole into water, when the DDT can be recovered and recrystallized from ethanol. Various syntheses and reaction conditions are discussed by Müller.[61]

In the years between the original Zeidler synthesis and the work by the Geigy group, other papers described the synthesis of various unsymmetrical DDT-analogues by the condensation of a particular intermediate carbinol with a differently substituted benzene. As an example, Chattaway and Muir slowly added benzene (1 mol to vigorously stirred chloral hydrate (3 mol) in concentrated sulfuric acid (350 ml), and after 3 hr poured the dark green emulsion onto ice.[10] Steam distillation then gave the intermediate low melting carbinol (1-phenyl-2,2,2-trichloro-ethanol) which was then condensed in the same manner with toluene, chlorobenzene, bromobenzene, and iodobenzene to give the corresponding unsymmetrical derivatives. The steam distillation which removes the carbinol from the reaction mixture leaves the bis-condensation product (1,1,1-trichloro-2,2-diphenylethane) behind as a solid which can be purified by recrystallization from ethanol. Various 4-halophenyl,4'-alkylphenyl-derivatives were also prepared as well as some symmetrical 4,4'-dialkyl-compounds. Such compounds have become of interest recently in relation to studies on biodegradable DDT analogues.[52] It is also possible to prepare intermediate alcohols for the synthesis of asymmetrical DDT analogues by the condensation of chloroform with an appropriate aromatic aldehyde:

$$ArCH=O + CHCl_3 \rightarrow ArCH(OH)CCl_3 \quad (Ar = 4\text{-chlorophenyl})$$

In another convenient preparation, anhydrous aluminum chloride is employed as the condensing agent.[38] To prepare methoxychlor, for example, anisole (0.1 mol) in chloroform (250 ml) is condensed with anhydrous chloral (0.05 mol) in the presence of anhydrous aluminum chloride (0.055 mol) at 4°C. After complete addition of the anhydrous aluminum chloride, stirring is continued for 30 min at 20°C, and then for 8 hr at ambient temperature before isolation of the product. To prepare the intermediate carbinol (1-p-methoxyphenyl-2,2,2-trichloroethanol), a mixture of chloral (0.05 mol) and anisole (0.05 mol) is added to anhydrous ether (100 ml) cooled at 5°C and containing anhydrous aluminum chloride (0.05 mol). After 20 hr at room temperature following the addition, the mixture is poured onto ice and HCl and the product isolated. The corresponding unsymmetrical 4-methoxy-4'-hydroxy-DDT analogue (which is a biological metabolite of methoxychlor) is prepared by reacting the carbinol (0.01 mol) with phenol (0.01 mol) in ethanol free chloroform at 4°C with addition of anhydrous AlCl₃ (0.01 mol). The mixture is stirred for 12 hr, diluted with water, the organic layer separated, and the chloroform and some unchanged phenol distilled off under reduced pressure to leave a residue of the desired compound.[38]

To prepare 2,4'-DDT (o,p'-DDT), an excess of chlorobenzene is condensed with 1-(2-chlorophenyl)-2,2,2-trichloroethanol in the presence of a mixture of 96% sulfuric acid and 25% oleum (3:1, v/v) at 60°C, and 3,4′-DDT (m,p′-DDT) is made by condensing 1-(3-chlorophenyl)-2,2,2-trichloroethanol (obtained in 70% yield from chloral and 3-chlorophenyl magnesium bromide) with a 50% excess of chlorobenzene using 100% sulfuric acid at room temperature for 6 hr:[62]

CCl₃CH=O + 3Cl–PhMgBr ⟶

3Cl–PhCH(OH)CCl₃

3Cl–PhCH(OH)CCl₃ + PhCl ——H₂SO₄——▶

Ar(3Cl–Ph)CHCCl₃

Close analogues of DDT such as the well-known 4,4′-difluoro-compound (DFDT) are also made by the Baeyer condensation, and these preparations serve to illustrate the general applicability of this reaction to the synthesis of DDT analogues. The reduction of DDT by its reduction with zinc and concentrated hydrochloric acid in boiling alcoholic solution gives 4,4′-DDD (Figure 2, structure 15; TDE) and also causes further reductive dechlorination resulting in compounds such as 1,1-bis(4-chlorophenyl)ethane, as well as molecular rearrangement to trans-4,4′-dichlorostilbene.[63] Other reducing agents such as acidic chromous chloride also produce 4,4′-DDD initially, followed by a range of further products.[64] A better method of preparation for 4,4′-DDD is the usual Baeyer condensation of dichloroacetaldehyde with chlorobenzene, or of 1-p-chlorophenyl-2,2-dichloroethanol with excess chlorobenzene under the conditions indicated above for 2,4′-DDT; in the latter case, the intermediate carbinol may be made by the Baeyer condensation or by the reaction of dichloroacetaldehyde with p-chlorophenyl magnesium bromide. For the preparation of 2,4′-DDD, 1-o-chlorophenyl-2,2-dichloroethanol, made by Grignard reaction, and an equimolar amount of chlorobenzene are condensed at 60°C for 1 hr in the presence of concentrated sulfuric acid. Other routes to the intermediate carbinols are available, as, for example, by the

Meerwein-Ponndorf reduction of appropriately halogenated acetophenones:

ArCOCHCl₂ ——reduction——▶ ArCH(OH)CHCl₂

Besides the DDT analogues referred to previously, Müller described the preparation of a number of moderately insecticidal propylene derivatives, which were prepared by the Friedel-Crafts reaction, or by a condensation of the Baeyer type, from pentachloropropene.[11] As an example of the method, aluminium chloride (5 g), added to a mixture of pentachloropropene (100 g) and benzene (500 g), cooled to maintain the temperature around 5°C, gave 1,1-bis(4-chlorophenyl)-2,3,3-trichloropropene-2:

2C₆H₆ + CHCl₂CCl=CCl₂ ——AlCl₃——▶ Ph₂CHCCl=CCl₂

The various 4,4′-dihalophenyl-derivatives were similarly made from the appropriate substituted benzenes. For the preparation of the 1,1-bis(4-methoxyphenyl)-2,3,3-trichloropropene-2, sulfuric acid monohydrate (35 ml) was added dropwise to an ice-salt cooled, stirred mixture of trichloroacrolein (11 g), anisole (22 g) and acetic acid (22 ml):

2(CH₃OPh) + OCHCCl=CCl₂ ——H₂SO₄——▶

(4–CH₃OPh)₂CHCCl=CCl₂

One of the most interesting aspects of the DDT investigations of the early 1950's was the work done on DDT-isosteres; that is, compounds related to DDT in which chlorine atoms are replaced by other groups of similar size so that the overall shape of the molecule remains the same. Notable among these investigations are the reports of Rogers et al.[65] and Skerrett and Woodcock,[66] which describe much synthetic work in this area. Attention centered upon the replacement of the trichloromethyl-group of DDT by tertiary butyl, and certain difficulties attend the preparation of such compounds. Three possible synthetic routes are

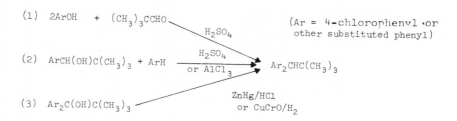

With the exception of (3), these syntheses proceed through the mono- or diaryl neopentyl ions $[ArCHC(CH_3)_3]^+$ or $[Ar_2CC(CH_3)_3]^+$, and molecular rearrangement results rather than formation of the desired product on the right. Thus, reaction (1) fails because of the tendency of pivalaldehyde to rearrange to methyl isopropyl ketone, although 1,1-bis(4-hydroxyphenyl)-2,2-dimethylpropane can be obtained from it by condensation with phenol in the presence of sulfuric acid. Attempts to prepare 1,1-bis(p-chlorophenyl)-2,2-dimethyl-propane from the corresponding alcohol and chloro-benzene via route (2), using either sulfuric acid or alu-minum chloride, were also unsuccessful. The copper chromite reduction appears to be the best method, but it cannot be used when the para-aromatic substituents are halogens or nitro-groups. Thus the method for the preparation of 1,1-bis(4-methoxy-phenyl)-2,2-dimethylpropane (dianisyl neopentane; DANP),[65] the earliest reported chlorine free isostere of DDT, involves first the condensation of ethyl pivalate with 4-anisyl mag-nesium bromide to give the intermediate tertiary alcohol:

$$(CH_3)_3COOC_2H_5 \xrightarrow{4-CH_3OPhMgBr}$$

$$(CH_3)_3C(OH)(PhOCH_3)_2$$

The alcohol was then reduced as in equation (3) above by high pressure hydrogenation (6000 lb/in.^{2}) at 250°C in the presence of the copper chromite catalyst. In a recently described synthesis of the corresponding 4,4'-dichloro-compound, the intermediate 4,4'-dichloro-alcohol is converted into the corresponding tertiary bromide, the bro-mine atom being subsequently removed by reduc-tion.[67] A very recent report describes the use of sodium in liquid ammonia for the reduction of intermediate alcohols from pivalaldehyde.[68]

A similar general route leading to intermediate tertiary alcohols which can then be reduced to the hydrocarbons has been used to produce other variants of the trichloromethyl group.[69] Thus,

ethyl trifluoroacetate condenses with an aryl-magnesium bromide to give a diaryl trifluoro-methyl carbinol from which the hydroxyl group may be removed by reduction, and various diaryl-methylcyclopropanes were made via the aryl-magnesium bromides and ethyl cyclopropane-carboxylate, followed by reduction of the inter-mediate alcohol. DDT analogues in which the aliphatic portion of the molecule is entirely replaced by the cyclopropane ring have assumed considerable interest recently. Thus, 1,1-bis(4-chlorophenyl)cyclopropane is prepared from the corresponding ethylene by its reaction with diazo-methane at 25°C to give 2,2-bis(4-chlorophenyl) pyrazoline, which decomposes into the cyclo-propane on heating.[67] Compounds containing halogenated cyclopropane rings are made by the addition of halocarbenes to the appropriate ethylenic derivative. In a preparation described by Wiles, 1,1-bis(4-chlorophenyl)ethylene in bis(2-methoxyethyl)ether is treated at 100 to 105°C in nitrogen with sodium trichloroacetate (as dichloro-carbene generator), until CO_2 evolution ceases, to give a moderate (27%) yield of the dichlorocyclo-propane derivative:[70]

$$Ar_2C=CH_2 + :CCl_2 \xrightarrow[CCl_3COONa]{N_2,105°} Ar_2C \overset{CCl_2}{\underset{CH_2}{\diagup\diagdown}}$$

Various other systems may be used to generate the necessary dichlorocarbene. Thus, Holan prepared various symmetrical and asymmetrical 4,4'-dialkoxy-analogues of DDT by refluxing the corresponding ethylenes in benzene with bromo-dichloromethylphenyl mercury as the dichloro-carbene source, or at 0° by treating them with chloroform and potassium t-butoxide, another useful dichlorocarbene generating system. The yield using the mercury derivative was almost quantitative after the reactants were refluxed for 10 hr.[71]

The synthesis of insecticidal nitro-aliphatic analogues of DDT was reported by Müller.[72] Aliphatic nitro-groups such as nitromethane contain reactive methylene groups which form sodio-derivatives in the presence of sodium methoxide; at low temperature (below 0°C) 1 mol of this nucleophilic reagent condenses with 1 mol of carbonyl compound such as 4-chlorobenzaldehyde to give the salt of a substituted ethanol without loss of water:

$$ArCH=O + CH_3NO_2 \xrightarrow{NaOCH_3}$$

$$\left[ArCH\!-\!CH_2NO_2 \atop \quad O^- \right] \rightarrow ArCHCH_2NO_2 \atop \quad\quad O^-Na^+$$

The sodium salt of 4-chlorophenyl nitroethanol separates as a white powder and can be filtered off, washed with ether and air-dried. The alcohol is liberated by acidification of an aqueous suspension of the salt with acetic acid and can be recovered by ether extraction; it undergoes the normal Baeyer condensation with chlorobenzene in the presence of sulfuric acid in the cold:

$$ArCH(OH)CH_2NO_2 + PhCl \xrightarrow{H_2SO_4} Ar_2CHCH_2NO_2$$

The original report also describes the preparation of 1,1-bis(p-tolyl)-2-nitroethane and 1,1-bis(3,4-xylyl)-2-nitroethane, as well as the asymmetrical derivative obtained by condensing benzaldehyde with nitromethane and the resulting alcohol with chlorobenzene. Products of this type may also be prepared by a reaction directly analogous to the Baeyer condensation:

$$NO_2CH_2CH(OC_2H_5)_2 + PhCH_3 \xrightarrow[-5°C]{H_2SO_4/oleum}$$

$$NO_2CH_2CH(PhCH_3)_2$$

Similar condensation products may be prepared from higher nitroalkanes;[73] the compounds Bulan (from nitropropane) and Prolan (from nitroethane) are the constituents of the commercial insecticidal mixture known as "Dilan", introduced in 1948 by the Commerical Solvents Corporation.[40]

Among the series of biodegradable DDT analogues described by Holan are compounds analogous to Prolan and Bulan, in which the 4,4'-chlorine atoms are replaced by ethoxy-groups, and also the 2-methyl-2-nitropropane analogue of Prolan.[74] These compounds were prepared by condensing 4-ethoxybenzaldehyde at ambient temperature with nitroalkane in dimethylsulfoxide containing a catalytic amount of 1,5-diazo-bicyclo[4.3.0] non-5-ene; the intermediate 1-(4-ethoxyphenyl)-2-nitroalkanol is then condensed with 4-ethoxyphenol as indicated previously. Another lightly chlorinated analogue of DDT which has found some practical application is 1,1-dichloro-2,2-bis(4-ethylphenyl)ethane (Perthane), a relative of DDD which, like the latter, may be prepared by the Baeyer condensation of dichloracetaldehyde, with the use of ethylbenzene instead of chlorobenzene as the second reactant.

Dehydrochlorination products of DDT and its relatives (that is, DDE and analogues) are fairly easily prepared by treating the parent compounds with a small molar excess of potassium hydroxide in 1% (w/v) ethanolic or aqueous ethanolic solution:

$$Ar_2CHCCl_3 \xrightarrow{OH^-} Ar_2C=CCl_2 + HCl \atop \quad\quad\quad\quad (DDE)$$

The speed of the E_2 elimination reaction depends on the nature of the nuclear substituents and those on the substituted methyl group, being accelerated by electron withdrawing groups and retarded by electron donating ones. Thus methoxychlor (Figure 2, structure 13; $R,R_2=OCH_3$) is dehydrochlorinated about 200-fold less rapidly[67] than DDT itself and a period of heating under reflux is required to complete the dehydrochlorination in such cases; with DDT the dehydrochlorination occurs at room temperature.

Some analogues of DDT are of interest because, although having little insecticidal activity, they act as inhibitors of DDT-dehydrochlorinase in vivo and hence as DDT synergists.[41] One or two of these are also useful acaricides and one (dicofol), is a DDT metabolite. A number of less closely related compounds used as acaricides at one time or another are of interest in view of their connection with DDT history mentioned earlier. The synthesis of these nonsynergist acaricides is outlined below for interest, but they are otherwise omitted from detailed discussion in this book. One of the best known acaricides which is also a synergist is 1,1-bis(4-chlorophenyl)ethanol (Chlorfenethol), also called 4,4'-dichloro-α-methyl benzohydrol or di-(4-chlorophenyl)methylcarbinol

(DMC), which is made by the Grignard reaction between 4,4'-dichlorobenzophenone and methyl magnesium bromide:[75]

$$Ar_2C=O + CH_3MgBr \longrightarrow Ar_2C(OH)CH_3$$
$$(DMC)$$

The corresponding trichloromethyl-compound (Figure 2, structure 19; R=CCl$_3$), dicofol (trade mark Kelthane® is an effective acaricide, as well as a DDT-synergist, and was reported to be formed by the high temperature chlorination of DMC. However, Bergmann and Kaluszyner found that this method gave the related acaricidal DDD-analogue (Figure 2, structure 19; R=CHCl$_2$), DDE, 4,4'-dichlorobenzophenone and oily material probably containing compounds having the >CClCHCl$_2$ moiety, there being only traces of dicofol present. They developed an indirect method of synthesis from DDE:[76]

$$Ar_2C=CCl_2 \xrightarrow{Cl_2} Ar_2CClCCl_3 \xrightarrow[\substack{boiling \\ acetic\ acid}]{CH_3CO_2Ag}$$

$$Ar_2C(OCOCH_3)CCl_3$$

$$Ar_2C(OCOCH_3)CCl_3 \xrightarrow[\substack{acetic\ acid}]{H_2SO_4} \underset{dicofol}{Ar_2C(OH)CCl_3}$$

Dicofol could not be prepared by direct hydrolysis of the tetrachloroethane intermediate, but the latter gave the acetate in 71% yield when heated under reflux for 90 min with a slight molar excess of silver acetate in glacial acetic acid; hydrolysis of the acetate proceeded in 70% yield in boiling 75% (v/v) aqueous acetic acid containing about 5% (v/v) of concentrated sulfuric acid. Other compounds, such as 1,1-diphenyl-2,2,2-trichloroethanol and 1,1-bis(4-fluorophenyl)-2,2,2-trichloroethanol were prepared from the corresponding DDE analogues by the same route. In this report, dicofol and its 4,4'-difluoro-analogue were claimed to be more effective DDT-synergists than the trifluoromethyl-analogue (Figure 2, structure 19; R=CF$_3$) of DMC. Various acaricidal esters of 4,4'-dichlorobenzilic acid, such as chlorobenzilate (Figure 2, structure 19; R=COOC$_2$H$_5$) and chloropropylate (Figure 2, structure 19; R=COOCH(CH$_3$)$_2$) are made by the esterification of this acid with the appropriate alcohol, and are the subject of a recent review.[14]

Acaricidal esters of derivatives of benzenesulfonic acid are prepared by the general reaction:[77]

$$ArONa + Ar'SO_2Cl \longrightarrow Ar'SO_2OAr + NaCl$$

(Ar and Ar' are benzene or various halogenated benzenes)

Thus, 4-chlorophenyl 4-chlorobenzene sulfonate (Ovex, trade mark Ovotran®) is prepared in this way. One mol of 4-chlorobenzenesulfonyl chloride and 1 mol of 4-chlorophenol are heated together with stirring at 60 to 70°C, while one mol of aqueous sodium hydroxide (0.7 N) is added during 30 min. The ester crystallizes from the reaction mixture during this time and the mixture is held at 60 to 80°C for another hour, after which the product may be recovered by filtration, washed with water, and recrystallized. These sulfonic acid esters are sparingly soluble in water so that the isolation process is easy. In a similar manner Genite (2,4-dichlorophenyl benzenesulfonate) is prepared from 2,4-dichlorophenol and benzenesulfonyl chloride, and fenson (4-chlorophenyl benzenesulfonate; trade mark Murvesco®) from 4-chlorophenol and benzenesulfonyl chloride in alkaline solution.[40]

Various sulfones have been used as acaricides. Sulfenone® (4-Chlorophenyl phenyl sulfone) is prepared by the reaction between chlorobenzene and benzenesulfonic acid at 250°, while tetradifon (2,4,4',5-tetrachlorodiphenylsulfone) is prepared by the Friedel-Crafts reaction between 2,4,5-trichlorobenzenesulfonyl chloride and chlorobenzene. Thus, Huisman et al.[78] prepared tetradifon and a number of similar compounds by reacting a stirred mixture of equimolar quantities of the appropriate aryl sulfonyl chloride and a substituted benzene at 80 to 90°C with slowly added aluminum chloride in slight excess. After the addition, heating was usually continued for some time at a higher temperature; in the tetradifon preparation, heating the mixture at 110°C for an additional 3 hr gave the product in 63% yield. The related sulfide, called tetrasul, in which the substituted benzene rings are linked by a sulfur atom, is also acaricidal, as are a number of similar sulfides.[40] Tetrasul is prepared by reaction between the sodium salt of 4-chlorothiophenol and 1,2,4,5-tetrachlorobenzene, which results in replacement of the 1-chlorine by the 4-chlorothiophenyl moiety:

1,2,4,5-tetrachlorobenzene + NaSAr $\longrightarrow$

2,4,5-trichloro-PhSAr
   (tetrasul)

Aralkyl sulfides are similarly prepared by reacting a benzyl halide with a suitable mercaptan:

ArCH$_2$Cl + NaSAr $\longrightarrow$ ArCH$_2$SAr

            chlorbenside (Chlorparacide[R])

Another group of acaricides are the aryloxy-alkanes, of which bis(4-chlorophenoxy)methane (Neotran), no longer of much commercial interest, is a good example:

2ArONa + ClCH$_2$Cl $\longrightarrow$ ArOCH$_2$OAr
                    Neotran[R]

A number of compounds of this type were prepared by heating (130 to 140°) a 2:1 molar ratio of the phenol and methylene chloride with a small excess over the theoretical amount of sodium hydroxide in methanol at 100 to 200 lb/in^2 for 12 hr. In a typical preparation, dichloromethane reacted with 2-allyl-4,6-dichlorophenol to give bis(2-allyl-4,6-dichlorophenoxy)methane in 64% yield.[79]

Radiolabelled DDT and its analogues have been valuable aids in metabolic studies on compounds of this series and labeling may be effected in various ways. When tritium labeling is acceptable for the experimental work in hand, aromatic ring hydrogen may be directly substituted by ^{3}H in the unlabeled molecules using the method of Hilton and O'Brien.[80] Although tritium labeling is not suitable for all purposes, the convenience of this method lies in its avoidance of total skeletal synthesis from labeled intermediates. Thus, recent metabolic studies by Metcalf[38,80a] employ methoxychlor and related compounds labeled by the above method to give the ^{3}H-compounds having specific activities 3.3 to 9.3 mCi/mM. DDT and analogues generally labeled with ^{14}C in the benzene rings are most widely employed for metabolic studies and are made by suitable adapta-

tion of the methods outlined previously. For example, ^{14}C-DDT, is made from chloral and a 10% excess of ^{14}C-chlorobenzene in 40 to 50% overall yield using chlorosulfonic acid as condensing agent, the chlorobenzene arising either by catalytic chlorination of ^{14}C-benzene or via the decomposition of benzenediazonium chloride (derived from ^{14}C-aniline) with cuprous chloride (Sandmeyer reaction).[81] Various ways have been described of incorporating ^{14}C into the aliphatic part of the molecule and routes will be evident from the reactions discussed earlier in this section. ^{36}Cl may be incorporated by using this isotope in the catalytic chlorination of benzene or in the preparation of chloral. Mention should also be made of the preparation by Hennessy and his colleagues of DDT analogues having the tertiary hydrogen atom replaced by deuterium; such compounds have proved valuable for investigations of structure-activity relationships.[82]

2. Technical Materials and Physical Properties

The commercial preparation of DDT is based upon the original Zeidler synthesis, and much effort has been directed toward improving the purity of the technical product, since only rather pure DDT can be used in the preparation of formulations such as wettable powders having a high DDT content. Since the melting point of pure 4,4'-DDT (108.5 to 109°C) is already rather low, the less pure technical products have melting characteristics such that they grind poorly in ball mills. The two problems that occur in manufacture concern (a) the preparation of high quality chloral, and (b) adjustment of the condensing agent and conditions of condensation of chlorobenzene with chloral so that the formation of by-products (Figure 3) is minimized.

Chloral may be produced by the direct chlorination of acetaldehyde, or more usually by the chlorination of alcohol, the last method being a complex sequence of reactions culminating in the formation of chloral hydrate and chloral alcoholate (a hemi-acetal):

$$C_2H_5OH + Cl_2 \longrightarrow [C_2H_5OCl] \longrightarrow CH_3CH=O \;(+ \;2HCl) \tag{1}$$

$$CH_3CH=O + C_2H_5OH + Cl_2 \longrightarrow CH_2ClCHClOC_2H_5 + H_2O \tag{2}$$

$$CH_2ClCHClOC_2H_5 + C_2H_5OH \longrightarrow CH_2ClCH(OC_2H_5)_2 + HCl \tag{3}$$

$$CH_2ClCH(OC_2H_5)_2 + Cl_2 \longrightarrow CHCl_2CH(OC_2H_5)_2 + HCl \tag{4}$$

$$CHCl_2CH(OC_2H_5)_2 + H_2O + Cl_2 \longrightarrow CCl_3CH(OH)OC_2H_5 + HCl + C_2H_5OH \qquad (5)$$
$$CCl_3CH(OH)(OC_2H_5) + H_2O \longrightarrow CCl_3CH(OH)_2 + C_2H_5OH \qquad \searrow (C_2H_5Cl) \qquad (6)$$

This process may be conducted batch-wise or continuously, and it is seen from the reaction sequence that some water should be present in the mixture to effect the hydrolysis of the di-acetal of equation (5) and the hemi-acetal of equation (6) to chloral alcoholate and chloral hydrate, respectively. The early stages of the chlorination are carried out at 50 to 60°C, and the later stages at 90°C. Treatment of the reaction mixture with concentrated sulfuric acid liberates chloral from the hydrate and alcoholate. If the temperature of chlorination is held below 35°C, dichloroacetaldehyde becomes the major product and is used to produce DDD by the usual Baeyer condensation. By-products include ethyl chloride, from the reaction between ethanol and hydrogen chloride. In the Brothman continuous process for DDT production described in 1944,[12] chloral alcoholate was formed by the chlorination of ethanol in the presence of ferric chloride as catalyst, the chloral being purified by fractionation following its liberation by acid treatment; excess chlorine was recovered as sodium hypochlorite. In this process, oleum was used as the condensing agent, but a number of other agents have been tried, such as chlorosulfonic acid, fluorosulfonic acid, hydrogen fluoride, and anhydrous aluminum chloride. The problem with concentrated sulfuric acid or oleum, which are most widely used, is that there is a sharp increase in the sulfonation of chlorobenzene at temperatures above 20°C. Although the product, p-chlorobenzene sulfonic acid (Figure 3, stucture 5) is a useful intermediate for the production of certain acaricides mentioned earlier and can be reconverted into chlorobenzene in the presence of sulfuric acid at 200 to 400°C, its excessive formation is clearly undesirable and can be minimized by the use of chlorosulfonic acid as condensing agent. In a preparation of DDT described by Sumerford,[83] the use of this acid gave a final yield of 65 to 70% of 4,4'-DDT melting at 107 to 108°C, but recrystallization was necessary for purification of the product, as it was in the method of Rueggeberg and Torrans,[84] who also used chlorosulfonic acid for the condensation and achieved a 77% yield of technical material operating at 20°C with a reaction time of 14 hr.

The reagent is not favored for commercial scale preparations because of its cost. The use of hydrogen fluoride as a condensing agent afforded a 58% yield of refined DDT on the basis of chloral and chloral alcoholate used, but a very large excess of the condensing agent was required to complete the reaction and there was some corrosion of the apparatus; copper and stainless steel vessels were used successfully but iron vessels were unsatisfactory.[12]

Mosher et al.[85] conducted a thorough investigation into the preparation of DDT in regard to optimal conditions for the Baeyer reaction using sulfuric acid as condensing agent. The conditions investigated included effect of reaction temperature, reaction time, concentration of the sulfuric acid employed, ratio of reactants, and purity of the chloral used. In a typical run using 1 mol of purified chloral and 2.2 mol of chlorobenzene with 104% sulfuric acid as condensing agent, reaction time 2.5 hr and temperature 20°C, 84% of DDT was obtained (based on chloral) having a setting point of 86.2°C; recrystallization from hexane gave a 57% recovery of 4,4'-DDT melting at 106.5 to 107.5°C. Although such conditions are not commercially applicable, best results were obtained when the reaction was effected at 15°C for 5 hr with 4 mol of chlorobenzene per mol of chloral in 98 to 99% sulfuric acid, 97 to 98% yields of crude DDT of setting point 90°C being obtained in this manner. Low boiling impurities in the chloral used did not affect the yield of crude material appreciably, but did lower the setting point of the product.

In the early days of manufacture in England, the DDT produced was required to contain not less than 70% of p,p'-DDT and have a setting point of not less than 89°C, with a hydrolyzable chlorine content of 9.5 to 11.0% The 1967 specification of the World Health Organization is very similar;[86] the minimum setting point is required to be 89°C and the product should have a total chlorine content of 49 to 51% by weight (wt) with hydrolyzable chlorine 9.5 to 11.5% wt and content of not less than 70% of p,p'-isomer of minimum melting point 104°C. The chloral hydrate content should not be greater than 0.005%, and the acidity (calculated as sulfuric

53

acid), insoluble material and water contents not greater than 0.3, 1.0, and 1.0% wt, respectively. For formulations of various types, a tolerance of ±5% is usually permitted for nominal contents of *p,p'*-DDT up to 50%, and of ±5% of (100-nominal) for contents of more than 50%.

In the U.S., three grades of DDT were recognized in the early days of production, namely, technical DDT containing not less than 70% of *p,p'*-DDT as above, partially refined DDT having a melting point of not less than 103°C, and the pure DDT of melting point 108.5 to 109°C used for research purposes. A detailed analysis of various samples of technical DDT was conducted by Haller,[62] who showed that up to 14 compounds may be present (Figure 3 and Table 7). These products may obviously arise from impurities in the chloral (as in the case of *p,p'*- and *o,p'*-DDD; 13 and 14, respectively (Figure 3), through the sulfonation of chlorobenzene, or through side reactions of the carbinols that are intermediates in the DDT synthesis. The formation of some of them, such as the amides, depends on the mode of neutralization of the sulfuric acid condensing agent. All of the products isolated and some of the oils obtained by chromatographic fractionation were tested for insecticidal activity, but none were as toxic as DDT, although some had insecticidal properties. Forrest and colleagues[63] also examined technical DDT and reached similar conclusions regarding its composition. Other workers identified *o,o'*-DDT, also indicated by Gunther to be present to the extent of 6% in technical DDT prepared by the Brothman process, along with compounds 9, 15, and 16 (Figure 3) derived from the intermediate carbinol (Figure 3, structure 3).[87]

DDT is a white, crystalline solid and its crystallographic properties were studied by Gooden,[88] who found that it formed more or less flattened needles or long prisms (from ethanol) with refractive indices (sodium lamp) α, β, γ; 1.618, 1.626, and 1.755, respectively. Other investigators determined from X-ray diffraction experiments that the substance crystallizes in the orthorhombic system with dimensions of a_0 = 19.25Å, b_0 = 10.04Å and c_0 = 7.73Å for the unit cell. This is in agreement with the results of Wild and Brandenberger,[89] who assumed four molecules per unit cell according to a density of 1.54 (1.62). DeLacy and Kennard have recently re-

ported on the crystal structures of *p,p'*-DDT, *o,p'*-DDT, and two DDT analogues.[713,713a]

Various melting points have been given for DDT. The setting point is a measure of the purity of the technical material, containing, as indicated above, up to 20% of the *o,p'*-isomer, less than 1% of the *o,o'*-isomer and the various other compounds shown in Table 7 and Figure 3. It is directly proportional to the content of p,p'-DDT and lies around 90°C for a good preparation. Müller has given a table relating the composition of various mixtures of *o,p'*- and *p,p'*-DDT to their observed setting point and melting point.[61] Melting points for the pure substance of 108 to 108.5°C, 109°C, 109.5 to 110°C and 110 to 110.5°C have been reported. The last mentioned melting point was obtained by diluting a thick paste, obtained from 100 g of a technical product of mp 60°C and 95% ethanol (50 ml), with water (300 ml), filtering the solid and then recrystallizing it several times alternatively from 95% ethanol and light petroleum (bp 30 to 60°C).[12] In this manner, the melting point was raised successively to corrected values of 109.5 to 110°C (60 g) and 110 to 110.5°C (46 g), the latter value being unchanged by further purification. The vapor pressure of DDT is reported to be 1.5 x 10^{-7} and 3 x 10^{-7} mm Hg at 20°C and 25°C, respectively, and the boiling point 185 to 187°C/0.5 mm.[61,90] Spencer and Cliath recently obtained closely similar values for the vapor pressure of DDT using a different technique, and for 2,4'-DDT, 4,4'-DDE, and 4,4'-DDD give values (at 30°C) of 55.3 x 10^{-7}, 64.9 x 10^{-7} and 10.2 x 10^{-7} mm Hg, respectively.[90a]

Fleck determined the volatilization rate of DDT by sieving a purified sample through a 325-mesh to give a coating of dust (63.36 mg) on a 5 x 10cm glass plate which was then kept at 45°C in air moving at 10 1/hr.[12] Weighing the plate at 4-day intervals revealed that 4.22 mg had evaporated in 37 days, corresponding to an average loss of 0.11 mg per day and a life of 18 months for the whole deposit. However, the loss was about three times higher than average during the first 4-day period and declined progressively thereafter, presumably due to a reduction in the surface area of the exposed dust particles. The material left on the plate at the end of the observation period was unchanged DDT, and it was concluded that the rate of loss of DDT from spray residues would not be of any practical significance. Another com-

TABLE 7

Composition of Technical DDT

Compound	Constitution (%) of sample of given setting point		
	88.6°C	91.2°C	91.4°C
p,p'-DDT(1)	70.5[b]	66.7[a] / 72.9[b]	72.7[a] / 76.7[b]
o,p'-DDT(2)	20.9	19.0	11.9[c]
o,o'-DDT	—	—	0.011
p,p'-DDD(13)	4.0	0.3	0.17[d]
o,p'-DDD(14)	—	—	0.044
2,2,2-Trichloro-1-o-chlorophenyl ethyl-p-chlorobenzenesulfonate(8)	1.85	0.4	0.57
1-p-Chlorophenyl-2,2,2-trichloroethanol(3)	—	0.2	—
1,1,1,2-tetrachloro-2-p-chlorophenylethane(9)	—	—	—
Bis(p-chlorophenyl)-sulfone(7)	0.1	0.6	0.034
o-chlorophenyl chloracetamide(11)	0.007	—	—
p-Chlorophenyl chloracetamide(10)	0.01	—	0.006
Sodium p-chlorobenzenesulfonate(5)	—	0.02	—
Ammonium p-chlorobenzenesulfonate(5)	—	—	0.005
Inorganic compounds	0.04	0.1	0.01
Losses or unidentified compounds	5.1	6.5	10.6

(Data adapted from West and Campbell.[12])

[a] By direct isolation from technical product.
[b] Recovered from technical product by recrystallization from aqueous ethanol saturated with DDT.
[c] Minimum value as some residues were not examined.
[d] Includes some of the corresponding olefin.

parison of the volatility of different DDT analogues at 130°F during a period of 18 days showed that of DDT, 4,4'-dichlorobenzophenone, 2,4'-DDT, and DDE, DDT was least volatile, its deposit losing 3.5% of the original weight in this time, whereas about 60% of the DDE volatilized. These differences may have an important influence on the loss of technical DDT and its degradation products from soils.[90a,90b]

DDT is highly insoluble in water, a property which makes the solubility difficult to define and the amount actually present in solution difficult to determine. According to measurements made in 1944, the "true water solubility of DDT" is 2×10^{-4} ppm.[24] Other values recorded range from 1 or 2×10^{-3} ppm up to 0.1 ppm, but the higher values undoubtedly represent colloidal solutions.[91] As indicated previously when discussing formulations, DDT is more soluble in olefins, cyclic and aromatic hydrocarbons than in paraffinic solvents; naphthalene derivatives such as tetralin and mixtures of various methylnaphthalenes being among the best solvents that are not halogenated hydrocarbons. A selection of representative solubilities obtained using DDT of melting point 107.5 to 108°C, either by the incremental addition of solvent to solid DDT or by comparison of the refractive indices of saturated solutions with those of solutions of known concentration in the same solvent, is given in Table 8, in which the most notable features are the superior solvent properties of aromatic and chlorinated aromatic solvents, and of most of the chlorinated alkanes or alkenes tested.[12] Besides being good solvents for DDT, the methylated naphthalenes can be used as extenders for pyrethrum and have insecticidal properties of their own (some are used as mosquito larvicides).

One of the problems early encountered with practical DDT solutions was the tendency of the toxicant to crystallize out in sub-zero temperatures. Thus, a 5-gal container of DDT solution containing 5 g of DDT/100 ml of kerosene would deposit several inches of crystalline DDT in winter.

An investigation of the effect of the addition of auxiliary solvents to prevent this phenomenon showed that solvents of equal dissolving power had differing efficiencies for this purpose. Thus, xylene and monomethylnaphthalene have about equal solvent power for DDT but monoethylnaphthalene and alkylated naphthalenes generally are far more effective than xylene in preventing the crystallization of DDT from kerosene solutions at −30°C, the effect lasting for several weeks with 10% or more of the former included in the solvent, in contrast to several days when xylene was added at the same rate. Table 8 shows that cyclohexanone is an excellent solvent for DDT, and its solubility in most solvents increases markedly with increase in temperature. Extensive references to solubility data are available in the books by West and Campbell and by Müller.[12,61]

A number of measurements have been made of the dipole moments of DDT and its analogues in relation to theories of toxic action; reported values for DDT are 0.93D (in heptane; o,p'-DDT gave 2.12D), 1.12D (CCl$_4$) and 1.07D (benzene).[61] Riemschneider[92] used a comparison of calculated and measured dipole moments of various analogues to show that with some types of aromatic substitution free rotation of the benzene rings is so restricted that the molecules exist in one preferred orientation of these rings that is not interchangeable with other possible configurations. The various measurements indicate that the distance between the 4 and 4'-chlorine atoms is 11.0 Å and the valence angle between the phenyl rings lies between 110° and 120° (123° according to X-ray diffraction experiments).[61] Some investigations indicated an inverse relationship between measured dipole moment and toxicity, but the value of such correlations is extremely limited.

One of the best known close analogues of DDT is 1,1,1-trichloro-2,2-bis(4-fluorophenyl)ethane (DFDT; Figure 2 structure 13; R$_1$, R$_2$ = F) developed in Germany during World War II under the name "Gix." The main constituent of the technical product, which is a colorless, viscous liquid, is the 4,4'-isomer of mp 45°C and bp 177°/9mm, but up to 10% of the 2,4'-isomer bp 135 to 136°C/9mm is also present. No. 2,2'-isomer could be isolated and with fluorobenzene the orientation appears to be predominantly para. The yield of 2,4'-isomer is greatest when the reaction is conducted for a short time at low temperatures. The 4,4'-isomer is very soluble in organic solvents; typical solubilities at 27°C in g/100 ml of solvent are cyclohexanone 850, o-dichlorobenzene 700, xylene 670, carbon tetrachloride 650, methylated naphthalene 460, dibutylphthalate 260, refined kerosene 140, and mineral oil 83. Although DFDT is more expensive to make than DDT, the solubility properties make it easier to formulate and the economies possible in the use of solvents probably contributed to its choice for development in Germany in World War II.[41,93]

In principle, methoxychlor (1,1,1-trichloro-2,2-bis(4-methoxyphenyl)ethane) can be made by the condensation of anisole with chloral in the presence of sulfuric acid, in a procedure similar to that used for DDT.[7] In this case, however, the use of 100% sulfuric acid or oleum leads to excessive formation of sulfonation products, and is not recommended; besides which, the reactivity of anisole is greater than that of chlorobenzene and the use of such strong acids is not necessary. In some of the earlier procedures, commercial concentrated sulfuric acid was added to a mixture of chloral and anisole, with excess of anisole present as diluent and a reaction temperature of 50°C. This led to a pasty product which had to be triturated with a solvent to induce solidification. The problem is to obtain a product which can be washed without the use of expensive solvents so that the final solid has a setting point reasonably close to the melting point (89°C) of the pure compound, which is considerably lower than that of DDT. In the procedure described by Schneller and Smith,[94] sulfuric acid was added to anisole and chloral with vigorous agitation and external cooling to maintain a reaction temperature of 25 to 30°C, the reaction time being less than 1 hr in these circumstances. Towards the end of the addition, when the reaction mass began to thicken, carbon tetrachloride was added to permit the continuation of stirring for an additional hour. The fluid reaction mixture could then be pumped into cold water for washing to remove acid and water soluble reaction products, and the carbon tetrachloride removed by steam distillation. Subsequent patents describe various other ways of effecting this condensation, as by the use of finely powdered absorbent clay impregnated with 96% sulfuric acid instead of the free acid, or using alkane-sulfonic acids as condensing agents.[95,96]

The technical product currently prepared is described by the manufacturers as a pale yellow solid, d^{25} 1.41, having a setting point of 77°C and

TABLE 8

Solubility of DDT in Various Solvents at 27 to 30°C

Solvent	Solubility of DDT	
	g/100 ml[a]	g/100 g[a]
Acetone	58	74
Acetophenone	67	65
Benzene	78	89
Carbon tetrachloride	45	28
Chlorobenzene	74	67
Cyclohexanol	10	11
Cyclohexanone	116	122
Diethyl ether	28[b]	39[b]
1,4-Dioxan	92	89
Ethyl alcohol	2[b]	2[b]
Ethylene dichloride	59	47
Mesityl oxide	65	76
Methyl isobutyl ketone	47	59
Methylene chloride	88[b]	66[b]
2-Nitropropane	40	40
Safrole	39	36
Tetrachloroethane	61	38
Tetrachloroethylene	38	23
Tetrahydronaphthalene	61	63
Tributyl phosphate	50	51
1,2,4-Trichlorobenzene	44	28
Trichloroethylene	64	44
o-Xylene	57	66
Kerosene (refined)	4	5
Gasolene	10	13
Velsicol AR-40 (mainly monomethylnaphthalenes)	48	51
Velsicol AR-50 (mainly mono- and dimethylnaphthalenes)	55	56
Velsicol AR-60 (mainly di- and trimethylnaphthalenes)	57	58
Velsicol NR-70 (mainly tetramethyl naphthalene)	52	50

(Data adapted from West and Campbell.[12])

[a]Of solvent.
[b]Approximate values only.

containing at least 88% of the 4,4'-isomer (mp 89°C), the remainder being mainly the 2,4'-isomer, as in the case of other technical DDT analogues.[40] The pure compound appears to exhibit dimorphism, since a form melting at 78°C gave the same ethylenic derivative, mp 108.5 to 109°C as the higher melting material.[94] The compounds were physically identical and the one of lower melting point exhibited the higher melting point when melted, allowed to solidify, then remelted. The vapor pressure (not given) is described as "low."

Like DDT, methoxychlor is not very soluble in aliphatic solvents such as kerosene, and the marked temperature coefficient of solubility in such solvents means that solid will be deposited when such solutions are frozen. Some typical solubilities (g/100 g of solvent at 22°C) are acetone (60), methylene chloride (60), dichlorobenzene (40), cyclohexanone (50), xylene (40), methylated naphthalenes (35), carbon tetrachloride (16), methanol (<5), and odorless kerosene (2). It is practically insoluble in water (about 0.1

ppm at 25°C).[39] Methoxychlor is stable to heat, oxidation, and moisture under normal conditions. Methoxychlor is also stable indefinitely when exposed indoors and relatively resistant to ultraviolet irradiation. Although it is less easily dehydrochlorinated by alcoholic alkalis than DDT, methoxychlor resembles this compound in being susceptible to dehydrochlorination catalyzed by heavy metals.[41] Methoxychlor can be formulated in the same way as DDT for practical use. It is normally marketed as a 25 or 50% wettable powder which may be diluted in the usual way with carriers for field use, or as the solid 90% technical Methoxychlor Oil Concentrate (90% MOC; equivalent to 79.2% of the active ingredient and 10.8% of other isomers and related materials) for conversion into emulsifiable concentrates either alone or with other pesticides such as malathion. Thus "90% MOC" may be dissolved in an aromatic solvent along with a suitable emulsifier to give a solution of 24% or 2 lb/gal of technical methoxychlor (equivalent to 21% of the pure compound). This solution, called Marlate® 2-MR, can be diluted with water for various applications; technical malathion may also be coformulated to give a concentrate containing 2 lb of each technical toxicant/gal.[39,40]

The commercial preparation of 1,1-dichloro-2,2-bis(4-chlorophenyl)ethane (DDD, Rhothane®; Figure 2, structure 15) from alcohol chlorinated at low temperature (that is, from the ethyl acetal of dichloroacetaldehyde) and chlorobenzene gives a technical product of setting point not less than 86°C consisting of the usual mixture of mainly 4,4'-isomer (mp 110 to 110.5°C and bp 185 to 193°C/1 mm), with 7 to 8% of the 2,4'-isomer (mp 76°C). Its physical properties such as solubility and friability resemble those of DDT, enabling it to be marketed (from 1945) as a 25% solution in an aromatic solvent (Rhothane S-215), as a 50% wettable powder (Rhothane WP-50), or as 5% and 10% dusts. Perthane, or 1,1-dichloro-2,2-bis(4-ethylphenyl)ethane, is another product of the Rohm and Haas Company analogous to DDD, which was introduced in 1950 under the code number Q-137. The technical material, obtained in a similar manner to DDD by the acid catalyzed condensation of chlorinated ethanol with ethylbenzene, is a waxy solid of melting point not less than 40°C, which consists of the pure compound, mp 60 to 61°C, admixed with position isomers, as with other DDT analogues.

Like the other compounds, it is virtually insoluble in water but soluble in most aromatic and chlorinated hydrocarbons. Formulations include a 75% solution in methylene chloride, a 4 lb/gal emulsifiable concentrate and a 50% wettable powder.[7,12,24,40,41]

The other compounds of interest in this series are "Prolan" and "Bulan": 1,1-bis(4-chlorophenyl)-2-nitropropane (mp 80 to 81°C) and 2-nitrobutane (mp 66 to 67°C), respectively, familiar as the commercial mixture "Dilan" introduced by the Commercial Solvents Corporation in 1948. The technical material Dilan is a brownish waxy semi-solid, density 1.28 which liquefies completely above 65°C and consists of 53.3% wt of Bulan, 26.7% wt of Prolan and 20% of related compounds, mainly the corresponding 2,4'-isomers. It is practically insoluble in water and sparingly so in aliphatic hydrocarbons, but soluble in more polar solvents and aromatic solvents. The electron withdrawing nitro-group renders it unstable to alkalis and oxidizing agents convert it into noninsecticidal compounds, so that although the spectrum of activity is somewhat similar to that of DDT, its persistence is much lower. Formulations consist of a liquid concentrate (80% Dilan, 20% xylene), used to prepare 25% emulsifiable concentrates, a 50% wettable powder and 1 to 2% dusts.[24,40,41]

The compound 1,1-bis(4-chlorophenyl)ethanol, common name chlorfenethol, also known as DMC, was introduced for use as an acaricide under the trade mark Dimite® in 1950, following the first description of its acaricidal properties in that year.[97] The product of the reaction between 4,4'-dichlorobenzophenone and methyl magnesium bromide melts at 69.5 to 70°C but may contain 2,4'- and 2,2'-isomers arising from corresponding impurities in the dichlorobenzophenone used. It loses water when heated, to give the corresponding ethylene derivative, and is decomposed by strong acids, but is alkali stable and compatible with most agricultural chemicals. Although virtually insoluble in water, it is readily soluble in the more polar organic solvents, and is formulated as a 25% emulsifiable concentrate.[40,75,97]

Technical dicofol is a viscous brown oil, d^{25} 1.45, containing about 80% of 1,1-bis(p-chlorophenyl)-2,2,2-trichloroethanol, mp 78.5 to 79.5°C. The material was introduced, in 1955, by Rohm and Haas as FW-293, trademark Kelthane®, its acaricidal properties being first described in detail by Barker and Maugham.[98] Like most DDT analogues it is highly insoluble in water but soluble in most of the regular aromatic and aliphatic solvents. Since alkaline hydrolysis converts it into the corresponding benzophenone and chloroform (the basis of certain analytical methods for the compound), it is incompatible with highly alkaline materials, but otherwise compatible with most agricultural chemicals. Various formulations have been described, such as 20% or 42% emulsifiable concentrates, 18.5% wettable powders and a 30% dust concentrate.[40]

3. Constitution and Chemistry

The chemistry of 4,4'-DDT and the other constituents of technical DDT (Table 7 and Figure

3) was examined in extensive programs of degradation and synthesis by Haller and by other investigators.[62,63] Dehydrochlorination of DDT or 2,4'-DDT with alcoholic potassium hydroxide gave the corresponding ethylenic derivatives (Figure 4), which were then oxidized to the corresponding chlorinated benzophenones, all six of the isomeric dichlorobenzophenones (2,2'; 2,3'; 2,4'; 3,3'; 3,4'; 4,4') with one chlorine atom in each benzene nucleus having been prepared by other synthetic routes. This established the location of the chlorine atoms in the 4,4'-positions of the benzene nuclei. In a similar manner 4,4'-DDD and 2,4'-DDD (Figure 3, structures 13 and 14) from technical DDT were dehydrochlorinated to their ethylenic derivatives which were oxidized to benzophenones identical

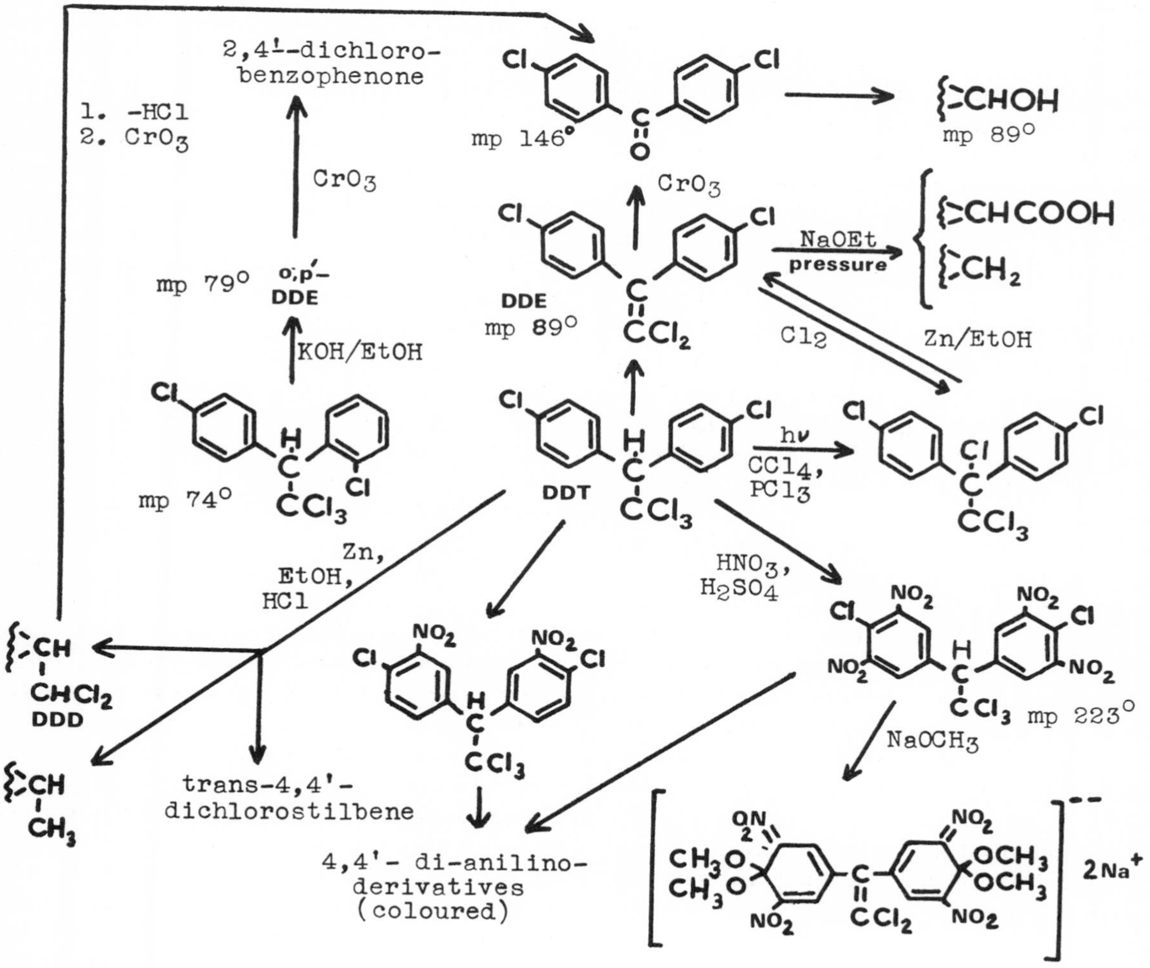

FIGURE 4. Some chemical reactions of DDT and relatives.

59

with those obtained from DDT and 2,4'-DDT, respectively. The ketone, 4,4'-dichloro-benzophenone (DBP) was reconverted into the ethylenic derivative from DDD (called DDMU) by the following sequence:

DDD and its 2,4'-analogue were also synthesized from 2,2-dichloro-1-p-chlorophenyl ethanol and 2,2-dichloro-1-o-chlorophenylethanol, respectively, via the Baeyer condensation with chlorobenzene. Another constituent,

$$(Ar)_2C=O \xrightarrow{CH_3MgI} Ar_2C(OH)CH_3 \xrightarrow[\text{in vacuo}]{\text{sublimation}} Ar_2C=CH_2 \xrightarrow[-HCl]{\dfrac{Cl_2}{CCl_4}} Ar_2CClCH_2Cl$$

$$Ar_2C=CHCl \ (DDMU)$$

2,2,2-trichloro-1-o-chlorophenylethyl-p-chlorobenzene-sulfonate (Figure 3, structure 8) was converted into 2-chlorobenzoic acid and 4-chlorobenzene sulfonic acid and the original compound was synthesized by condensing 2,2,2-trichloro-1-o-chlorophenylethanol with 4-chlorobenzenesulfonyl chloride. 4-Chlorophenyl chloracetamide (Figure 3, structure 10) gave 4-chlorobenzoic acid and ammonia on degradation, and was synthesized by forming the related chlorophenyl chloroacetyl chloride from phosphorus pentachloride and 4-chloromandelic acid; the acid chloride gave the required amide (Figure 3, structure 10) by ammonolysis. The ortho-amide (Figure 3, structure 11) was similarly prepared from o-chloromandelic acid. 1,1,1,2-Tetrachloro-2-p-chlorophenylethane was prepared by treating 2,2,2-trichloro-1-p-chloro-phenylethanol (Figure 3, structure 3) with phosphorus pentachloride and gave a nitro-derivative identical with that formed from the compound obtained from technical DDT. This compound was not found in the samples of technical DDT listed in Table 7 but is present in others.

The replacement of the benzylic hydrogen atom of DDT by chlorine, either via reversible addition of chlorine to DDE or directly by the light catalyzed chlorination of DDT with phosphorus trichloride (Figure 4) shows the relationship between these three compounds, and the relationship between DDT and DDD is shown by the analogous syntheses and by formation of DDD when DDT is reduced with zinc dust. The nature of the aliphatic moiety is shown by the conversion of DDE into the corresponding benzophenone on the one hand and into the corresponding carboxylic acid, together with some of the decarboxylated derivative bis(4-chlorophenyl) methane, on the other (Figure 4). The stability of DDT towards oxidizing agents

is shown by its recovery unchanged after 20-hr heating with excess potassium permanganate in 1N alkali or 8 hr in chromium trioxide/acetic acid mixture. DDE is unaffected by boiling sodium chromate in acetic acid, but when boiled for 1 hr with 3 mol of chromium trioxide in acetic acid, DBP is formed.

Of special interest is the formation of nitro-derivatives from DDT. Mild nitration gives the 3,3'-dinitro-compound (mp 143°C),[99] while more energetic treatment with a 1:1 mixture (v/v) of 30% fuming nitric acid and concentrated sulfuric acid gives the 3,5,3',5'-tetranitro-derivative (mp 223°C) which is useful from an analytical point of view.[100] In this compound, the 4- and 4'-chlorine atoms flanked by nitro-groups are labile; they can be replaced, for example, by aniline to give an orange-red 4,4'-bis-anilino-derivative (mp 231°C). The 3,3'-dinitro compound and its corresponding ethylenic derivative (DDE analogue) also react with aniline with replacement of the labile chlorine atoms, and with pyridine to give corresponding water soluble 4,4'-bis(pyridinium)-salts. 2,4'-DDT and 4,4'-DDD give analogous di-and tetranitro-compounds.[61]

One of the most important chemical changes undergone by DDT is the loss of one molecule of hydrogen chloride by various means to give the corresponding ethylenic derivative known as DDE. Since this conversion results in the loss of toxicity, its occurrence under various conditions has been thoroughly explored. Thermal decomposition with elimination of hydrogen chloride occurs under various conditions, as, for example, when attempts are made to distill DDT at atmospheric pressure. According to one source, little decomposition of DDT occurred when it was heated at 150°C for 24 hr. Other investigators heated it slowly in a tube immersed in a liquid bath held successively at 115°C, 120°C, 125-130°C, 140°C and 145°C for about 1 min. A rapid current of air was passed through the melt and the vapors were examined for hydrogen chloride content; its evolution was first detected at 140 to 145°C. In a similar experiment, Balaban and Sutcliffe determined the decomposition of pure DDT, mp 108.5 to 109.5°C (corrected), to begin at 195°C,[101] and a later thorough investigation by Scholefield et al.[102] indicated a decomposition temperature of 200°C, in good agreement with Balaban and Sutcliffe. The liberation of one molecule of hydrogen chloride is completed in about 160 min at 236°C and in 30 to 40 min at 260°C.[61]

A number of compounds catalyze the liberation of hydrogen chloride from DDT, and Balaban and Sutcliffe showed that the addition of 0.1% of ferric chloride in their thermal decomposition experiments markedly lowered its decomposition temperature. The reaction is catalyzed by anhydrous ferric oxide, ferric chloride or aluminum chloride, iron and a number of other substances such as Fuller's earth and various minerals.[103] With the last materials, the catalytic effect may have been due to the presence of traces of iron compounds. In aqueous suspension catalyzed decomposition will take place at 100 to 110°C, and at even lower temperatures in nitrobenzene, naphthalene, chlorobenzene, or 1-chloronaphthalene. On the other hand, decomposition is inhibited in solvents such as n-octadecyl alcohol, stearic acid, and petroleum fractions. Clearly, this phenomenon questions the compatibility of DDT in admixture or in contact with other compounds. No hydrogen chloride was liberated in the presence of sodium fluosilicate, sodium fluoride, cryolite, calcium arsenite, Paris green, lead arsenate, pyrethrum, or rotenone at 115 to 120°C; nicotine caused some loss of hydrogen chloride which was probably due to the base catalyzed reaction rather than to catalytic decomposition of the type under discussion. Fungicides such as 2,3-dichloro-1,4-naphthaquinone and commercial lime-sulfur are not catalytic, but there is some decomposition in the presence of Bordeaux mixture, sulfur, or dolomite limestone used as a fertilizer. This fertilizer liberated nearly the theoretical amount of hydrogen chloride in 1 hr, but a number of others such as ammonium sulfate, ammoniated superphosphate, manure salts, dicalcium phosphate, and steamed bonemeal, showed no catalytic action.[12]

Ferric chloride catalyzed decomposition of DDT proceeds to completion in solvents such as chlorobenzene, o-dichlorobenzene, nitrobenzene, naphthalene, and 1-chloronaphthalene, but is retarded in many other solvents. Various metals appear to be particularly good catalysts of DDT composition. Thus, aluminum (but apparently not aluminum powder), chromium, stainless steel, iron powder, and iron filings liberate 1 mol of HCl/mol of DDT at 115 to 120°C. This finding is rather critical in connection with the formulation of DDT aerosols in metal bombs; an investigation of the stability of DDT when heated at 60°C with strips of various metals, especially under pressure, showed that the amount of decomposition occurring depended strongly on the type of solution used in the formulation.[61] The iron catalyzed thermal decomposition of DDT may be inhibited by the addition of compounds which combine with ferric iron, such as salicyl aminoguanidine and picolinic acid.[12] Thus, a sample of technical DDT of setting point 90°C containing 1.5% of ferric nitrate and 2% of either of these inhibitors, when heated to 110 to 120°C for 24 hr, decomposed to the extent of 1% when picolinic acid was added and 2.5% when the inhibitor was salicylaminoguanidine. With ferric nitrate and no inhibitor, complete decomposition occurred. Stabilization is also achieved by the addition of alkali carbonates or alkaline earth carbonates, phosphorus pentoxide, or neutral esters of phosphoric acid. The use of esters such as triethyl or tributyl phosphate is advantageous since they form homogeneous solutions with anhydrous DDT and in other nonaqueous organic systems. Thus, 0.25% of either of these esters in technical DDT strongly inhibits its decomposition catalyzed by a 0.1% content of ferric chloride.[104]

There are strong indications that technical DDT is more stable than the purified material, due to the presence of impurities which inhibit the catalytic decomposition. For example, technical DDT (setting point 88°C) did not decompose in 24 hr at 115°C whereas this treatment caused evolution of hydrogen chloride from a sample that had been purified to a melting point of 105 to 107°C. It seems that the removal of decomposition inhibitory substances by recrystallization, if conducted without parallel steps to remove iron, can lead to an apparent instability of the purified DDT. Ferric chloride is able to effect changes other than dehydrochlorination in various DDT analogues. When the DDT-derivative in which the benzylic hydrogen is replaced by chlorine is heated to 210°C with a trace of ferric chloride, molecular rearrangement occurs to give 1,1,2,2-tetrachloro-1,2-bis(4-chlorophenyl)-ethane, and DDT is said to give the same compound under similar conditions:[61]

$$Ar_2CClCCl_3 \xrightarrow[FeCl_3]{210°C} ArCCl_2CCl_2Ar \ (mp \ 193°C)$$

Other reports indicate that when DDD is heated with 6% of anhydrous ferric chloride at its melting point, the molecule rearranges to give a derivative of 1,2-dichloro-ethane, whereas at 300°, two molecules of hydrogen chloride are lost:[61]

$$Ar_2CHCHCl_2 \xrightarrow[FeCl_3]{112°C} ArCHClCHClAr$$
DDD
$$(mp \ 227 \ to \ 228°C)$$

$$\xrightarrow[FeCl_3]{300°C} ArCH_2CH_2Ar$$

DDT reacts with 1 mol of anhydrous aluminum chloride in excess of dry benzene to give 10% of symmetrical tetraphenylethane, together with a little DDE:[105]

$$Ar_2CHCCl_3 \xrightarrow[C_6H_6]{AlCl_3} Ph_2CHCHPh_2$$

Like DDT, methoxychlor is also subject to catalytic dehydrochlorination in the presence of heavy metals.

The other well-known dehydrochlorination reaction of DDT is that effected by alkalis. DDT is said to withstand boiling with 20% aqueous sodium hydroxide solution for 24 hr, but dehydrochlorination occurs very rapidly in alkaline alcoholic solution. If an alcoholic solution of DDT is treated with 0.1 to 0.2 M alcoholic potassium hydroxide at room temperature, the theoretical amount of hydrochloric acid is liberated in 15 to 20 min and can be titrated with silver nitrate solution. As in the case of the catalytic decomposition this reaction has an important influence on the stability of DDT under conditions of practical use; DDT preparations applied, for example, to lime washed walls which are quite common in houses in the tropics have a much shorter effective life than when they are applied to neutral surfaces. The exposure of DDT to heat and sunlight under these alkaline conditions may be expected to destroy it rather quickly.

When DDT is boiled with an alcoholic solution of an alkali hydroxide, more than the theoretical amount of hydrogen chloride is liberated, and, in addition to DDE, bis(4-chlorophenyl)acetic acid can be detected in the reaction mixture due to the liberation of three chlorine atoms per molecule of DDT. This reaction can be envisaged as a direct hydrolysis of the -CCl₃ moiety, or as a further step in the hydrolysis of the major product, DDE:

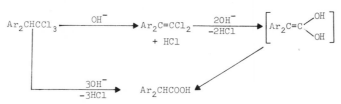

When this reaction is used as an analytical method for the estimation of DDT, it is advisable to conduct it at room temperature to avoid any elimination of HCl beyond the DDE stage.[12] It may be that this additional hydrolytic reaction, involving the liberation of two more molecules of HCl, contributes to the corrosive effects of technical DDT on metals under storage conditions when moisture is present.

The dehydrochlorination is an elimination of the E_2-type; its rate depends on the availability of electrons at the benzylic carbon atom, and is therefore influenced by the aromatic substituents,

as well as those on the terminal aliphatic carbon atom. Cristol[106] examined the dehydrochlorination rates of eight DDT analogues and demonstrated a linear relationship between the summation of Hammett sigma values for the aromatic substituents and the logarithm of the dehydrochlorination rate constant, which is in accordance with the Hammett equation:

$$\log \frac{k}{k_o} = \sigma\rho$$

in which k is the rate constant for the dehydrochlorination of a compound with aromatic sub-

TABLE 9

Chemical Dehydrochlorination Rates of Some DDT Analogues and Resistance Ratios Measured for Houseflies

$$R_1\text{—}\underset{R_2}{\overset{R_1}{\bigcirc}}\text{—}\underset{\underset{R_4}{\overset{|}{\underset{C}{C}}}\text{—}R_5}{\overset{H}{\underset{|}{C}}}\text{—}\overset{R_6}{\bigcirc}$$

R_1	R_2	R_3	R_4	R_5	R_6	Dehydrochlorination rate ($K \times 10^5$; 1/min/mol; 30°C)	Σ^a ($R_1 + R_6$)	Resistance ratio[b]
F	H	Cl	Cl	Cl	F	1.03×10^3	+0.124	24
Cl	H	Cl	Cl	Cl	Cl(DDT)	6.1×10^3	+0.454	276
Br	H	Cl	Cl	Cl	Br	8.9×10^3	+0.464	515
I	H	Cl	Cl	Cl	I	1.04×10^4	+0.552	–
NO_2	H	Cl	Cl	Cl	NO_2	6.9×10^6	+1.556	–
H	H	Cl	Cl	Cl	H	1.22×10^2	0.0	–
CH_3	H	Cl	Cl	Cl	CH_3	36.2	−0.34	1.65
C_2H_5	H	Cl	Cl	Cl	C_2H_5	27.0	−0.302	4.0
SCH_3	H	Cl	Cl	Cl	SCH_3	6.1×10^2	0.0	–
OCH_3	H	Cl	Cl	Cl	OCH_3	29.5	−0.536	5.0
Cl	Cl	Cl	Cl	Cl	Cl(o-Cl-DDT)	1.07×10^3	–	–
H	Cl	Cl	Cl	Cl	Cl(o,p'-DDT)	1.10×10^2	–	–
Cl	H	Cl	Cl	H	Cl(DDD)	2.98×10^3	–	460
C_2H_5	H	Cl	Cl	H	C_2H_5	16.7	–	–

Data adapted from Metcalf and Fukuto.[67]

[a] Summation of Hammett's sigma values for the para-substituents.
[b] Ratio of LD50 for the DDT-resistant Bellflower strain to LD50 for the susceptible NAIDM strain.

stituents having a summation of Hammett sigma values equal to σ, k_0 is the rate constant for the unsubstituted parent and ρ is a measure of the susceptibility of the reaction to the influence of polar effects. A similar result is demonstrated for a further 16 compounds by Metcalf and Fukuto,[67] who obtained a value for ρ of 2.71, in good agreement with the value obtained by Cristol. Electron-withdrawing substituents in the 4- and 4'-positions of DDT increase the rate of dehydrochlorination of DDT and its analogues that can be dehydrochlorinated, while electron-donating groups retard it (Table 9).

A detailed examination of the mechanism of this reaction suggests that although the mechanism is E_2, the removal of the benzylic proton by the base in the transition state occurs well in advance of the heterolysis of the C-Cl bond.[107] Nuclear magnetic resonance studies (NMR) show that the values (δ) for the chemical shifts of the benzylic proton are linearly related to Hammett values for the 4,4'-substituents, the shift increasing as the electron-withdrawing power of these substituents increases. As might be expected, substituents on the aliphatic group adjacent to the benzylic carbon also influence the dehydrochlorination rate. In this case the effect is found to be related to the values of Taft's polar constants (σ^*) for aliphatic substituents. Electron-withdrawing groups again increase the rate of dehydrochlorination and also the δ values for the chemical shifts of the benzylic proton, which are linearly related to the values of σ^*.[67]

With the symmetrical 4,4'-dihalogen substituted compounds, the dehydrochlorination rate decreases in the order I, Br, Cl, F, while the rate for the 4,4'-dinitro-analogue of DDT is about 600 times greater than that for the most reactive halogenated derivative, the 4,4'-diodo-compound. The dehydrochlorination rate is considerably reduced (50 times compared with DDT) in the DDT-analogue lacking any aromatic substituents, and 4,4'-dialkyl or dialkoxy-substituents greatly reduce the rate, that for methoxychlor being 200 times lower than is found with DDT.[67] The dehydrochlorination rate of 2,4'-DDT is 50 to 60 times lower than that of DDT, and, since the other position isomers are also dehydrochlorinated more slowly, the alkaline dehydrochlorination of a technical DDT mixture at room temperature favors selective decomposition of the 4,4'-isomers; Cristol claimed that by conducting the dehydrochlorination at 20.1°C with sodium hydroxide in 92.6% ethanol, conditions could be arranged so that the dehydrochlorination rate of 2,4'-DDT was nearly 70-fold lower.[108]

The effect of alteration of the trichloromethyl group on the dehydrochlorination rate is seen in the change from DDT to DDD; replacement of one chlorine atom by hydrogen halves the rate of dehydrochlorination, since the electron-withdrawing power of the aliphatic moiety is reduced. If the 4,4'-chlorines are further replaced by alkyl-groups in the DDD-structure, as in Perthane (the 4,4'-diethyl-analogue), the dehydro-chlorination rate falls about 180 times, this rate being similar for the symmetrical compounds having methyl- and isopropyl- 4,4'-substituents. The reduction in dehydrochlorination rate which occurs when the 4,4'-chlorine atoms of DDT are replaced by alkyl, or especially alkoxy-groups, is of significance in relation to insect resistance when this is caused by the enzymatic dehydrochlorina-tion of DDT. It is found that whereas DDT becomes ineffective, compounds having 4,4'-alkoxy-substituents frequently retain considerable activity toward these resistant insects.[52,67]

DDT undergoes some interesting reactions in the presence of reducing agents.[61] Catalytic hydrogenation at 45 to 50°C in alcoholic solution with a palladium-calcium carbonate catalyst gives a 36% yield of a double compound formed by the reductive elimination of two chlorine atoms be-tween two DDT molecules:

$$Ar_2CH\underset{\underset{Cl}{|}}{\overset{\overset{Cl}{|}}{C}}Cl \quad + \quad Cl\underset{\underset{Cl}{|}}{\overset{\overset{Cl}{|}}{C}}CHAr_2 \xrightarrow[-2HCl]{H_2;Pd/CaCO_3}$$

$Ar_2CHCCl_2CCl_2CHAr_2$ (mp 271.5°C)

The same compound is formed, along with DDD and DDE, by the reaction of DDT with methyl-magnesium iodide. Continual hydrogenation under the same conditions results in a further reductive dechlorination reaction:

$$Ar_2CHCCl_2CCl_2CHAr_2 \xrightarrow[-2HCl]{H_2;Pd/CaCO_3}$$

$Ar_2CHCCl=CClCHAr_2$ (mp 229°C)

Both types of double molecule evidently undergo further elimination of chlorine atoms, either under the conditions of catalytic reduction or with zinc

dust and acetic acid to give the corresponding allenic or acetylenic derivatives:

$Ar_2C=C=C=C\ Ar_2$ $Ar_2CHC\equiv CCHAr_2$

mp 288°C   mp 174°C

The acetylenic compound is also said to be formed in modest yield (13%) by direct electrochemical reduction of DDT in boiling ethanol acidified with HCl, some of the chlorinated ethylene derivative, mp 229°C, being formed at the same time.

The reduction of DDT with zinc dust and hydrochloric or acetic acid affords *trans*-4,4'-dichlorostilbene and compounds, such as DDD, derived by progressive reductive dechlorination of the trichloromethyl-group (Figure 4).[63] No forma-tion of unsaturated compounds has been indi-cated. Cochrane has investigated the action of aqueous chromous chloride solution (pH 3.9) on DDT and some of its relatives, mainly in connec-tion with the development of methods for the identification of residues in agricultural products.[64] Chromous chloride is a powerful reducing agent that is capable of effecting the protonolysis, hydrogenation, and deoxygenation of various structural types. In a typical experi-ment, DDT (2 g) in acetone (250 ml) treated with chromous chloride solution (125 ml) in nitrogen at 55-60°C for about 15 hr gave a 45 to 55% yield of *trans*-4,4'-dichlorostilbene (DCS, mp 177-8°C), the remainder being a mixture of 4,4'-dichloro-benzophenone, 1,1-bis(4-chlorophenyl)-2-chloroethylene (DDMU), and 1,1-bis(4-chlorophenyl)-ethylene (DDNU), to-gether with unchanged DDT, DDD and 2,4'-DDT (the last three compounds totaling about 5% of the reaction mixture). DDE is apparently not formed in this reaction; it is not itself changed by a 24-hr treatment with chromous chloride under these conditions. DDD appears to be the first product of the reaction, and then slowly re-arranges to DCS. The formation of DDMU may result from the dehydrochlorination of DDD, while DDNU might arise either by reductive dechlorination of DDMU or by dehydrochlorina-tion of 1,1-bis(4-chlorophenyl)-2-chloroethane (DDMS), a further reduction product of DDD, which was not detected, however.

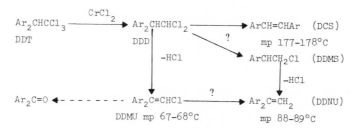

The mechanisms leading to these products have not been investigated in detail. 4,4′-dichlorobenzophenone might arise by a hydrolytic replacement of the single chlorine atom in DDMU; the resulting compound is the enol of an aldehyde which is known to convert rather readily into the benzophenone. Thus, attempts to prepare 2,2-bis(*p*-chlorophenyl)acetaldehyde by acidic hydrolysis of the 1,3-dioxolane derivative from DDD always give a mixture of the aldehyde and the benzophenone, and the aldehyde deformylates readily upon heating or under acidic conditions:[109]

$$Ar_2CHCHCl_2 \xrightarrow[(CH_2OH)_2]{NaOH} Ar_2CHCH \overset{O}{\underset{O}{<}}$$

$$\xrightarrow[H_2O/HCl]{dioxan} Ar_2CHCH{=}O \; + \; Ar_2C{=}O$$

$$(Ar = 4\text{-chlorophenyl-})$$

The photochemistry of DDT is of considerable interest, both in connection with its influence on the efficiency of DDT deposits under conditions of practical use and because it is necessary to determine the nature of the residues left on crops following DDT applications. Of equal or greater interest is the fate of DDE, which is considerably less toxic than DDT to many insects and has long been regarded as a detoxication product. Whether this is entirely true for all other living organisms is the subject of much speculation at present, and the question is of some importance in view of the persistence of DDE in living organisms and in the environment generally.

Both DDT and DDE are photooxidized in methanol to give a large number of products. Free radicals generated by the process may abstract hydrogen from the solvent, react with oxygen in solution, or remove hydrogen from the parent molecules. Short-lived intermediates decompose to give many other compounds. When DDT is irradiated in methanol with light of wavelength 260 nm in a nitrogen atmosphere, chlorine is progressively removed from the trichloromethyl group as the radicals produced from this group abstract hydrogen from the solvent to form DDD and DDMS:[110]

$$Ar_2CHCCl_3 \xrightarrow{h\nu} Cl^{\bullet} + Ar_2CHC^{\bullet}Cl_2 \quad (Ar = 4\text{-chorophenyl}) \quad (1)$$

$$Ar_2CHC^{\bullet}Cl_2 + CH_3OH \xrightarrow{h\nu} {}^{\bullet}CH_2OH + Ar_2CHCHCl_2 \;(DDD) \quad (2)$$

$$Ar_2CHCHCl_2 \xrightarrow{h\nu} Cl^{\bullet} + Ar_2CH\overset{\bullet}{C}HCl \quad (3)$$

$$Ar_2CH\overset{\bullet}{C}HCl + CH_3OH \xrightarrow{h\nu} Ar_2CHCH_2Cl + {}^{\bullet}CH_2OH \quad (4)$$

$$(DDMS)$$

At shorter wavelengths, the aromatic rings are dechlorinated, and extensive molecular disruption occurs to give a variety of products. When oxygen is present, products arise from its reaction with radical intermediates. Thus, 4,4′-dichlorobenzophenone is produced from both DDT and DDE and the following sequence is suggested for its formation from DDT:

$$Ar_2CHCCl_3 + Cl^{\bullet} \xrightarrow{h\nu} Ar_2\overset{\bullet}{C}CCl_3 + HCl \quad (5)$$

$$Ar_2\overset{\bullet}{C}CCl_3 + O_2 \longrightarrow Ar_2C\overset{\bullet}{O}_2CCl_3 \quad (6)$$

$$2Ar_2C\dot{O}_2CCl_3 \longrightarrow 2Ar_2C\dot{O}CCl_3 + O_2 \qquad (7)$$

$$Ar_2C\dot{O}CCl_3 \longrightarrow Ar_2CO + \dot{C}Cl_3 \qquad (8)$$

A reaction sequence leading from DDT to the formation of 2,2—bis(4-chlorophenyl)-acetyl chloride is also postulated to account for the eventual formation of methyl-2,2-bis(4-chlorophenyl)acetate;

$$Ar_2CHCCl_3 \xrightarrow{h\nu} Ar_2CH\dot{C}Cl_2 + Cl\cdot$$

$$Ar_2CH\dot{C}Cl_2 + O_2 \longrightarrow Ar_2CHCCl_2(\dot{O}_2)$$

$$2Ar_2CHCCl_2(\dot{O}_2) \longrightarrow 2Ar_2CHCCl_2\dot{O} + \dot{O}_2$$

$$Ar_2CHCCl_2\dot{O} \longrightarrow Ar_2CHCOCl + Cl\cdot$$

$$Ar_2CHCOCl + CH_3OH \longrightarrow Ar_2CHCOOCH_3 + HCl$$

The irradiation of DDE in methanolic solution in nitrogen results in reductive removal of the olefinic chlorine atoms. When oxygen is present, photocyclization also occurs with wavelengths above 260 nm to give various derivatives of fluorene and fluorenone, as indicated in Figure 5. Both DDE and 4,4'-dichlorobenzophenone are obviously well suited structurally for such cyclization reactions. The reaction with solvent of a radical such as that postulated in equation (7) would account for the formation of chlorobenzoic acid, methylchlorobenzoate, various chlorophenols and other fragmentation products, while the formation of compounds with additional nuclear chlorine substituents shows that the fragments produced can react with free chlorine radicals. DDT dissolved in γ-valerolactone and irradiated for 4 hr in an open dish in air gives a mixture of unchanged DDT and 4,4'-dichlorobenzophenone. When irradiated in alcohol in the absence of air, the reductive dechlorination of two molecules occurs:

$$2Ar_2CHCCl_3 + 2C_2H_5OH \xrightarrow{h\nu} 2CH_3CHO$$

$$+ Ar_2CHCCl=CClCHAr_2$$

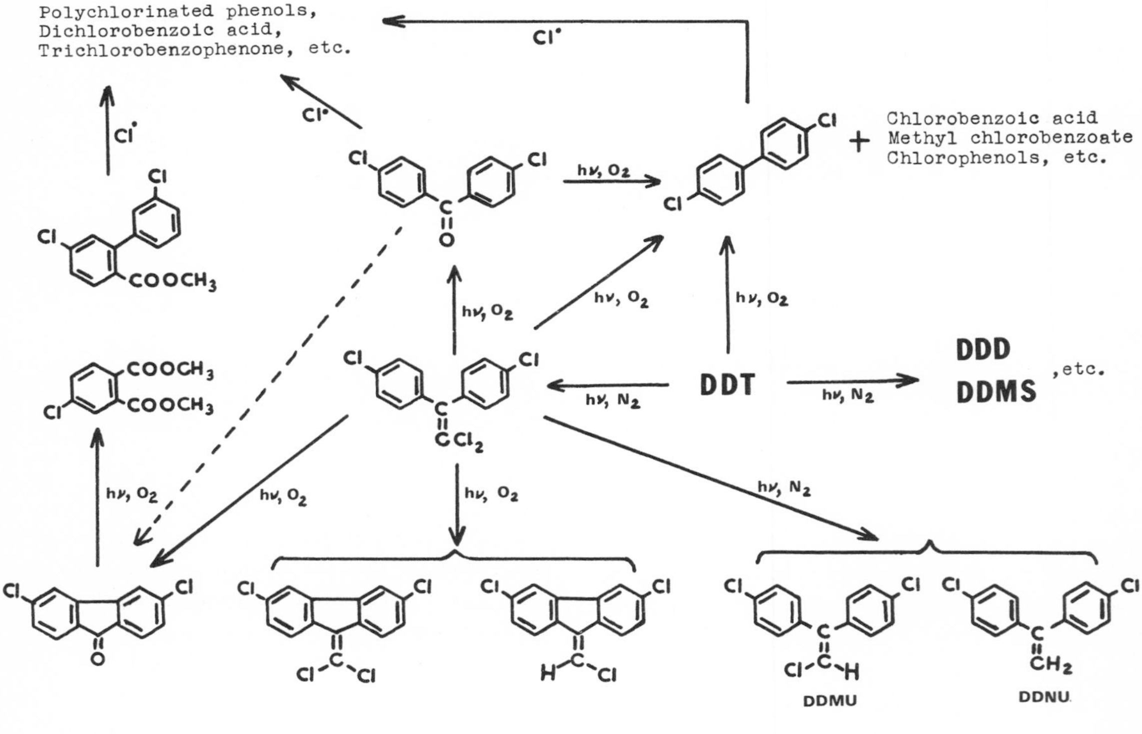

FIGURE 5. Photolysis of DDT and DDE in methanol.[110]

This product was identified by dehydrochlorinating it to the allenic compound discussed previously in connection with reductive processes, and the allene was oxidized to 4,4'-dichlorobenzophenone. It has been suggested that the last sequence of events from the allene also occurs in light and air, to account for the formation of the benzophenone:

$$Ar_2CHCCl=CClCHAr_2 \longrightarrow [Ar_2C=C=C=CAr_2]$$

$$\longrightarrow 2Ar_2CO + 2CO_2$$

DDD and DDE were said to be unchanged under the conditions of irradiation of DDT in these experiments. Work published in 1948 suggests that in sunlight and under field conditions, DDT residues give DDE, 2,2-bis(4-chlorophenyl)acetic acid (DDA), DBP and bis(4-chlorophenyl) methane, and that the last two compounds react together to give symmetrical tetra-(4-chlorophenyl)ethylene.[41]

No formation of double molecules has been mentioned by Korte and his colleagues, who showed that DDE irradiated in the solid phase with a pyrex filter to remove wavelengths below about 300 nm is progressively decomposed into 4,4'-dichlorobenzophenone (DBP) and the monodechlorination products (1) and DDMU:[111]

$$Ar_2C=O \quad (DBP)$$

$$Ar_2C=CHCl \quad (DDMU)$$

$$DDE \xrightarrow{h\nu} Ar(x,4\text{-dichloro-Ph})C=CHCl \qquad (1)$$

$$Ar_2C=CH(n\text{-hexyl}) \qquad (2)$$

$$Ar_2C=CCl(n\text{-hexyl}) \qquad (3)$$

After 3 days' irradiation the mixture consisted of DDE (40%), DBP (20%), DDMU (25%) and 15% of photoproduct (1). After 10 days only 12% of DDE remained, the major products being DBP (40%) and DDMU (45%) with less than 1% of photoproduct (1). With the addition of chlorophyll as a photosensitizer, a 90% yield of photoproduct (1) having an additional aromatic chlorine atom is obtained after 3 days of irradiation, the remaining material consisting of a little unchanged DDE, together with small amounts of DBP and DDMU. Irradiation in hexane affords the same products with additional ones, (2) and (3), derived by reaction with the solvent. With long irradiation times the major products are the mono-dechlorination products (1) and DDMU, and DDE completely disappeared after 8 days. If a quartz filter is used, so that the full range of wavelengths is available, decomposition occurs more rapidly, a 92% yield of DDMU being obtained after 11 hr irradiation. In acetone or acetone-water mixtures, the decomposition is very slow, and although the reaction proceeds in dioxan-water mixtures to give DBP, DDMU, and photoproduct (1), it is slower than when hexane is the solvent. DDMU and photoproduct (1) are also produced in up to 20% yield when an airstream saturated with DDE is irradiated. In addition to the reactions with solvent to give the products (2) and (3) above, DDT in solution in unsaturated hydrocarbons is said to react slowly with them to give products of telomerization, a reaction which is accelerated by heat and light:

$$Ar_2CHCCl_3 + n\ CH_2=CH_2 \longrightarrow Ar_2CHCCl_2(CH_2CH_2)_nCl$$

This change results in a loss of activity when DDT solutions in unsaturated hydrocarbons are allowed to stand for a long time.

DDT-analogues of the nitroalkane type (Prolan) are unstable to alkali because of the presence of the activating nitro-group, are destroyed by oxidizing agents, and lose the nitro-group when heated in light, especially if acids are present:

$$2Ar_2CHCH(NO_2)CH_3 \longrightarrow 2Ar_2CHCOCH_3 + H_2O + N_2O$$

Compounds like dicofol, having the benzylic hydroxyl group adjacent to an electron deficient carbon atom (in this case the CCl_3 group), readily eliminate this carbon atom under a variety of conditions to give the benzophenone. Thus, dicofol gives DBP and chloroform, which is the basis of its colorimetric determination. This occurs, for example, under conditions (zinc powder in dimethylformamide at 125 to 140°C) which convert DDT into DDE as well as DDD and several other products of reductive dechlorination. The same treatment converts the acetate derived from dicofol into DDE (with a little DDMU and other compounds), thus providing an interesting route, via initial hydroxylation and final reduction, from DDT to DDE:[112]

$$DDT \xrightarrow{\text{O}} Ar_2C(OH)CCl_3 \xrightarrow[\text{HAc}]{CH_3CO_2Ag} Ar_2C(OCOCH_3)CCl_3 \xrightarrow{\text{reduction}} Ar_2C=CCl_2$$

C. PRINCIPLES OF ANALYSIS

1. Gravimetric, Total Chlorine, and Colorimetric Analysis

Analytical methods available for DDT and its relatives can be divided into macro-methods, applicable when there is no restriction on the amount of materials available, and micro-methods which are more generally applicable but especially valuable in residue analysis. DDT is formulated as dusts, dispersible powders, solutions, emulsifiable concentrates, aerosols, and smoke-generating formulations. Full procedural details for the analysis of most of these are given in the book series, *Analytical Methods for Pesticides, Plant Growth Regulators and Food Additives,* edited by Zweig,[113] the review series *Residue Reviews* edited by Gunther,[114] and the *CIPAC Handbook*[115] and references cited therein; only principles are discussed here.

For the analysis of technical DDT,[115] the simplest method involves treatment of the sample with active carbon (a sample which does not absorb DDT) in acetone to remove impurities, removal of the solvent, and recrystallization of the residue from ethanol that has been saturated with pure DDT at say 25°C. The sample is heated under reflux with this saturated solution and the resulting solution cooled to 25°C, so that the DDT which crystallizes out is that contained in the original technical material. The determination of pure DDT content is then a simple matter of weighing the crystalline material. Volatile substances (including water) in technical DDT are estimated from the loss in weight when a sample is oven dried at 60°C for 1 hr, and insoluble material by weighing the residue when a sample is dissolved in acetone. Technical DDT may contain chloral hydrate which may be distilled off from a sample that has been diluted with water. The organic layer in the distillate is treated with 40% aqueous sodium hydroxide and pyridine and heated at 100°C, the red color which develops being compared with a standard prepared at the same time from pure chloral hydrate. Since the preparation of DDT is conducted under acidic conditions, it is frequently necessary to determine the acidity of the material, and this is done by diluting an acetone extract with water and titrating the solution with sodium hydroxide using methyl red as indicator, or measurement of the end-point electrometrically.

For the determination of total chlorine in DDT samples, the organic chlorine is converted into inorganic chlorine by refluxing the material with potassium in xylene or with sodium in dry isopropanol (the Stepanov method). The inorganic chlorine in the diluted aqueous reaction mixture is then precipitated with a known excess of silver nitrate solution, and the unused silver nitrate determined by titration with potassium thiocyanate. Inorganic chloride may also be determined electrometrically, but the organic solvents present may sometimes foul the electrodes. The amount of DDT present may be calculated from the fact that the molecule contains nearly exactly 50% of total chlorine, but it must be remembered that this method releases all the organic chlorine, including that which is contained in DDT-isomers and any other organochlorine compounds present. Various modifications are available for removing possible interferents, such as pretreatment with activated charcoal to remove pigments and the inclusion of a treatment with hydrogen peroxide to remove organic thiocyanates. The modifications used depend on the type of formulation examined, and the CIPAC handbook should be consulted for details.[115] A semi-micro method is also under consideration for the estimation of total chlorine when this is greater than 1% but would be unsuitable for formulated products having a high content of inorganic filler. In this method 15 to 100 mg of the sample (depending on the chlorine content) is wrapped in a piece of filter paper and burned in a combustion flask containing oxygen and neutral hydrogen peroxide solution, the liberated chloride ion being determined as in other cases.

A similar method in principle is the Parr bomb technique in which the sample is oxidized with sodium peroxide, containing sodium nitrate to accelerate the reaction, in the presence of added sucrose to supply oxygen.[113] The platinum-lined bomb is pressurized to about 25 atm with oxygen and the sample, contained in a small cup, is ignited electrically. The combustion products are absorbed in aqueous sodium hydroxide solution which is placed in the bomb with the sample before combustion takes place. After the reaction, the bomb contents are diluted with water, acidified with nitric acid and boiled to destroy any remaining peroxide; the liberated halide may then be determined by the usual titration methods. This method, like the previous one, is applicable for the determination of chlorine content in samples such as animal tissues, provided the moisture content is reduced before combustion. Freeze drying is a convenient way of doing this.

Another nonspecific method for total chlorine that has found favor for organochlorine pesticide formulations generally, as well as DDT, is the use of the sodium biphenyl reagent. The sodium in this reagent is highly reactive and reacts almost instantaneously with most halogen-containing compounds, although the reaction is slower with cyclodienes such as dieldrin. To avoid excessive consumption of the reagent, the sample should be dry and free from extraneous compounds containing labile hydrogen. Liquid formulations may be dissolved directly in toluene for determination; the toxicant is extracted from solid formulations with organic solvent and the residue after evaporation dissolved in toluene. The toluene solution is treated with excess of the reagent in a separating funnel and the mixture is allowed to stand until a dark green color develops (frequently within 5 min). Excess reagent is then destroyed with water, the inorganic halide extracted with dilute nitric acid, and the usual halide determination conducted on the total aqueous extract.[113,115,116]

If a total chlorine analysis gives approximately the 4,4'-DDT content expected, the actual DDT content may than be determined more accurately by controlled dehydrochlorination of a suitable sample. Samples are equilibrated in ethanol at 25°C, then dehydrochlorinated with normal ethanolic sodium hydroxide for 15 min, hydrolysis then being arrested by the addition of nitric acid. Under these conditions, all of the 4,4'-DDT is hydrolyzed, but only 2 to 10% of any 2,4'-isomer

present. If other chlorinated constituents are present the 4,4'- and 2,4'-DDT may be separated from these by column chromatography on silica gel carrying nitromethane as the stationary phase, with hexane saturated with nitromethane as mobile phase. The total chlorine content in the fraction containing the mixed isomers is then determined by the potassium in xylene method, and the 4,4'-DDT by selective dehydrochlorination, whence the content of 2,4'-isomer is given by difference.[115] The influence of chemical structure on dehydrochlorination rates of DDT analogues is discussed in the section on the chemical reactions of the DDT-group compounds (Vol. I, Chap. 2 B.3).

The nonspecific nature of the total chlorine analysis has been emphasized, but it is clear that its appeal lies in a simplicity which does not require sophisticated equipment, and for this reason, the Stepanov method (alkali metal dechlorination) appears to be a favored official method for formulation analysis. The lack of specificity means that, with the reservations already made, the method can be used to determine methoxychlor, DDD, dicofol, Perthane, chlorobenzilate, chloropropylate, and any other halogen containing DDT analogues. In appropriate cases, the process involving measurement of hydrolyzable chlorine can be used if steps are taken to make the reaction conditions selective for the compound investigated. In residue analysis, a selective dehydrochlorination process is commonly used as a prelude to some other reaction that can be carried out on the ethylenic product to give a measurable, colored reaction product. In the case of chlorobenzilate and chloropropylate, which are esters, alkaline saponification provides a method of estimation while acetylation of the free hydroxyl group is the prelude to an infrared procedure using the strong carbonyl stretching frequency for analysis.[14,117]

Dicofol is a good example of a compound which lends itself to analysis by several of these methods. Total chlorine may be determined by the Parr bomb, Stepanov, or sodium biphenyl methods. Following hydrolysis with boiling 0.5N ethanolic potassium hydroxide for 1 to 2 hr, the chloroform first formed gives chloride ion which may be measured in the usual way. Alternatively, either the chloroform or the benzophenone liberated in the reaction may be determined by colorimetric methods, as mentioned in the next section.[113]

The methods discussed above are mostly applicable to technical materials and formulations for which sizeable samples are available. Methods to be discussed in the following sections are predominantly semi-micro- or micro-methods which can be applied to bulk materials but find their main application in qualitative and quantitative analysis related to the determination of pesticide residues in foodstuffs or other commodities, and in metabolic and related investigations.

A number of colorimetric procedures have been described for the quantitative determination of DDT.[61] One method involves heating a mixture of DDT, glacial acetic acid, and concentrated sulfuric acid for 10 min on a boiling water bath and measuring the absorbance of the resulting yellow solution at 435 nm. The concentration and reaction time are rather critical for the success of the reaction and other isomers and chlorinated impurities also react. This method was said to detect 50 to 500 μg of 4,4'-DDT with an accuracy of ±1%. Various reactions with anhydrous aluminum chloride have also been utilized for colorimetric analysis. In one method, DDT is heated with an excess of anhydrous aluminum chloride at 66°C for 1 hr in dry, thiophene free benzene. Excess aluminum chloride is destroyed with water and the precipitate allowed to settle when the solution has an orange color in transmitted light and is greenish orange by reflected light. By this method 0.1 to 1 mg of DDT may be estimated by absorbance measurement at 420 nm. Ethylene chloride may also be used as solvent, in which case a brownish yellow color is obtained, without fluorescence.

Other procedures have been described, of which the best known are those devised by Stiff and Castillo,[117] and by Schechter and Haller.[100] In the first procedure, DDT is heated in water-free pyridine with xanthydrol (9-hydroxyxanthene) and solid potassium hydroxide to give a red color of intensity proportional to the content of DDT. The lower limit of detection by this method is about 75 μg. A critical examination of this method as applied to various DDT analogues shows that all compounds having the group -CHCX$_3$ or -C=CX$_2$ (where X = Cl or Br) give a positive reaction. The Stiff-Castillo method was claimed to be ideal for the detection and estimation of DDT residues on sprayed surfaces, including those of plants. Claborn evaluated the method and confirmed its value for estimation and detection of larger amounts of DDD as well as for DDT determination. The pyridine used need not be absolute, but must have a constant water content. Although the method is not particularly exact, its simplicity enabled it to be used for the determination of residues under difficult conditions, as for example, in the field.[61,118]

In a method described by Bradbury et al.,[119] a solution of DDT in diethylsulfate was heated for 3 min in a boiling water bath with a freshly prepared solution of 0.5% hydroquinone in concentrated sulfuric acid. The resulting wine-red solution has a characteristic absorption curve with a strong maximum at 480 to 500 nm. This method is sufficiently specific to enable it to be used for rapid measurement of the purity of technical DDT preparations; in an artificial mixture of 2,4'- and 4,4'-DDT, the content of 4,4'-isomer could be measured to within 0.5% and in the technical product to within ±1%. However, both DDE and DDD give a similar color reaction and are therefore not distinguished from DDT. An obvious method of estimation which was used at an early stage in the development of DDT involves its dehydrochlorination to DDE, which is then oxidized to DBP for estimation as the colored 2,4-dinitrophenylhydrazone.[61]

The most commonly used color reaction is that of Schechter and Haller,[100] which utilizes the blue color produced by the reaction between the tetranitro-derivative of DDT (Vol. I, Chap. 2 B.3) and sodium methoxide. When treated in benzene with sodium methoxide in methanol, this tetranitro-compound is converted into a blue, cross-conjugated di-anion (Figure 4). In the original form of the test, 20 μg of DDT could easily be detected and the limits can be improved by reducing the volume of benzene used as solvent. Traces of water must be avoided. Any aromatic compound that can be nitrated is a potential source of interference. Thus, chlorobenzilate, dicofol, DDE, DDD, methoxychlor, Perthane, tetradifon, and 2,4'-DDT give colors. However, in all cases except DDD and 2,4'-DDT, the absorption maxima lie below that for DDT, being usually in the 530 to 540 nm range. Compounds such as chlorobenzilate can be removed as salts following hydrolysis before DDT is estimated. The absorption spectrum of the colored product has only a single maximum and minimum in the case of DDT or DDD (maxima at 600 nm and 598 nm, respectively, according to Müller), whereas 2,4'-DDT gives a red-violet color reaction with maxima at 590 and 511 nm. DDE may be estimated in the presence of DDT by measuring absorbance due to the former at 540 nm. According to further work carried out on this reaction, the color due to DDD may be distinguished by its persistence when the solution is diluted with water; the colors due to DDT or 2,4'-DDT are said to disappear under these circumstances. DDT in a sample of unknown content is determined by reference to a standard curve of absorbance versus concentration obtained using the pure compound. When the DDT content of residues is determined, the limit of sensitivity of the reaction is considered to be 0.1 ppm. Disadvantages of the method are that it requires 2 or 3 hr for completion, and cleanup is required for oily or waxy samples. When, for example, DDT residues are determined in milk, a chloroform solution of milk fat is extracted successively with concentrated sulfuric acid, sodium sulfate in concentrated sulfuric acid, 1:1 concentrated sulfuric acid-oleum, and sodium sulfate-sulfuric acid, a procedure which removes fat but not DDT from the chloroform. After being neutralized with sodium bicarbonate solution, the chloroform is evaporated, and the residue nitrated for the Schechter-Haller determination.[120]

Although the Schechter-Haller method may be used for DDD determination, it is said not to be adequately sensitive. Instead, another method is suggested which involves its dehydrochlorination to produce DDMU, followed by treatment of the latter to give a colored carbonium ion complex which is then measured spectrophotometrically.[120] The crop or other sample to be analyzed is macerated and thoroughly extracted with an organic solvent, which may be hexane or one which can later be replaced by hexane. Use of the latter permits a cleanup process in which partitioning of the hexane solution with acetonitrile removes the pesticide and leaves many interfering materials in the hexane. The acetonitrile is then dried with sodium sulfate containing some celite and other adsorbents, filtered, and evaporated nearly to dryness in nitrogen at 50 to 55°C. Dimethylformamide (3 ml) is added and the dissolved residues are treated with sodium ethoxide (0.1N in ethanol; 1 ml) with

swirling for exactly 1 min. Water (25 ml) is then added and the products immediately extracted into hexane. Finally, the hexane solution is evaporated at room temperature in nitrogen and treated with concentrated sulfuric acid. The absorbance is read at 502 nm using concentrated sulfuric acid in the reference cuvette. DDT, methoxychlor, or Perthane do not interfere because the dehydrochlorination reaction conducted in this manner is too mild to dehydrochlorinate them; the more vigorous conditions used in their dehydrochlorination destroy DDD. The DDD content of the sample is determined from a standard curve obtained with pure DDD, and recoveries estimated using crop samples, etc. containing added known amounts of it. Reproducibility is said to be good (average deviation of ±5%) and 1 μg of DDD can be detected in 5 to 15 g of crop sample.

Methoxychlor has been determined colorimetrically by the pyridine-xanthydrol method of Stiff and Castillo, the Schechter-Haller procedure, and a technique involving dehydrochlorination followed by sulfonation. According to Claborn and Beckman,[121] a modification of the Schechter-Haller procedure in which methoxychlor is treated with fuming nitric acid for 15 min is specific for this compound, although the blue color formed from the product with sodium methoxide is not distinguishable from that given by the tetranitro-derivatives obtained from DDT and DDD in the normal nitration procedure. The fuming nitric acid procedure does not give the tetranitro-derivatives of these other compounds and so, no color reaction is obtained with sodium methoxide. When these authors treated 100 μg of DDT (75% p,p'- and 25% o,p'-DDT) with fuming nitric acid and the products with sodium methoxide, the color produced was only equivalent to that given by 6 μg of methoxychlor, so that small amounts of the other compounds in methoxychlor are not expected to interfere. The absorbance of the colored solution is measured at 590 nm.

For determination by the sulfonation procedure, the crop is extracted, the extract subjected to a cleanup process, and the methoxychlor in the residue dehydrochlorinated by boiling (30 min) with 4% ethanolic potassium hydroxide solution (50 ml). The organic products are picked up in organic solvent, which is then evaporated and the residue treated with 85% sulfuric acid (20 ml) containing 10 mg of ferric chloride per liter; the pink color produced is regarded as fully developed after the mixture has been allowed to stand for 90 min, and the absorbance is then read at 550 nm using the sulfuric acid reagent as reference solution. The method is sensitive to 2 μg of methoxychlor, and DDT does not respond. Cleanup from waxes and oils is required because these are blackened by sulfuric acid. An 85% recovery of quantities of methoxychlor from a few hundredths of a ppm to 10 ppm is claimed.[113]

The Schechter-Haller type procedure is not readily applicable to Perthane because nitration is difficult and there is danger of a high level of interference from similar chlorinated insecticides that react more readily. One method originally used involved oxidation to the corresponding 4,4'-dicarboxylic acid, followed by dehydrochlorination to the ethylenic derivative and absorbance measurement at 264 nm. A recommended colorimetric method which is highly specific for Perthane involves extraction of the crop and cleanup of the extract by the acetonitrile partition method, dehydrochlorination of Perthane at 100°C in nitrogen with 1% ethanolic potassium hydroxide, and treatment of the recovered organic material with concentrated sulfuric acid for a short time. A characteristic peach color develops and the absorbance is measured at 495 nm, using concentrated sulfuric acid as reference. DDT, methoxychlor and dicofol do not interfere and the method is applicable to the detection of 0.05 to 0.1 ppm of Perthane on fruits and vegetables, as little as 1 μg being detectable in most cases. The limit of detectability in milk is 0.02 ppm and in fatty tissues, 0.5 ppm. Recoveries from crops range from 75 to 100%, while those from milk and fatty tissues average about 83% and 70%, respectively.[113]

Besides lacking specificity, the Schechter-Haller method is not highly sensitive for difocol but the compound may be quantitively determined, following its alkaline hydrolysis to 4,4'-dichlorobenzophenone (DBP) and chloroform, by the analysis of either of these fragments. DBP may be determined either by measurement in the ultraviolet at 268 nm, or colorimetrically by conversion

into its 2,4-dinitrophenylhydrazone and measurement of the absorbance at 510 nm.[113] The other method makes use of the colorimetric determination of chloroform by means of the well-known Fujiwara reaction which involves the formation of a red compound of undefined structure when chloroform reacts with pyridine in the presence of a strong base. Any compound which can liberate chloroform on hydrolysis is a potential interferent in this reaction, but in the DDT-group, only dicofol and its acetyl-derivative give a response; 100-fold greater levels of DDT and DBP cause no interference. The dicofol analogue (called FW-152) derived from DDD by hydroxylation at the benzylic carbon atom also gives a small response, but this amounts to about 5% of the absorption obtained from the equivalent amount of dicofol at the same wavelength.[122]

In the basic method, the dicofol is extracted from the sample and heated in xylene with 50% aqueous sodium hydroxide solution at 100°C, while nitrogen is bubbled through to expel the liberated chloroform. The latter is collected in a mixture of pyridine (9.0 ml), water (0.6 ml), and 50% aqueous sodium hydroxide (0.4 ml), which is then heated to 100°C for 5 min, cooled to 0°C for 5 min, and the absorbance due to the red color measured at 525 nm. In this way, 10 μg of dicofol at a level of 0.1 ppm can be determined in a crop. The red color developed with low concentrations of base fades quickly on contact with carbon dioxide and atmospheric exposure should be avoided on this account.[113] Improvements of this procedure said to give greater sensitivity involve measurement of the absorbance of the colored complex at 366 nm in 50% aqueous pyridine or at 365 nm after dilution of the pyridine solution to a 10% concentration in methanol. In the latter case, the solution is yellow and lacks the higher wavelength peak.[122]

Amounts of nitroalkane analogues of DDT of the Prolan or Bulan type (constituents of the technical mixture called Dilan) as low as 10 μg can be detected by a procedure that is specific for such compounds and depends on their existence as an equilibrium mixture of the normal and aci-forms:

$$Ar_2CH\ NO_2 \rightleftharpoons Ar_2C=NO(OH) \xrightarrow{\ FeCl_3\ } colored\ complex.$$

Under basic conditions, the equilibrium is shifted to the right owing to the formation of the aci-form salt, which reacts with ferric chloride to give a

colored complex useful for colorimetric analysis. The sample in ethanol containing 16% of water is treated with 0.5N methanolic sodium hydroxide and then with ferric chloride in hydrochloric acid to give the red complex, which is stable for about 1 hr. Color development is rapid and should be complete in 10 min. According to Jones and Riddick who developed the method, 50 μg of the nitro-compounds can be estimated with an error not greater than 10 μg, with decreasing error as the amount of material present increases to 100 μg.[123]

The Schechter-Haller procedure can be used for determination of compounds such as chlorobenzilate or chloropropylate, since hydrolysis followed by nitration gives tetranitrodichlorobenzophenone which reacts with sodium methylate to give a red complex with maximum absorption at 418 and 538 nm. Several variants of the process have been described, depending on the likelihood of other compounds being present in the material analyzed. If DDT is present, direct nitration is possible, followed by absorbance measurements at different wavelengths on the colored complex formed with base; the different absorption maxima produced by the two compounds make this possible. Alternatively, alkaline hydrolysis will convert these compounds to the benzylic acid derivative; extraction of the alkaline solution removes any DDT analogues produced, following which acidification and re-extraction recovers the acid. This can then be oxidized to the benzophenone-derivative for direct estimation in the ultraviolet (264 nm) or colorimetrically via the 2,4'-dinitrophenylhydrazone. The Shechter-Haller method is said to detect 2 μg of chlorobenzilate and the benzophenone method is sensitive to 15 μg of chlorobenzilate recovered from admixture with 3 g of citrus extractives.[14,113]

2. Spectrophotometry, Polarography, Isotope Techniques

The term colorimetry generally implies the measurement of the quantity of a material in a solution absorbing light in the visible region of the spectrum by comparison with color standards at a given wavelength, whereas "spectrophotometry" implies that a photoelectric cell is used to convert the changes in absorbance into an electrical response that can be measured on some kind of recording instrument. The use of such instruments permits the continuous measurement of absorbance as the wavelength range is scanned, and the absorption spectrum consists of a continuous

plot of absorbance versus wavelength (λ). In such a plot, the maxima and minima correspond to the absorption maxima and minima of the substance whose spectrum is being measured. The three main wavelength regions employed for analytical measurements are the ultraviolet (UV) lying between 185 and 400 nm, the visible, lying between 400 and 800 nm, and the infrared (IR), lying between 800 and 16,000 nm. The region below about 120 nm is called the far UV, and 800 to 3,500 nm the near IR. Much of the residue analysis is conducted in the 400 to 800-nm region.[113]

When light passes through a transparent medium, the intensity of the light transmitted at a given wavelength is given by Beer's Law, according to which each successive layer of the medium absorbs an equal fraction of the light passing through it and,

$$\log \frac{I_0}{I} = kcd$$

where I and I_0 are the intensities, respectively, of the transmitted and incident light, c the concentration of the solution and d the thickness of the solution through which the light has passed. When c is in mol/l, and d in cm, k is the molar extinction coefficient (ϵ). Log I_0/I is the absorbance at the particular wavelength, and is equal to log $(1/T)$, where the transmittance (T) is equal to I/I_0. As indicated in the previous section, many colorimetric methods will routinely detect as little as 10 μg of organochlorine insecticides, and smaller amounts can be detected by micro-methods. In general, measurements in the UV region are more sensitive than those in the visible by a factor of ten or more, depending on the wavelength used for measurement. The sensitivity of measurement is greater when the maxima observed are near the lower wavelength end of the UV scale than when they are near the visible region. However, increased sensitivity applies also to measurement of impurities and rigorous cleanup is required for best results with UV-analysis.[113]

Since members of the DDT group are aromatic, they exhibit UV absorption spectra of which selected wavelengths may be used for direct analysis of samples. Maxima are reported for DDT in ethanol at 240 and 270 nm (log ϵ, 4.3 and 2.8, respectively), for methoxychlor in benzene at 230, 238, 270, and 275 nm (log ϵ, 4.3, 4.2, 3.9, and

3.8, respectively) and for DDD in ethanol at 233 nm (log ϵ, 4.3). Gillett has given a useful account of the use of both spectrophotometric and colorimetric methods for the determination of various DDT-analogues in chromatographic column effluents in relation to insect metabolism studies.[122] In this method the chromatographic column was eluted with isooctane (2,2,4-trimethylpentane) saturated with 3-methoxypropionitrile, and the absorbances of fractions (solution versus pure solvent) compared at 235 and 268 nm, the absorbance ratios characteristic of the various compounds at these two wavelengths having been observed previously. In this report, dicofol and its acetate were also determined colorimetrically by measurement of the colored Fujiwara complex in methanol at 365 nm, as described in the section on colorimetric methods.

Other methods of UV-analysis depend on the chemical transformation of the compound into a product having a functional group with suitable UV absorbance maxima.[124] As an example, chlorobenzilate may be converted successively into the corresponding acid and then into DBP:

$$Ar_2C(OH)COOC_2H_5 \xrightarrow{OH^-} Ar_2C(OH)COO^- \xrightarrow{CrO_3} Ar_2C=O$$

Chlorobenzilate   DBP

The absorbance of the DBP produced is then measured at 268 nm or that of its derived 2,4-dinitrophenylhydrazone at 510 nm. This method can, of course, be applied to any DDT-derivative which can be converted into DBP, and will determine any impurities that are similarly converted along with the substance being determined. However, the fact that these acaricides are derived from benzilic acid enables this method to be used specifically, as indicated in the previous section.

Absorption measurements in the infrared are based on Beer's Law, as in the case of the other spectral measurements, and the same principles are involved. In IR measurements, the positions of absorption maxima are recorded in wavelengths (micrometers; μ) or as the corresponding frequencies (wavelengths per cm; cm^{-1}). The region of most interest for analytical purposes lies between about 2.5 μ (4,000 cm^{-1}) and 15 μ (667 cm^{-1}). IR-analysis has become a widely used technique for the analysis of both technical pesticides and the formulated products. Particularly useful applications arise in the field of process control and in other cases where the history of the sample in known, in which case analysis can frequently be based on a measurement made at one convenient wavelength. The technique has not been used as frequently as UV-analysis for pesticide residue work, because the sensitivity is lower, but it is valuable for the identification of a few μg of purified, or partly purified substances isolated in metabolic or residue studies.[113,115,124]

Cell path lengths of 0.1 to 0.5 mm are recommended for quantitative analysis, and the windows may be of sodium chloride for measurements between 2 and 15 μ; beyond 15 μ sodium chloride is insufficiently transparent and potassium bromide must be used. For quantitative estimations, an absorption band is selected that is unique for the compound being determined. If the material being examined is a mixture, it is necessary to ensure that absorption bands due to contaminants do not fall in the same position as the chosen band of the compound to be analyzed. Another method is to cross check the results by carrying out the analysis for a particular component using more than one of its unique absorption bands. Using standard solutions of the substance to be determined, the absorbance at the chosen wavelength is determined from the ratio of the distance between the zero transmission line to the peak minimum (I_0) to that from zero transmission to peak maximum (I). Modern spectrophotometers have charts calibrated either in transmittance or absorbance, according to choice; if the absorbance is available directly, it gives a linear plot against concentration. If only linear transmittance (I/I_0) charts are available, a plot of linear concentration against (I_0/I) on semilogarithmic paper gives a linear relationship from which unknown concentrations can be determined.

In practice, a strong, sharp and well-separated band is chosen and a baseline is drawn between two wavelengths on either side of the band and as parallel as possible to the curve obtained in the spectrum of any impurities in this region. I_0 is then taken as the distance between the zero line and this baseline. A similar technique may be employed using potassium bromide pellets. Double beam instruments are most convenient for this sort of work because they permit solvent compensation in the reference beam and various other manipulations that are valuable in microanalysis.[113,115]

An absorption maximum at 1,020 cm^{-1} (9.8 μ) has been used to determine 4,4'-DDT, and maxima at 834 cm^{-1} and 750 cm^{-1} have been recommended for the determination of 4,4'- and 2,4'-DDT in a mixture. The IR method is recommended for the analysis of methoxychlor in complex mixtures, using plots of absorbance versus concentration at 8.0, 8.48, 9.60, 12.54, and 13.30 μ for the determination. Absorption maxima and baseline data used for the determination of some other DDT relatives and acaricides are given in the *CIPAC Handbook.*[115]

When conducting IR analyses, advantage can be taken of the chemical properties of the substances under investigation, as in the case of UV measurements.[124] Tetradifon, for example, has a strong sulfone absorption band at 1,160 cm^{-1} (6.25 μ) and is resistant to many oxidation procedures. Accordingly, oxidation with chromium trioxide in acetic acid can be used to destroy many interfering substances and the recovered tetradifon may then be measured at 1,160 cm^{-1} in CS$_2$. The same cleanup procedure can be used for the nonspecific determination of DDT, DDD, dicofol, chlorobenzilate, chloropropylate, and DMC, which are converted into the benzophenone DBP during the oxidation process. To promote the conversion, oxidation is preceded by a dehydrochlorination step (for DDT and DDD) or a hydrolytic step (for chlorobenzilate and chloropropylate). DBP is then determined at 930 cm^{-1} (10.75 μ) in CS$_2$. An analytical method for chlorbenside provides another example of a useful derivatization procedure; the compound is completely oxidized to the sulfone, which may then be determined at 1,155 cm^{-1} (8.65 μ) in CS$_2$:

$$\text{ArCH}_2\text{SAr} \xrightarrow{\;O\;} \text{ArCH}_2\overset{O}{\overset{\uparrow}{\text{S}}}\text{Ar} \xrightarrow{\;O\;} \text{ArCH}_2\overset{O}{\underset{O}{\overset{\uparrow}{\underset{\downarrow}{\text{S}}}}}\text{Ar} \quad (\text{chlorbenside sulfone})$$

Chlorbenside

When mixtures of pesticides are present in the sample to be examined, the main maxima of one component frequently are overlapped by subsidiary absorption from other components, in which case calculations of the absorbance due to one compound must allow for that due to others.[115] In such cases, a method of successive approximations is used to arrive at the individual absorbances. Dieldrin and DDT present together in dusts have been determined by extracting the dust with carbon disulfide and measuring the absorbances at the dieldrin band (912 cm^{-1}; 10.96 μ) and at the DDT band (711 cm^{-1}; 14.06 μ). Some band overlap requires that the absorbance at each wavelength is corrected for absorbance due to the other compound. At the 95% confidence level, an accuracy of $\pm 0.34\%$ for DDT and $\pm 0.10\%$ for dieldrin has been claimed for this method.[113,116] In quantitative IR analysis, care is needed to ensure that the temperature conditions are kept constant, especially when spectra are measured in solution, as is often the case. The volatility of the solvents used and the need to wash the cells out with them mean that time must be allowed for the cells to attain ambient temperature again before fresh spectra are run. Analysis using fluorescence spectra is not normally applicable to organochlorine insecticides, although an indirect method has been described which relies on the ability of these compounds to inhibit pig pancreatic lipase, thereby preventing it from liberating highly fluorescent 4-methyl umbelliferone by hydrolysis of non-fluorescent 4-methyl umbelliferone heptanoate.[125] The principle is interesting but the method seems unlikely to compete with established spectrophotometric methods.

Polarography is a rapid, sensitive, and relatively specific technique that is used for the detection, identification and quantitation of less than μg amounts of trace components of mixtures. The technique is fully described by Allen,[126] and in the CIPAC handbook in connection with the analysis of lindane.[115] It depends on the sudden increase in residual current measured by a polarizable electrode (usually a dropping mercury electrode) and a nonpolarizable reference electrode when the gradually increasing potential applied to a solution containing a reducible ion reaches the reduction potential of this ion. The difference between the new steady current achieved (the limiting current) and the original residual current is the diffusion current, which is proportional to the concentration of the reacting ion present. The method has been applied in the determination of a large number of pesticides, including the organochlorine compounds, in which reductive cleavage of carbon-chlorine bonds can occur. DDT has been determined both directly and following its conversion into the tetranitro-derivative, and the method has also been used for cyclodiene insecticides and for the hexachloro-cyclohexane isomers.[127] DDT can be detected at 0.5 μg/ml in solution and at 0.5 ppm in crop residues and the method is also suitable for the analysis of technical DDT and for DDT in emulsions and other formulations. Compounds containing a trichloromethyl group are found to be reducible, whereas those containing dichloro-groups, including the olefins from DDT analogues, are not.[128] A series of papers by Cisak presents a detailed analysis of the polarographic reduction of HCH isomers, some analogues, and various cyclodiene insecticides,[129,130] while reviews are available on the general subject of pesticide analysis by this method.[126,131,132]

In the technique of neutron activation analysis,[113] some particular element in a molecule is converted by neutron bombardment either into another radioactive element or into a radioisotope of the same element, so that the level of the original element present can be determined by radiometric methods. Crops, or crop extracts, in the form of solid or liquid samples, are irradiated together with appropriate reference standards in polythene vials. When the half-lives of the radio-isotopes to be measured are less than 30 min, irradiation times are chosen to be approximately equal to the half-lives. If the resulting radio-isotopes have half-lives greater than this, the irradiation period may be up to 30 min. Compared with the result of an irradiation time of one half-life, exposure for two half-lives increases the induced radioactivity by 50% and for about six half-lives by about 100%, but the statistics of counting are not improved to the same extent (22% and 41%, respectively). The method has found some application in the determination of organochlorine compounds. For example, various crops that had been grown in soil containing the nematicide, 1,2-dibromo-3-chloropropane (DBCP), were subjected to suitable thermal neutron fluxes ($\sim 10^{12}$ n/cm^2/sec) to convert the bromine into ^{80}Br (half-life 18 min) and ^{82}Br (half-life 36 hr) for radiometric assay. In this way, bromine levels

in the crops from 0.01 to 448 ppm were detectable. Chlorine may be determined with slightly lower sensitivity by counting the ^{38}Cl (half-life 37 min) generated at the same time.

The analysis of pesticides is greatly facilitated if the compounds are available in radioisotope labeled form, and the use of radiolabeled insecticides in a variety of investigations, especially those involving metabolic studies, has become commonplace during the last 20 years.[133] In such investigations, the experiments are conducted with starting material of known history, namely, the labeled compound. However, the availability of radiolabeled DDT or other pesticide permits the specific determination of the corresponding unlabeled materials, in a mixture of unknown content by the elegant isotope dilution technique.[134] This simple technique makes it unnecessary to isolate all the material of interest from the mixture, and is often capable of a precision and sensitivity unobtainable by any other method. Suppose, as an example, that a sample of ^{14}C-DDT (or ^{36}Cl-DDT) of activity A_0 and mass W_0 (and therefore of specific activity $S_0 = A_0/W_0$) is added to and thoroughly incorporated into a mixture containing DDT of unknown mass W_u. By mixing the two samples of DDT, the total mass of insecticide in the mixture becomes $W_0 + W_u$ but the activity A_0 remains the same, so that the specific activity of DDT in the mixture is now $A_0/(W_0 + W_u)$.

It is now only necessary to isolate a sample of the DDT from the mixture in order to determine the new specific activity S_1; that is, $A_0/W_0 + W_u$). The unknown DDT content of the mixture is then given by $W_u = W_0 [(S_0/S_i)-1]$. For the success of the method it is essential (a) that the ^{14}C-DDT, or other compound, should be chemically and radiochemically pure to start with, (b) that it should be thoroughly mixed with the unknown, so that the sample isolated for purification is representative, and (c) that the sample isolated for the determination of S_1 is exhaustively purified by some suitable method or combination of methods. Since the final yield of pure material is unimportant, wasteful techniques that result in rapid purification may be used (the final mass may be determined by modern microtechniques if necessary), and derivatization may be resorted to if it assists purification. One way of checking the purity of the labeled compound used is to dilute it with inactive material; if the specific activity of this mixture remains unchanged on purification, then the original sample was pure. The techniques used to handle and measure radiolabeled chlorinated insecticides are the same as those used with other labeled compounds and have been extensively described in the literature.

3. Analysis and Structure Determination of Organochlorines by Chromatography, NMR and Mass Spectrometry

The increasingly stringent requirement for the location and identification of minute amounts of drugs and pesticides in biological materials has led to remarkable advances in analytical technique during the last decade, and it may be fairly stated that in the pesticide area the modern development began with the introduction about 1960 of GC-analysis with electron capture detection for the analysis of organochlorine compounds.[135],[136] Broadly speaking, the steps involved in identifying a trace pesticide present in crude biological material include (1) cleanup by adsorption or partition chromatography on columns to remove most of the biological debris present, frequently accompanied by a rough or complete separation of pesticides present together, silica gel, alumina, and florisil being frequently used for this purpose; (2) more exact separation identification and quantitation using various combinations of paper (PC) or thin-layer (TC) chromatography with appropriate detection methods (color reactions, radiotracers, etc.); (3) the use of gas-liquid chromatography either instead of or subsequent to the operations in (2), which frequently permits simultaneous and rapid separation and quantitation of individual components; (4) the application of modern methods of physical organic chemistry to the unequivocal identification of compounds separated in the previous stages. When (3) and (4) can be combined in an "on line" system (for example, gas-liquid chromatography with infrared or mass spectrometry) powerful facilities for rapid identification result. Such systems have recently been discussed by Widmark.[137]

A great deal has been written on this subject in relation to organochlorine insecticides, which are considered together in this section as is appropriate in view of their frequent simultaneous occurrence in residues. Some additional information is given in the analytical sections relating to organochlorines other than DDT. Space permits only a brief account; the subject is being contin-ually updated and the reader is referred particularly to the series, *Analytical Methods for Pesticides, Plant Growth Regulators and Food Additives,* edited by Zweig,[113] and the series *Residue Reviews,*[114] edited by Gunther, for theoretical aspects, the last source being especially useful for appraisal of current techniques.

a. Column, Paper, Thin-layer, and Gas-liquid Chromatography

Before subjection to column chromatographic separation, organochlorine residues may be recovered from agricultural crops or other biological samples by exhaustive extraction with a variety of organic solvents, for example, hexane, ether, acetone, alcohols, and their combinations. Liquid-liquid partitions are then frequently used to further the cleanup. Thus, hexane extracts may be partitioned with acetonitrile, in which organochlorine insecticides are more soluble than most coextracted lipid materials; the latter therefore remain largely in the hexane phase. Treatment of extracts with sulfuric acid is another way to remove fats. Assuming that several organochlorines are present together, a crude separation into groups may be effected using various types of column chromatography. As an example, aldrin is eluted, while dieldrin and endrin are retained on a florisil column eluted with 6% ether in petroleum ether. These compounds can be eluted, as can dicofol, with benzene or benzene containing 1% acetone or acetonitrile, after DDD, *p,p'*-DDT, *o,p'*-DDT, DDE, methoxychlor, Perthane, chlordane, heptachlor and its epoxide, toxaphene, and HCH (isomeric mixture or lindane) have been eluted with the ether-petroleum ether mixture. The elution pattern of components from such columns is followed ideally using TC- or GC-analysis of the fractions.

Groups of compounds obtained in this manner may be further separated by paper chromatography, and the available methods have been reviewed by McKinley;[139] the method was at one time widely used for organochlorine analysis but has now been largely superseded by TC-analysis. Since organochlorines are highly lipid soluble, "reversed phase" systems are used in which chromatographic paper is impregnated from a 1% solution in ether, with refined soybean oil, vaseline, or liquid paraffin as stationary phase, the mobile phase being acetone:water (3:1), methanol:water (85:15), pyridine:water (3:2), or other aqueous organic solvents. These systems depend on partition of the insecticides between nonpolar stationary phase and mobile polar phase, so that the less polar components (for example, aldrin; R_F typically about 0.2) run more slowly than the more polar ones (dieldrin; R_F ca 0.6). Alternatively, a polar stationary phase such as dimethyl formamide or phenoxyethanol is combined with a nonpolar mobile phase such as iso-octane, in which case the above order of R_F values is reversed. Solvent systems for separating mixtures of various organochlorines have been extensively described by Mitchell.[138] The most popular way of visualizing the spots involves spraying the chromatogram with silver nitrate solutions in ethanol or acetone, when exposure to UV-light gives distinct reddish-purple spots on a white background if the paper was properly washed beforehand to remove traces of materials likely to decompose silver nitrate. For quantitation, various amounts of standards are run with the unknown to provide a calibration curve and spot areas are measured.[712] Use of paper impregnated from a 10% (v/v) ethereal solution of liquid paraffin with suitable tank design gives a linear relation between spot area and quantities of 2.5 to 14 μg of various cyclodiene insecticides with an average deviation of ±4%.[139] If radiotracers are used, the compounds can be detected by radiometric scanning and measurement of areas of peaks of radioactivity.[140]

Thin-layer chromatography is easy to use, more versatile than the above technique, and offers great scope for imaginative innovation.[141] Plates of metal or glass may be coated with various thicknesses of silica gel or alumina or even spread with "loose layers" of these adsorbents, permitting rough separations on "thick" layers for cleanup or micro-separations on thin layers (usually 0.25 mm) for analysis. Miniature plates made by dipping microscope slides in adsorbents suspended in an organic solvent are valuable for the rapid analysis of organochlorines from column effluents, etc., but greater reproducibility is achieved by the use of commercially prepared deposits on aluminum foil, which can be cut into any desired shape and run in small jars as solvent tanks. Organochlorines may be detected in amounts down to fractions of a μg using the silver nitrate spray, with UV-exposure as for paper chromatography. Alternatively, the fluorescence quenching of various dyes incorporated in the adsorbent or applied as sprays after the plate is run, affords "dead" spots against a fluorescent background. Quantitation by area measurement or radiometric scanning may again be used, but the great value of the technique lies in the fact that the material can be extracted from the inorganic adsorbent without the contamination with a stationary phase which is usually a problem with paper chromatography. Thus, the components can be rapidly separated by TC, extracted from the adsorbent with solvent, and assayed by GC-analysis or dissolved (or slurried along with the adsorbent) in a suitable scintillation medium for scintillation counting if radioactive.[142] Location of the spots, if not radioactive, can be effected by analyzing sections of the plate using GC, or reference to the position of a reference spot on the plate. It turns out that the silver nitrate reagent usually decomposes organochlorine compounds superficially, so that the material beneath the color may be extracted largely unchanged, especially if exposure to UV-light is as brief as possible. In the absence of contaminating stationary phase, it is also a relatively simple matter to recover sufficient pure material for IR-analysis (10 μg being suitable for measurement in many cases). In normal systems, solvents such as hexane or its mixtures with more polar solvents constitute the mobile phase so that more polar insecticides have lower R_F values; reversed phases may again be employed by impregnating normal plates with a lipophilic stationary phase, as in paper chromatography. The techniques used have been extensively discussed by Abbott and his colleagues and typical R_F values obtained with organochlorine insecticides in several solvent systems are given in Table 10.[141]

In gas-liquid chromatographic analysis, the mixture of organochlorine insecticides is carried onto an inert carrier with a suitable stationary phase, so that the materials to be separated are partitioned between the gas phase and the stationary phase.[113,143] The separation has similarities with fractional distillation and the performance of the column in separations is defined in terms of theoretical plate content (n) by the equation $n = 16(R_t/W)^2$, in which R_t is the retention time (distance between solvent peak and sample peak expressed as time) and W, the sample peak base width expressed as time. The materials used to carry the stationary phases are by no means always inert and usually have to be treated in various ways to remove reactive centers which may

TABLE 10

Thin-layer Chromatography of Some Organochlorine Compounds (15 cm Development in a 28 x 26.5 x 21 cm Tank)

Compound	R_F Value x 100 in system no:[a]							
	1	2	3	4	5	6	7	8
p,p'-DDT	42	89	69	57	39	54	52	91
o,p'-DDT	50	89	73	59	46	58	50	90
p,p'-DDE	65	95	78	65	57	62	74	98
Methoxychlor	–	–	–	–	10	36	28	–
p,p'-DDD	25	71	57	52	26	46	67	77
p,p'-DDD ethylene derivative	53	93	75	49	53	62	67	98
p,p'-dichlorobenzophenone	14	31	55	59	27	48	–	–
Heptachlor	58	95	78	65	53	62	48	98
Heptachlor epoxide	17	49	57	39	–	–	–	–
Aldrin	70	95	82	67	64	69	58	98
Dieldrin	12	37	52	65	48	48	30	58
Endrin	13	51	61	49	26	52	–	–
Endosulfan-α	17	65	64	58	35	52	–	–
Endosulfan-β	2	4	9	12	–	–	–	–
α-HCH	34	87	63	52	28	43	–	69
γ-HCH	21	78	55	46	18	37	–	58

(Data adapted from Abbott and Thomson.[141])

[a]System no/adsorbent/mobile solvent:
1/silica gel/hexane
2/alumina/hexane,
3/alumina/petroleum ether (40 to 60°C): liquid paraffin: dioxane; 94:5:1
4/silica gel/petroleum ether (40 to 60°C): liquid paraffin: dioxane; 94:5:1
5/silica gel/petroleum ether (40 to 60° C): liquid paraffin; 4:1
6/silica gel/cyclohexane: liquid paraffin: dioxane; 7:2:1
7/silica gel/cyclohexane: benzene: liquid paraffin; 9:9:2
8/silica gel: alumina; 1:1/cyclohexane:silicone oil; 92:8

decompose the samples in one way or another. Many methods have been reported for the preparation of column-packing materials, and a large number of stationary phases suitable for various situations in organochlorine pesticide analysis have been described in the pesticide literature. Details of the technique are to be found in the sources cited at the beginning of this section and will not be considered here.

Although excellent separations can be achieved by this method, its value as a tool for quantitative analysis is clearly limited by the sensitivity of the device used to detect the material of interest in the gaseous effluent, and there is also the problem that extraneous biological material in the solutions to be analyzed can interfere with the detection system. Attempts to achieve both sensitivity and specificity in the GC-analysis of organochlorine compounds met with success almost simultaneously with the development of the microcoulometric detector of Coulson et al.[144] and the electron-capture detector of Goodwin et al. in 1960.[135] In the microcoulometric method, the sample leaving the chromatographic column is combusted in oxygen and the halide (or SO_2 content) of the combustion gases is determined in a coulometric titration cell arranged to determine either halide or sulfur specifically. In the case of the halide cell, silver ions are generated to restore any instantaneous off-balance due to the presence of halogen halide in the cell and the consequent increase in current taken by the cell appears as a peak on the current-time curve for silver ion generation. For quantitation, the percentage of chlorine in the sample must be known since it relates the amount of pesticide present to the amount of silver ion generated (proportional to the peak area) according to Faraday's Law. The method, which permits the detection of nanograms of compounds, is specific for either chlorine or sulfur, depending on the titration cell employed and is much less sensitive to interference than detection methods that depend on ionization. An early review of the technique was given by Cassil in 1962, and it is also discussed by Bevenue in the analytical series edited by Zweig.[113]

The electron-capture method for organochlorine detection achieved great popularity in a very short time, partly due to the ease with which detectors having excellent sensitivity can be made easily and inexpensively in the laboratory. Originated by Lovelock and Lipsky,[146] the method involves the generation of a standing current between two electrodes in the gas effluent (carrier gas nitrogen, argon or helium) by the application of a small negative voltage to one of them which is a source, such as tritium or [63]Ni, of β particles. The electrons picked up by the other relatively positive electrode constitute the standing current which is recorded potentiometrically as the baseline. When a compound with high electron affinity enters the detector (conjugated unsaturated ketones, nitro-compounds and organic halides are examples) a sudden reduction in standing current occurs during the passage of sample through the detector and is recorded as a peak on the current versus time recording. The method is so sensitive that picogram (10^{-12}g) quantities of organochlorine compounds (and usually rather larger amounts of other electron-capturing substances) can be detected with present-day instrumentation, although careful sample cleanup is desirable if the detector is to retain its sensitivity for long periods without cleaning. The response curve for a particular organochlorine, derived by injecting increasing amounts of it, is usually nearly linear at first, then slowly becomes logarithmic and finally reaches a maximum value (R_o). This maximum is a characteristic of the particular detector which is controlled by the nature and flow rate of the carrier gas, the voltage applied to the detector (determining the standing current), and its state of cleanliness. For such detectors, the electron absorption characteristics follow laws similar to those of light absorption spectroscopy, a relationship between peak height and sample quantity analogous to Beer's Law being observed. Thus, kW = log $[R_0/(R_0-R)]$, where k is a constant, W the amount injected and R the observed peak

height.[147] In the first detailed paper on the electron-capture detection technique a 2.5% loading of silicone elastomer E301 on kieselguhr deactivated with Shell Epikote® resin 1001 was employed in a 2-ft column (at 163°C) with nitrogen as carrier gas to separate and detect a mixture of lindane, five cyclodiene insecticides and DDT in various crop extracts.[136] DDT had the longest retention time and was detectable down to 1 ppm, the others being detectable at 0.1 to 0.25 ppm under these conditions. The potential of this technique, which is the starting point of the environmental contamination controversy, is evident from this paper and those from many other laboratories which swiftly followed it. A review of the early developments was given in 1964.[148]

Table 11 lists some retention times relative to aldrin obtained for organochlorine pesticides using a mixed packing of QF1 and DC-200 on Gas Chrom Q in a 6-ft column at 200°C.[149] Except in the case of some of the technical materials, these are listed in order of retention time so that the possibilities for interference in residue analysis using this particular system can be seen. The microanalysis of DDT and its relatives by GC-methods has to be conducted with some care, since breakdown is apt to occur on columns that have not been properly conditioned. Ott and Gunther examined the breakdown of DDT on a packing of acid-washed GC-22 firebrick carrying 18.5% HiVac grease and provided evidence for the formation of DDE, DDD, and DDD-ethylene (the DDD dehydrochlorination product corresponding to DDE); DBP and 4,4′-dichlorodiphenyl acetic acid were also thought to be formed.[150] To prevent this decomposition, 2-biphenyl diphenyl phosphate has been recommended as a column conditioning agent. Dicofol and endrin provide other examples of decomposition on chromatographic columns; the former gives DBP and the latter the rearranged half-cage hexachloro-ketone (delta-keto-153, Table 11) together with the alternative aldehyde.

Besides the use in quantitation, comparisons of GC-retention times (and/or R_F values on TC) of unknowns with corresponding values for suspected authentic compounds are regularly made on different stationary phases to provide evidence of identity. There are pitfalls in these methods of identification, which are only valid if the retention volumes (gas flow rate x retention time) of a

TABLE 11

Retention Times Relative to Aldrin of Some Organochlorine Pesticides

Pesticide	Retention time relative to aldrin $(R_t = 3.5 \text{ min})^a$
Tech. DDT	0.39;1.88;2.48;2.70;*3.28*
Tech. HCH	*0.46;*0.58;0.68
a-HCH	0.46
Lindane (γ-HCH)	0.58
β-HCH	0.60
δ-HCH	0.68
Tech. methoxychlor	0.51;3.23;*4.8*
Tech. dicofol	0.58;0.74;1.09;*1.31;*1.86;2.21;2.37
Heptachlor	0.81
Dimite®	0.82;*1.31*
Aldrin	1.00
Isobenzan	1.14
o,p'-DDD-olefin	1.21
Isodrin	1.25
Perthane®-olefin	1.31
Dicofol	1.31
o,p'-DDE	1.46
Heptachlor epoxide (HE 160)	1.47
p,p'-DDD-olefin	1.50
o,p'-dichlorodiphenyl-monochloroethane	1.53
a-chlordane	1.57
β-chlordane	1.73
Tech. chlordane	0.72;*0.81;*1.07[b];155;*1.72;*2.78
Toxaphene	1.70;2.25;2.37[b];2.55;*3.03;3.66;*4.10;*4.70;*5.8
Strobane®	1.09;1.38;1.53;1.69[b];2.52;3.01;3.56;4.57
p,p'-DDE	1.88
a-Endosulfan	1.89
p,p'-Dichlorodiphenyl-monochlorethane	1.94
Perthane®	*2.02;*2.60
o,p'-DDD	2.04
Dieldrin	2.22
o,p'-DDT	2.48
Endrin	2.55
Sulphenone	2.62
Kepone®	2.67
p,p'-DDD	2.70
Chlorfenson	2.78
β-Endosulfan	2.92
p,p'-Methoxychlor-olefin	3.00
p,p'-DDT	3.28
Endrin aldehyde[c]	3.98
p,p'-Methoxychlor	4.80
Prolan®	4.90
Dilan®	4.40;4.90;*5.70*
Mirex®	5.15
Bulan®	5.70
Delta-Keto-153[c]	6.05
Tetradifon	8.75

(Data adapted from Maier-Bode,[248] Burke and Holswade.[149])

[a]Major peak(s) italicized when more than one present
[b]Additional small peaks not included
[c]Half-cage rearrangement products from endrin

compound on different columns or its R_F values in different TC systems are truly independent. When the number of components in a mixture is small, or following preliminary separations, derivatization techniques such as those described[64] by Cochrane provide valuable "fingerprint" chromatograms as aids to the identification, and examples are given in the sections on analysis of individual compounds. Another simple and sensitive confirmatory test is provided by the solvent partition method (determination of p-values) described in a series of papers by Bowman and Beroza.[152] Such complementary techniques are essential when electron-capture detection is used in residue analysis, because the detector is not completely specific for chlorine (a disadvantage as compared with microcoulometry), a fact which should be remembered when considering the analytical results. Bache and Lisk applied emission spectrometry to the determination of halogens in gas-chromatographic effluents.[153] Chlorinated compounds are excited and fragmented in a plasma generated in the helium carrier gas by an electrode-less microwave discharge at 5 to 10 mm Hg. Examination of the halogen line in the emission spectrum permits detection and quantitation of p,p'-DDE, lindane, and cyclodienes at the picogram level. The equipment used is costly, but the method is highly specific.

b. Modern Physical Methods

Infrared spectrometry was the first of the modern methods to be extensively used in residue analysis and pesticide metabolism studies and has been of great value in the organochlorine area because of the "fingerprint" character of the spectra, obtainable in favorable circumstances with a few μg of compound. The inclusion of proton chemical shifts from nuclear magnetic resonance (NMR) measurements is now standard practice in reports relating to organochlorine compounds, although such measurements require relatively large amounts of material (1 mg or more), even when time-averaging computer facilities are available to improve the spectra obtained. Apart from the information on proton chemical shifts available in the more recent papers on various aspects of DDT chemistry and metabolism, a recent paper by Keith and colleagues discusses the high resolution NMR spectra of some 26 DDT analogues and related compounds. One of the

matters discussed is the strong effect on the ortho-aromatic protons of the electro-negative substituent on the α- or benzylic carbon atom; these protons are most strongly deshielded when this substituent is the $-CCl_3$ group (β-carbon), the effect being relaxed when less highly chlorinated α-substituents such as $CHCl_2$ are present.[154]

Information regarding the application of NMR measurements to structural determination in the cyclodiene series is to be found in the publications of Robinson, Mackenzie, Cochrane, Keith, McKinney, Korte and their colleagues.[155-160] McKinney recently described the use of this method in the determination of the structures of cyclodiene metabolites,[158] and the use of the relatively recent NMR shift reagent Eu(DPM)$_3$ to effect the resolution of superimposed signals is particularly interesting. Another fairly recent paper describes the assignment of chlorine nuclear quadrupole resonance (NQR) frequencies to chlorine atoms in different environments in closely related compounds such as the HCH isomers and several cyclodiene insecticides[161] but this method is unlikely to present an immediate challenge to the currently established analytical methods.

The combination of GC-separation with mass spectrometric examination of the separated fractions[137] is undoubtedly the most valuable combination currently available to the residue analyst or toxicologist attempting to identify minute traces of organochlorine or other pesticides. In some cases, the mass spectrometer is able to provide analytical data from samples (peaks) not many orders of magnitude greater than the limits of comfortable detection afforded by the most sensitive GC-electron capture combinations. Recent publications on organochlorine compounds, especially those involving environmental studies, usually contain data regarding mass spectral fragmentation patterns, and the technique has been especially valuable for identification of some of the molecular rearrangement products of cyclodiene molecules. A detailed analysis of cyclodiene fragmentation for five methanoindene (chlordane) relatives and for aldrin, isodrin, dieldrin, and endrin, together with illustrations of the spectra, is given by Damico, Barron, and Ruth.[162]

The presence of ^{35}Cl and ^{37}Cl in natural chlorine results in a group of fragments separated by two m/e units for each chlorine containing moiety in the mass spectrum. Thus, the molecular ion (rather small for some cyclodiene molecules) which is the positively charged parent molecule, gives rise to a series of peaks of which that of lowest molecular weight arises from individual molecules containing only ^{35}Cl, while ions at M + 2 contain one ^{37}Cl atom, and so on. There are theoretically (n + 1) peaks for a fragment containing n chlorine atoms, the one of highest m/e value containing only ^{37}Cl and having lowest probability. The relative abundance of these combinations in a particular fragment is a characteristic of the number of chlorine atoms present so that the m/e value for the molecular ion and the ratios of peak heights for M, $M + 2$, $M + 4$, and $M + 6$ reveals the molecular weight of the compound and the number of chlorine atoms present. When M^+ is small, there is usually a prominent group corresponding to the combinations of $(M\text{-}Cl)^+$, from which similar information can be derived. For cyclodiene compounds, the remainder of the pattern consists of ionic fragments corresponding to reversal of the Diels-Alder reaction that produced the molecules, and successive loss of chlorine and/or HCl from the parent molecule. In the case of epoxides, ions are present whose formation is more specifically associated with the presence of this functional group. There are also fragments produced by combinations of these modes and by more complex processes.[162]

4. Bioassay

When the relationship between the dose applied under given conditions and the response shown by a particular insect is well known, then, in principle, this species can be used under the standard conditions to detect and determine insecticide residues. The housefly has been frequently used as a test insect for this purpose; other insects used include the fruit or vinegar fly (*Drosophila melanogaster*) and *Folsomia fimetaria*, a member of the springtail family.[163] The use of more than one species for bioassay is advantageous since species differences in susceptibility may permit determinations at different levels of sensitivity. The basis of the technique is that the inside of a jar, a filter paper, or some suitable carrier for the toxicant is impregnated with it in a standard manner from a solution in an organic solvent. Dosage-mortality curves are prepared from replicated observations of the response of groups of insects to a range of known doses applied in this way. The solution containing the unknown amount of toxicant is then used to impregnate the receptacle in the same way, the response of the insects in the standard time of observation is noted, and the amount of toxicant present is read off from the dosage-mortality curves already obtained. Although the presence of a different toxicant than the one whose measurement is desired might be evident from the nature of the poisoning process, the method clearly will not distinguish between different toxicants with closely similar action. Experiments with *F. fimetaria* and *D. melanogaster* have shown that the first insect is generally more susceptible to organophosphorus and carbamate insecticides than to aldrin and lindane (γ-HCH) while the reverse is the case with *D. melanogaster*. Both species are rather resistant to DDT. A 90 μg deposit of this toxicant was required to kill 60% of *Drosophila* (as compared with 0.43 and 0.54 μg, respectively, for lindane and aldrin) within the standard 8-hr exposure period, and this deposit killed none of the exposed *Folsomia* in a 24-hr period.[163] This illustrates the point made earlier regarding species differences. Sun has given a comprehensive account of the use of bioassay methods in pesticide analysis.[164]

Chapter 3

INSECTICIDES OF THE DIENE-ORGANOCHLORINE GROUP

A. INTRODUCTION

1. The Diels-Alder Reaction

The Diene-organochlorine or "cyclodiene" chlorinated insecticides constitute a remarkable series of compounds that arise or can be considered to arise from the elaboration of hexachlorocyclopentadiene or closely related dienes. The synthetic reaction used to prepare these compounds is the Diene-synthesis or Diels-Alder reaction, which consists in the addition of a compound containing a double or a triple bond (usually, but not necessarily activated by additional unsaturation in the adjacent α,β-positions) to the 1,4-positions of a conjugated system. Therefore, in its simplest form, the reaction involves the addition of a molecule of type (1) or (2) (Figure 6; R may be carbonyl, carboxyl, methoxycarbonyl, acetoxy, nitro, sulfonyl, cyano, amino, vinyl, or even hydrogen), designated the "dienophile," to a diene of type (3) or (4). Representative dienophiles include α,β-unsaturated acids and their anhydrides, esters, and halides; α,β-unsaturated aldehydes and ketones, quinones, and simple olefins.

The Diene-synthesis between components of the type shown results in the formation of a six-membered carbocyclic ring without displacement of hydrogen. Thus, (1) and (2) (Figure 6) react with (3) and (4) to give the ring systems (5), (6), (7), and (8). According to Alder's rules,[165,166] the addition of diene to dienophile is a purely cis-addition and the relative positions of substituents in the dienophile are retained in the adduct. Accordingly, a cis-dienophile will not give rise to a trans-adduct nor will a trans-dienophile give rise to a cis-adduct as the initial products. However, it is possible that the cis-adduct formed may isomerize to a trans-adduct under the conditions of experiment, especially if high temperatures are used to effect the reaction. From a thorough investigation of the reactions, all of which lead to methylene bridged molecules (these bridges were called endo-bridges, but the term is apt to cause confusion with other stereochemical nomenclature and is not much used now, especially in cyclodiene chemistry), Alder and Stein[165] proposed that when a

1,3-diene and a dienophile react, the molecules tend to assume that mutual orientation which corresponds to the greatest accumulation of unsaturation. In modern terms, the Diels-Alder reaction is governed by the Woodward-Hoffmann orbital symmetry rules and should be characterized by an endo-transition state, as is found.[166a] Thus, in the well-known reaction between cyclopentadiene and maleic anhydride, the isomer resulting from the maximal interaction between the unsaturated systems is the endo-isomer (Figure 6), in which the acid anhydride ring system is fused to the newly formed methylene-bridged cyclohexene nucleus (bicyclo[2.2.1]hept-2-ene; norbornene) through its endo-positions (which are necessarily cis- to one another).

The endo-positions are those which lie below the plane containing the methylene bridge (also called methano-bridge) carbon of norbornene and the two carbon atoms carrying the exo-hydrogens at the ring junction. Another way of expressing this is to say that the endo-positions lie on the side of the general plane of the norbornene nucleus opposite to that of the methano-bridge, a statement which has the advantage that it is independent of the way the molecule is drawn; the author prefers to draw these molecules so that the methano-bridge (the one which is chlorinated in the cyclodienes that have two) is uppermost and on the left of the molecule, and they will be so depicted as far as possible in this book. This means that either definition of the endo-positions of the norbornene system is valid as the molecules are normally drawn here.

Although the formation of endo-isomers is the general rule, the Diels-Alder reaction is subject to temperature effects which may result in isomerization. Thus, if furan, rather than cyclopentadiene, reacts with maleic anhydride (cf. Figure 6), the exo-isomer is the predominant product. The endo-isomer is also formed, but dissociates in solution and at moderate temperatures, with recombination to give the more stable exo-isomer.[166a]

The Diene-synthesis is an extremely versatile method for producing polycyclic systems and has been widely employed for this purpose. Simple reactions of the type just discussed are those

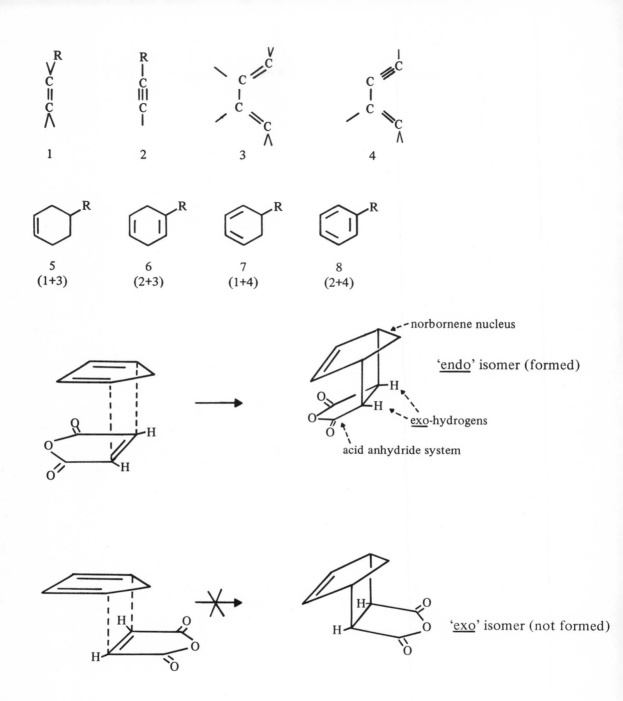

FIGURE 6. The Diels-Alder addition reaction with simple dienes and dienophiles.

between cyclopentadiene and ethylene, cyclopentadiene and cyclopentadiene, or cyclopentadiene and norbornadiene (bicyclo[2.2.1]hepta-2,5-diene). These are shown in both planar and three-dimensional formulae in Figure 7. The second reaction is a Diene-synthesis leading to the dimerization of cyclopentadiene; it occurs fairly readily at ambient temperature and can be reversed by heating the liquid dimer at 180 tc 200°C. The carbon skeleton of this dimer forms the basis of cyclodienes of the heptachlor-chlordane series. In the third reaction, norbornadiene (itself produced by the reaction between cyclopentadiene and acetylene) reacts with cyclopentadiene to give molecules which have the basic carbon skeleton of the dimethanonaphthalene series of cyclodienes. It is seen from Figure 7 that the planar representation is more than usually inadequate in this case; apart from the possibility of both *exo*- and *endo*-addition (considering the left-hand norbornene nucleus only *endo*-addition is shown), two possible orientations are available for each of these modes of fusion, depending on the final disposition of the methano-bridge of the right hand norbornene nucleus. As shown in Figure 7, the two possible orientations of the methano-bridge associated with *endo*-fusion correspond to the isodrin and aldrin series of cyclodiene insecticides. Evidence for the assignment of structures to the two series will emerge in later sections in which the general synthesis and chemistry of the cyclodienes is considered.

2. The Nomenclature of Cyclodiene Insecticides

The systematic naming of the cyclodiene insecticides has presented some difficulties for chemists and others who are familiar with the subject, and so it is not surprising that the nomenclature is confusing for those who are unfamiliar with these compounds. In fact, the various names used for the primary products of the Diels-Alder reaction are relatively easy to understand once the reader has become familiar with the convention used in a particular publication; the difficulty lies in the number of conventions that have been used. The simplest system is the one which presents the molecules as they arise from the Diels-Alder reaction, and the planar structures of Figure 8 are not difficult to understand. These then have to be elaborated to depict the true, three-dimensional stereochemistry, shown in Figure 8 for some compounds. The conversion of a written structure

into a recognizable three-dimensional one is easiest if the reader can first mentally relate the former to some familiar planar one; with some knowledge of the Diels-Alder reaction, it is then easy to convert from two to three dimensions. Clearly, however, such processes will be extremely difficult for those who lack a certain basic knowledge of organic chemistry, and, unfortunately, the situation becomes even more complex in the case of the molecular rearrangement products of cyclodienes, some of which have elaborate cage structures. For the benefit of the less chemically inclined readers and others who find the subject confusing, a more detailed discussion of nomenclature will be given here than is usually found in literature relating to cyclodiene compounds.

For the simple, primary adducts, the naming system that relates the molecules to indene, indane (for the heptachlor, chlordane series), or naphthalene (for the aldrin, isodrin series) is undoubtedly easiest to understand and will be considered first. In this system, the molecule heptachlor (Figure 8, structure 2), for example, is called 1-*exo*-4,5,6,7,8,8-heptachloro-3a,4,7,7a-tetrahydro-4,7-methanoindene. This molecule may be considered, for the purpose of name derivation only (since such a synthetic route is hardly practicable), as deriving from the aromatic molecule called indene (Figure 8, structure 1). This does not mean that heptachlor is aromatic, but only that indene is a very convenient mental starting point. The next operation is the substitution of the hydrogen atoms on positions 4- and 7- of the benzene nucleus by the bridging -CH_2 group (C_8 of heptachlor) to give the hypothetical 4,7-methanoindene (Figure 8, structure 4).

If one hydrogen atom is now added to each of the 4- and 7-positions (equivalent to a 1,4-reduction of the conjugated 4,5,6,7-double bond system) the remaining double bond necessarily appears in the 5,6-position to give the more feasible molecule 4,7-dihydro-4,7-methanoindene (Figure 8, structure 5). To obtain the Diels-Alder adduct dicyclopentadiene (the basic skeleton of the heptachlor-chlordane series) from structure 5, a hydrogen atom must now be added to each of the 3a and 7a-positions of (5) to give, proceeding in sequence round the ring system, 3a,4,7,7a-tetrahydro-4,7-methanoindene (dicyclopentadiene; Figure 8, structure 6). It is important to note that the name must be read in full because *tetrahydro* immediately shows that the molecule is *not*

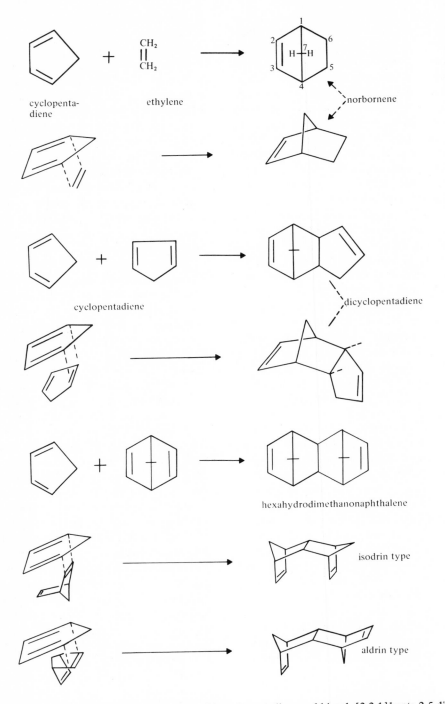

FIGURE 7. The Diels-Alder reaction with cyclopentadiene and bicyclo[2.2.1]hepta-2,5-diene (norbornadiene).

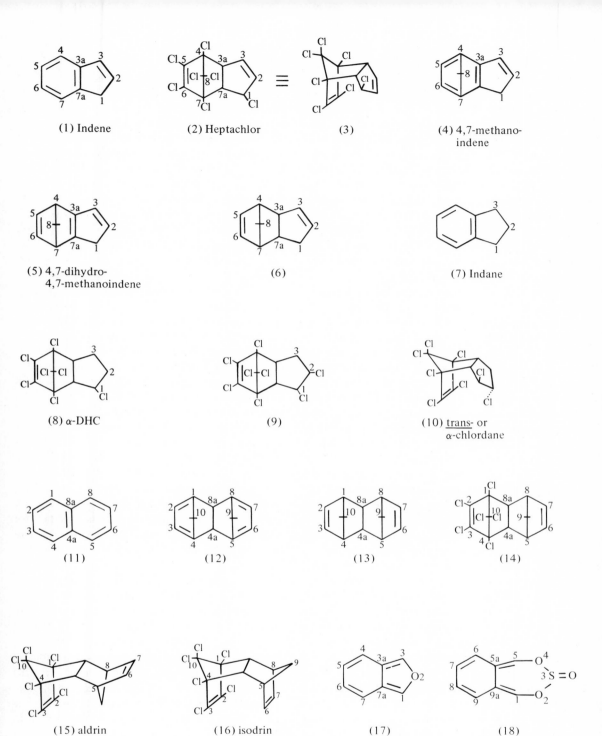

FIGURE 8. Numbering systems and derivation of systematic names, for cyclodiene compounds (methanoindane and dimethanonaphthalene nomenclature).

89

aromatic, a point which sometimes confuses people with no chemical training. Finally, the ring system hydrogens are substituted by chlorine atoms in the appropriate places to give the planar structure (Figure 8, structure 2) of heptachlor, which is 1,4,5,6,7,8,8-heptachloro-3a,4,7,7a-tetrahydro-4,7-methanoindene, a name which gives no indication of the three-dimensional structure of the molecule. The "left-hand" nucleus of heptachlor is clearly the hexachloro-derivative of the norbornene molecule discussed earlier and is called the hexachlorobicyclo[2.2.1]hept-2-ene (hexachloronorbornene) nucleus. What remains to be defined is the manner in which this nucleus is fused to the chlorocyclopentene ring, and the orientation of the chlorine atom at C_1.

There is much evidence that in cyclodiene compounds, the addendum (in this case the chlorocyclopentene nucleus) is fused to hexachloronorbornene through its *endo*-positions, and this appears to be taken for granted in naming simple molecules like heptachlor, so that the name contains no indication of the nature of the stereochemistry of this ring junction. When simple ring systems such as cyclopentene or cyclopentane are attached to hexachloronorbornene, substituents in these rings are called *exo*- or *endo*- according to whether they are on the same or opposite sides of the ring as the *exo*-hydrogens at the ring junction. Another way of expressing this is to use the terms *syn*- (on the same side of the ring as the *exo*-hydrogens) and *anti*- (on the opposite side of the ring to these hydrogens). This last usage has the advantage that it is convenient to use for designating individual substituents and avoids confusion with the terms *exo*- and *endo*- relating to the basic stereochemistry of the carbon skeleton. The terms *cis*- and *trans*- have also been used with the same meanings, but these are best reserved to designate the relationship between individual substituents when more than one is present in the ring (examples below). The C_1-chlorine of heptachlor is known to be in the *exo*-position (Figure 8, structure 3) and is so indicated by the expression 1-*exo*, ... in the full name of heptachlor, which as indicated above is only incomplete in respect to the lack of formal designation of the ring junction stereochemistry; this could, in fact, be rectified by writing -*endo*-4,7-methanoindene in the name, as is done in the case of aldrin and dieldrin (below).

In an exactly similar manner, the molecule 1-*exo*-4,5,6,7,8,8-hexachloro-3,4,7,7a,-tetrahydro-4,7-methanoindane (Figure 8, structure 8) can be regarded as being derived from indane (Figure 8, structure 7) and again the C_1-chlorine atom is in the *exo*-position. Since this molecule is a derivative of heptachlor in which the 2,3-double bond is reduced, it is convenient to call it α-2,3-dihydroheptachlor, or simply α-dihydroheptachlor, abbreviated further to α-DHC. Other molecules will be discussed later which have the cyclopentane chlorine atom at C_2; although α-DHC is the only one of these isomers which is strictly analogous to heptachlor, the others have also been called dihydroheptachlors and are designated β-DHC and γ-DHC (Figure 14) to distinguish them from the compound just discussed.[167] At this point, it may be mentioned that heptachlor and a number of the molecules in this series are asymmetrical and are capable, in principle, of being resolved into optically active enantiomorphs. This should be kept in mind but need not be considered for most purposes. If a chlorine atom is present at each of carbons C_1 and C_2 (Figure 8, structure 9) there are four possible isomeric forms (apart from optical isomerism) of the resulting molecule (the ring junction being assumed *endo*- in each case). The chlorine atoms may both be on the *exo*-side of the cyclopentane ring (with reference to the ring junction *exo*-hydrogens), in which case they are *exo*- (i.e., 1-*exo*, 2-*exo*-) and also *cis;* they may both be on the *endo*-side of the ring, in which case they are *endo*- (i.e., 1-*endo*, 2-*endo*-) and *cis,* or they may be on opposite sides of the ring, in which case they are *trans*. In the last case, the possibilities are 1-*exo*, 2-*endo*- and 1-*endo*, 2-*exo*-. Only two of these four possible isomers, namely, 1-*exo*, 2-*endo*- and 1-*exo*, 2-*exo*-, are actually known, and they are known as α- (better called *trans*-) and β- (better called *cis*-) chlordane, respectively. The 1-*exo*, 2-*endo*- isomer (α-) is shown three-dimensionally in Figure 8. The full chemical name for this compound is therefore 1-*exo*, 2-*endo*-4,5,6,7,8,8-octachloro-3a,4,7,7a-tetrahydro-4,7-methanoindane. When the *exo*, *endo*-terminology is used in this manner, it is not necessary to use the terms *cis*- and *trans*- as well, since this is fully indicated by the other terms, but it is convenient to refer to *cis*- and *trans*-chlordane for simplicity. There are, of course, other possible isomers of the chlordane structure in which the chlorine atoms are on C_1 and C_3 of the cyclopentane ring, and these are usually designated for simplicity by different Greek prefixes, as in the case of the dihydroheptachlors. Other groups attached to the cyclopentane ring are designated similarly; for example, the main biological metabolite of heptachlor is its 2,3-*exo*-epoxide, called 1-*exo*-4,5,6,7,8,8,-heptachloro *exo*-2,3-epoxy-3a,4,7,7a-tetrahydro-4,7-methanoindane (Figure 14, structure 4).

In a similar manner, the compounds of the aldrin and isodrin series may be regarded, although they are *not* aromatic, as deriving from naphthalene (Figure 8, structure 11). In this case, each of the fused benzene rings is -CH$_2$-bridged (hypothetically) to give 1,4;5,8-dimethanonaphthalene (Figure 8, structure 12). Two 1,4-additions of hydrogen (that is, one hydrogen atom added to each of the carbons 1,4,5 and 8) and one hydrogen atom added to each of the carbons 4a and 8a then give the basic carbon skeleton 1,4,4a,5,8,8a-hexahydro-1,4;5,8-dimethanonaphthalene (13) of these two series. Hydrogen atoms are then replaced by chlorine in the appropriate positions, and the planar structure 1,2,3,4,10,10-hexachloro-1,4,4a,5,8,8a-hexahydro-1,4;5,8-dimethanonaphthalene (Figure 8, structure 14) results.

As indicated earlier, the situation in three dimensions is more complicated than in the case of heptachlor, since not only is the stereochemistry of the ring junction at positions 4a and 8a undefined, but for each of the two possible configurations there (*endo*- or *exo*-), the methano-bridge of the unchlorinated norbornene nucleus may be in one of two different configurations (Figure 8, structures 15 and 16; only *endo*-fusion is shown) relative to the dichloromethano-bridge. Molecules of this type are named so that the stereochemistry is fully defined. Unfortunately, two systems are in use, based on dimethanonaphthalene, which lead to precisely opposite terminology. In the British system, the prefixes *exo*- or *endo*- are applied to the methano-bridges according to whether they are "outside" or "inside" the bent cage structure as a whole. The first prefix that appears in the name refers to the methano-bridge of the lowest numbered nucleus, which is the one carrying the chlorine atoms, while the second prefix applies to the methano-bridge in the highest numbered nucleus. In this system, the 1,4-dichloromethano-bridge is *exo*- and the 5,8-methano-bridge *endo*- in the case of structure 15, which is therefore called 1,2,3,4,10,10-hexa-chloro-1,4,4a,5,8,8a-hexahydro-1,4-*exo*, *endo*-5,8-dimethanonaphthalene (abbreviated to HHDN). In isodrin (16), both methano-bridges are outside the bent cage structure (*exo*-) and the molecule has the same name as HHDN, but with 1,4-*exo*, *exo*-5,8- instead of 1,4-*exo*, *endo*-5,8-. When the unchlorinated double bond in these structures is oxidized to give the corresponding epoxide, the resulting epoxide ring is generally accepted to be fused through the *exo*-positions of the norbornene nucleus. Such a ring would be *endo*- when added to HHDN, but could conceivably be either, although probably should be called *exo*- when added to isodrin, which is not very satisfactory.

In the American system, which is used in this book, the terms *exo*- and *endo*- have the different meaning which was defined in the preliminary discussion of the Diels-Alder reaction; they indicate which of the pairs of *exo*- or *endo*- bonds of each of the norbornene nuclei are used to fuse the two ring systems together and each prefix therefore defines the stereochemistry of the norbornene nucleus to which it applies.[168] As before, the first prefix appearing in the name refers to the lowest numbered and therefore the chlorinated nucleus. Accordingly, in HHDN (Figure 8, structure 15), the hexachloronorbornene nucleus contains the 1,4-methano-bridge and is fused to the attached norbornene system through its *endo*-positions so that it is designated 1,4-*endo*. On the other hand, the norbornene nucleus containing the 5,8-methano-bridge is fused to the hexachloronorbornene nucleus through its *exo*-positions and it is therefore designated 5,8-*exo;* the full name therefore becomes 1,2,3,4,10,10-hexachloro-1,4,4a,5,8,8a-hexahydro-1,4-*endo*, *exo*-5, 8-dimethanonaphthalene. In the case of isodrin (Figure 8, structure 16), both norbornene nuclei are fused through their *endo*-positions and the name is the same as for HHDN, except that -1,4-*endo*, *endo*-5,8- replaces -1,4-*endo*, *exo*-5,8-. Since this nomenclature describes the mode of fusion of addenda to the norbornene ring system, *exo*-fused epoxide rings are described as such in the name. Thus, the molecule HEOD (Figure 9) is 1,2,3,4,10,10-hexachloro-6,7-*exo*-epoxy-1,4,4a,5,6,7,8,8a-octahydro-1,4-*endo*, *exo*-5,8-dimethanonaphthalene in *this* naming system.

HHDN and HEOD are the names of the pure major components of the *technical* compounds known as aldrin and dieldrin, respectively, and this is another source of difficulty with regard to

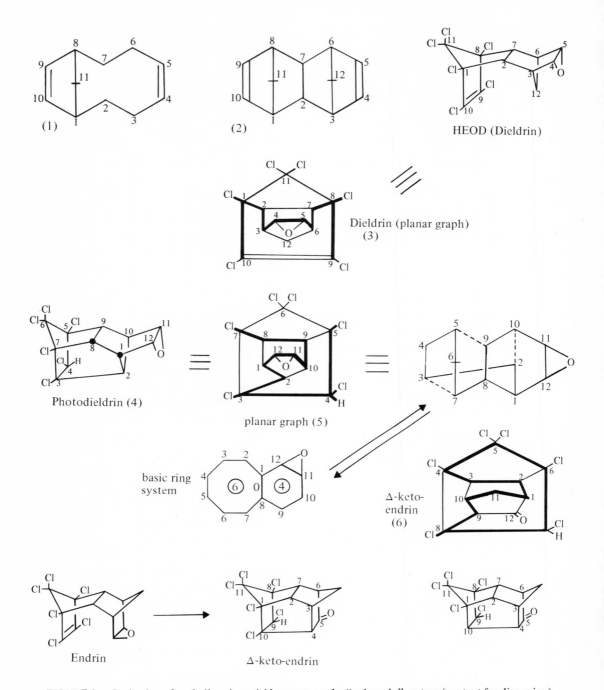

FIGURE 9.   Derivation of cyclodiene insecticide names on the "polycyclo" system (see text for discussion).

nomenclature. The product aldrin is a material which contains not less than 95% of HHDN (mp 104 to 104.5°C), together with a remainder of related compounds (Vol. 1, Chap. 3C.3a,b). Dieldrin contains not less than 85% of HEOD (mp 175 to 176°C). Much experimentation has been conducted with the pure compounds, and in these cases, the names HHDN and HEOD should have been used throughout or the definitions made clear to begin with. However, because of the complexity of the proper names, it has been highly convenient to use the names aldrin and dieldrin to denote the pure compounds. An additional advantage is that these names lend themselves readily to the convenient assignment of names to simple derivatives. For example, reduction of the 6,7-double bond in HHDN gives a 6,7-dihydro-HHDN which is very conveniently called 6,7-dihydroaldrin, and this name requires only a little elaboration to include other derivatives of this structure. Thus, the *trans*-diol produced by hydrolytic opening of the epoxide ring of dieldrin is a derivative of dihydroaldrin, and can conveniently be called *trans*-6,7-dihydroxydihydroaldrin. Isodrin, the aldrin stereoisomer, is not marketed as a technical material and the name is used to denote the pure compound. On the other hand, technical endrin contains 92% of the pure *endo, endo* stereoisomer of HEOD. While this compound might have been called *endo, endo*-HEOD (since HEOD is an abbreviation for the basic planar skeleton), or perhaps β-HEOD, it is simply called endrin, in analogy with the isodrin situation.

It is regrettable that the names derived from simple aromatic compounds are inadequate to describe accurately the more complex ring structures produced by the various molecular rearrangements and that some of the names in current use are difficult to translate into structures that are recognizable and understandable in terms of the origin of the molecules as outlined above. A useful discussion of the problem has been given by Benson,[169] who provides diagrams illustrating the way in which names are derived on the "polycyclo" system of nomenclature, although a more rigorous application of nomenclature rules leads to different names for photodieldrin and the related Klein ketone (see below and Figure 9).[169a] Further examples are to be found in the papers by Bird and colleagues,[170] and by Parsons and Moore.[171]

The planar structure (Figure 8, structure 14) shown for aldrin or isodrin (aldrin is used here to denote HHDN) consists of four fused five-membered rings and is regarded as deriving from the basic molecule (Figure 9, structure 1) in which a nine- and a five-membered ring are separated by the 1,8-methano-bridge. Starting from C_1 and considering the rings in order of size, there are *six* carbon atoms (2,3,4,5,6, and 7) between C_1 and C_8 to the right of the bridge, two to the left of it (9 and 10), and *one* (C_{11}) on the bridge itself. The situation is now similar to the simple case of norbornene, which has two five-membered rings and seven carbon atoms and is a *bicyclo* [2.2.1] heptene. The more complicated molecule under discussion contains eleven carbon atoms and is *bicyclo* [6.2.1] undeca-4,9-diene, which is converted into the desired planar cyclodiene structure by joining carbons 2 and 7, and placing a methano-bridge between positions 3 and 6. Therefore, there are *no* carbon atoms between positions 2 and 7 in the final structure and *one* carbon atom between positions 3 and 6. On this basis, the final basic structure contains 12 carbons and is tetracyclo $[6.2.1.1^{3,6}0^{2,7}]$ dodeca-4,9-diene (Figure 9, structure 2). The six chlorine atoms may now be inserted to give, for aldrin or isodrin, 1,8,9,10,11,11-hexachlorotetracyclo$[6.2.1.1^{3,6}0^{2,7}]$dodeca-4,9-diene.

The same result is obtained by drawing the molecule according to Eckroth's method for the derivation of correct von Baeyer names for the polycyclic hydrocarbons (based on IUPAC 1957 rules), as described by Benson.[169] In the derivation discussed above, the molecule was considered in "plan" view. In the Eckroth system,[172] the molecule is presented as a planar graph or Schlegal diagram in which the molecule viewed from one end is collapsed or projected so that all the ring systems are visible and the *largest possible* circuit of the planar graph is chosen for numbering according to IUPAC rules.[172a] The application of this principle to dieldrin results in the planar graph (3) shown in Figure 9, in which the same numbering system is used as before.

Both approaches lead to the same planar structure, but the stereochemistry of the ring junction between the norbornene systems and of the epoxide rings now requires definition. From this point, the stereochemistry is defined in the publications of Bird and colleagues and those of Parsons and Moore, by the assignment of an α-prefix to the substituent at position 1 and then relating the disposition of the substituents at C_2 and C_3 to the C_1-substituent according to whether they are on the same side (α) or different sides of the plane (considered to be flat for this purpose) of the largest ring. The stereochemistry of the epoxide ring was implied but not defined in these publications. The α-, β-system seems less satisfactory than the system suggested by Benson,[169] which is related to the American method of assigning the ring fusion stereochemistry.

As shown previously, the disposition of the two methano-bridges relative to each other is defined as soon as the mode of fusion (*endo*- or *exo*-) of the norbornene systems, each to the other, has been defined. The planar graph (3) (Figure 9) is drawn to look like dieldrin (used here to denote pure HEOD), but may equally well represent endrin once the mode of ring-fusion is defined. Thus, the complete description of dieldrin is given by saying that the 2,3- and 7,6-bonds are *endo*-, and the 2,1- and 7,8-bonds are *exo*- (using that order of numbers for the bonds that indicates which norbornene nuclei are being considered). Since the weight of evidence supports the *exo*-configuration for the 4,5-epoxide ring, the full name now becomes 1,8,9,10,11,11-hexachloro-4,5-*exo*-epoxy-2,3-7,6-*endo*-2,1-7,8-*exo*-tetracyclo[$6.2.1.1^{3,6}0^{2,7}$]dodec-9-ene. The corresponding name for endrin differs only in that the 2,1- and 7,8-bonds are *endo*-, and so we have -2,1-7,8-*endo*-tetracyclo- . . . for this molecule.

The photoisomer of dieldrin (photodieldrin; Figure 9, structure 4), in which a bridge is formed between the dieldrin-9 (or 10) position and the C_{12}-methano-bridge (Figure 9), presents a more complicated example. In this case, the full name is 3 *exo*-4,5,6,6,7-hexachloro-11,12-*exo*-epoxy-pentacyclo[$6.4.0.0^{2,10}.0^{3,7}.0^{5,9}$]dodecane, in which the fusion of the norbornene nuclei is fully defined by the presence of the C_2-C_3 bridge. The name is derived quite logically, as can be seen by working backwards with the aid of Figure 9. Unfortunately, it is quite difficult, without foreknowledge of the compounds involved, to arrive at a recognizable cyclodiene structure starting from this description. As in the case of dieldrin, the name is arrived at by considering the planar graph obtained by projecting the end on view of the molecule as shown in Figure 9. In this case, the presence of the additional bond between C_2 and C_3 makes possible a complete circuit called a Hamiltonian line which encompasses all of the

carbon atoms and passes through every vertex only once; it converts the molecule into a basic [6.4.0] ring system. In this molecule, the chlorine atom at C_4 may be either *exo*- or *endo*- relative to the hexachloronorbornane nucleus and has been assigned the *exo*-position, as signified in the name; the [6.4.0] ring system obtained when C_1-C_8 is the main bridge is preferred over the less symmetrical [7.3.0] system afforded if C_3-C_7 is regarded as the main bridge. One of the metabolites produced from both dieldrin and photodieldrin in rats is the ketone, called Klein's metabolite, which has been assigned the structure of photodieldrin in which C_4 has been converted into a carbonyl group with concurrent loss of the chlorine and hydrogen atoms. The correct name for this compound is therefore derived from that of photodieldrin by a simple modification to indicate this change; it is 3,5,6,6,7-pentachloro-11,12-*exo*-epoxy-pentacyclo[$6.4.0.0^{2,10}0^{3,7}.0^{5,9}$]dodecan-4-one. Since the two molecules just discussed lack any features of symmetry, each should be capable of resolution into two optically active enantiomorphs.

In the case of the "polycyclo" names for dieldrin and endrin, one can draw from the written structure something recognizable as a dimethanonaphthalene-derivative by a reversal of the above process for deriving the name. The [6.2.1] system is written down as a symmetrical ring system as shown (Figure 9, structure 1), with six- and two-membered chains fused to the one-membered bridge. One end of the bridge has to be C_1, and the rest of the system is numbered sequentially, starting with the largest ring as shown. The rest of the name within square brackets shows where a methano-bridge is to be inserted and a direct C-C link is to be made, the result being recognizable as the planar depiction of the dimethanonaphthalene structure because the system is symmetrical. All that remains is to add the epoxide ring, the double bond, and the chlorine atoms according to the numbers to complete the planar formula, which can then be written three-dimensionally by attention to the stereochemical instructions in the name.

Considering the planar graph for photodieldrin, the reader will see that the Hamiltonian line is that which remains when the 2-10, 3-7, and 5-9 carbon-carbon bonds are removed from the planar dimethanonaphthalene structure. If the latter structure, minus these bonds, is imagined in three dimensions and then flattened out, there is obtained the fused six- and eight-membered ring system shown, which is the [6.4.0] skeleton contained in the systematic name and is the starting point for construction of the three-dimensional molecule. The numbering is as before, proceeding round the large ring from C_1, which is one end of the main bridge. Insertion of the other links as indicated by the $0^{2,10}.0^{3,7}.0^{5,9}$ designations in brackets results in a depiction of the dimethanonaphthalene system that is complicated by the asymmetry due to the additional bridge; it becomes more recognizable when the epoxide ring, the keto-group (for Klein's ketone) and the chlorine atoms are inserted according to the numbers.

In the case of the molecular rearrangement products of endrin, prominent among which are ketones produced by cross-linking to form half-cage structures (Figure 9), widely used but systematically incorrect names simply adapt the polycyclo- name for endrin by accounting for the fact that certain other carbon atoms are now linked. Thus, the half-cage pentachloro- and hexachloro-ketone which have been obtained as ultraviolet irradiation products of endrin each have a new bond between C_4 and C_{10} of the dimethanonaphthalene skeleton and a keto-group at C_5; the pentachloro-ketone has lost a formerly vinylic chlorine atom from the carbon to which the new link was formed. The inclusion of the additional information '$0^{4,10}$' in the description of the ring system, which is now *pentacyclic* instead of *tetracyclic,* gives for the hexachloro-ketone the name 1,8,*exo*-9,10,11,11-hexachloropentacyclo[6.2.1.1^{3,6}0^{2,7}.0^{4,10}]dodecan-5-one, in which the C_9-chlorine is assigned the *exo*-configuration. Similarly, the pentachloro-ketone is 1,8,*exo*-9,11,11-pentachloropentacyclo[6.2.1.1^{3,6}.0^{2,7}.0^{4,10}]dodecan-5-one. The link between C_4 and C_{10} shows that both norbornane nuclei must be in the *endo*-configuration.

With the same notation, the full-cage molecule (Figure 21, structure 3) obtained by the rearrangement of isodrin under various conditions (called photoisodrin) becomes 1,8,9,10,11,11-hexachlorohexacyclo[6.2.1.1^{3,6}.0^{2,7}.0^{4,10}.0^{5,9}]dodecane, which shows that there are now bonds between the pairs of carbon atoms 4, 10 and 5, 9 in the dimethanonaphthalene nuclei, making a hexacyclic system, and this again means that both norbornane rings necessarily have the *endo*-configuration. Bird

et al.[170] arrived at other names for the half-cage ketones which are based on the [7.2.1] ring system. In fact, rigorous application of the IUPAC rules, taking the longest Hamiltonian line in an appropriate planar graph (Figure 9, structure 6), gives the name 4,5,5,6,*exo*-7,8-hexachloropentacyclo[7.2.1.0^{2,6}.0^{3,10}.0^{4,8}]dodecan-12-one for Δ-keto-endrin (Figure 9), the other ketone being its 4,5,5,6, *exo*-7-pentachloropentacyclo-relative. Photoisodrin becomes 1,7,8,11,12,12-hexachlorohexacyclo[5.4.1.0^{2,6}.0^{3,10}.0^{5,9}.0^{8,11}]dodecane, so that these names disguise the structural relationships in this series.

Another interesting series of compounds results from the rearrangement in ultraviolet light of heptachlor (photoheptachlor), or from the addition of two molecules of hexachlorocyclopentadiene to give mirex (Figure 10). By applying the process used for dieldrin, the planar graph shown is obtained for heptachlor which gives the basic ring system as tricyclo[5.2.1.0^{2,6}]deca-3,8-diene, and the complete molecule as 1,*exo*-5,7,8,9,10,10-heptachloro-2,3-6,5-*endo*-tricyclo[5.2.1.0^{2,6}]deca-3,8-diene, in which the *endo*- fusion of the cyclopentene ring to the hexachloronorbornene system is indicated. In the rearrangement product, photoheptachlor, it is seen that links are now present between the pairs of carbon atoms 3,9 and 4,8 in the first planar graph, and so the product may be simply related to heptachlor (as photodrin and the ketones from endrin were related to endrin) by modifying the above name to account for the new bonds and for the fact that two additional rings have been formed by the bridging process. Photoheptachlor is then described as 1,*exo*-5,7,8,9,10,10-heptachloropentacyclo[5.2.1.0^{2,6}.0^{3,9}.0^{4,8}]decane (since the double bonds are no longer present).

However, photoheptachlor is more correctly represented by the second planar graph of Figure 10, in which the C_1-C_7 bond forms the main bridge in the basic ring system, thereby permitting a complete circuit (heavy line) which includes every carbon atom. The basic system now becomes pentacyclo[5.3.0.0^{2,6}.0^{3,9}.0^{5,8}]decane and the complete molecule is 1,*exo*-4,7,8,9,10,10-heptachloropentacyclo[5.3.0.0^{2,6}0^{3,9}.0^{5,8}]decane. Mirex is completely chlorinated and is the related symmetrical molecule in which there is a bond between C_4 and C_8 instead of C_5 and C_8. In this case, therefore, the basic ring system is pentacyclo[5.3.0.0^{2,6}.0^{3,9}.0^{4,8}]decane and the full name is

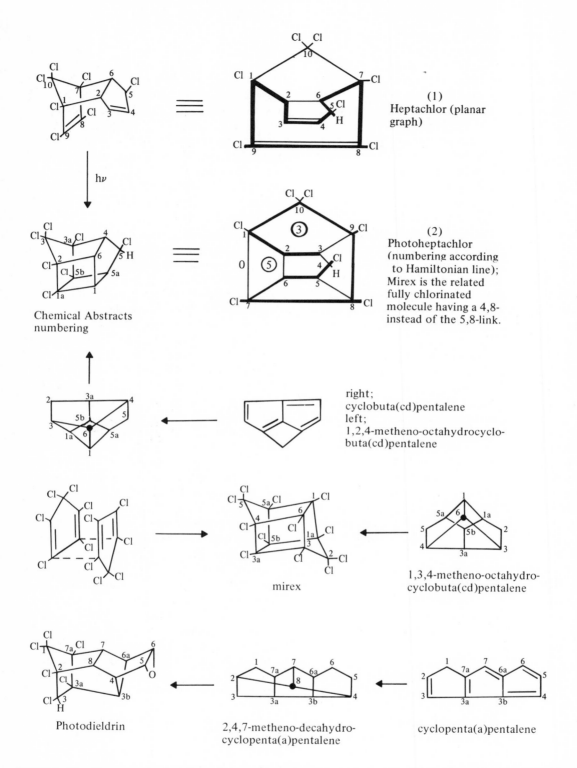

FIGURE 10. Planar graph depiction of heptachlor, photoheptachlor, and mirex. Lower half; *Chemical Abstracts* derivation of systematic names for mirex, photoheptachlor, and photodieldrin.

1,2,3,4,5,5,6,7,8,9,10,10-dodecachloropentacyclo [5.3.0.0^{2,6}.0^{3,9}.0^{4,8}] decane, which might be called dodecachloropentacyclo [5.3.0.0^{2,6}.0^{3,9}.0^{4,8}] decane or perchloropentacyclo [5.3.0.0^{2,6}.0^{3,9}.0^{4,8}] decane, since there is no ambiguity regarding the positions of the chlorine atoms in this molecule. The related ketone, called Kepone®, has the dichloromethano-bridge at C_5 replaced by a carbonyl group, and would be called decachloropentacyclo [5.3.0.0^{2,6}.0^{3,9}.0^{4,8}] decan-5-one. Names for these last two compounds may also be derived using the planar graph drawn for heptachlor in Figure 10, although the IUPAC rule requiring the largest possible complete circuit is thereby ignored; the extra bridges in these molecules are between C_3 and C_9, and between C_5 and C_8, so that with this numbering system, each is based on the [5.2.1.0^{2,6}.0^{3,9}.0^{5,8}] ring system. Mirex becomes dodecachloropentacyclo [5.2.1.0^{2,6}.0^{3,9}.0^{5,8}] decane, and Kepone, decachloropentacyclo [5.2.1.0^{2,6}.0^{3,9}.0^{5,8}] decan-4-one. Names for these two compounds derived by both methods are to be found in the literature. While use of the [5.2.1] ring system is helpful to the user in that it relates the cage structures to the well-known molecule heptachlor, it is essential for uniformity in documentation that the IUPAC Rules are rigorously applied; hence, the [5.3.0] system is preferred.

Chemical Abstracts uses the methanoindene or dimethanonaphthalene naming system for the simple cyclodiene structures, and a rather different polyclonomenclature from the one described above for the more complicated molecules. This can be seen in Figure 10, in which photodieldrin is shown to be derived from three linearly fused cyclopentane rings linked together by the carbon atom at position 8 (the metheno-group). Three cyclopentane rings can be fused together in a variety of ways, and for each mode, an unsaturated structure can be envisaged having a maximum of five double bonds. This hydrocarbon is one of the cyclopentapentalenes and the designation [a] in cyclopenta[a]pentalene means that the three five-membered nuclei are linearly fused. Reduction of all five double bonds requires ten hydrogen atoms, so the fully reduced, linearly fused system is decahydrocyclopenta[a]pentalene. The metheno-carbon (C_8) is then linked to positions 2, 4, and 7, and the basic skeleton becomes decahydro-2,4,7-metheno-1H-cyclopenta[a]pentalene. The symbol 1H means that the only CH$_2$

group in the original unsaturated molecule was at C_1 rather than C_3 and fixes the position of the original double bond in this terminal ring. When the chlorine atoms and the 5,6-epoxide ring are inserted, we derive the full name for the planar structure as 1,1,2,3,3a,7a-hexachloro-5,6,-epoxy-decahydro-2,4,7-metheno-1H-cyclopenta[a]pentalene.

If C_8 is now pulled out of the plane of the paper, the three-dimensional structure is obtained, and the stereochemical disposition of the methano-bridges relative to one another is fixed by the mode of linking (that is, 2,4,7-) of this carbon atom; if the linking is 2,5,7-, it will be seen that the bridged system becomes a half-cage structure derived from the isodrin series of compounds. In accordance with the previous procedure, the epoxide ring should be designated *exo-* in the above name. In this terminology, Klein's ketone becomes 1,1,2,3a,7a-pentachloro-*exo*-5,6-epoxy-decahydro-2,4,7-metheno-3H-cyclopenta[a]pentalene-3-one, in which 3H indicates that the CH$_2$ group in the original unsaturated molecule was at C_3. These molecules are derived from the fully saturated dechydrocyclopenta[a]pentalene structure and specification of the positions of the hydrogen atoms is consequently unnecessary. However, a double bond in the molecule must have its position indicated, and this is done by indicating all the positions which carry hydrogen, so that those not so indicated necessarily carry the double bond. Accordingly, photoaldrin, the photoisomer of aldrin and the unsaturated precursor of photodieldrin, has a 5,6-double bond, and is therefore 1,1,2,3,3a,7a-hexachloro-2,*exo*-3,3a,3b,4,6a,7,7a-octahydro-2,4,7-metheno-1H-cyclopenta[a]pentalene.

The series of compounds derived from heptachlor by molecular rearrangement may also be described by this system. Thus, photoheptachlor (Figure 10) is a derivative of cyclobutapentalene, two fused five-membered rings and a four-membered ring (the "base" of the cage as depicted in Figure 10) linked once again by a metheno-group (C_6), as is seen when the structure is flattened. In this case the original unsaturated molecule would consist of the two five-membered rings containing a maximum of four double bonds and having positions 1a and 5a linked by a single carbon atom (C_1, which forms the cyclobutane ring). The fully saturated molecule is therefore an octahydro-derivative, namely, octahydrocyclobuta[cd]pentalene, in which [cd] indicates the way in which the cyclobutane ring is incorporated into the structure. The metheno-group (C_6) is now linked to the 1,2 and 4-positions to give octahydro-1,2,4-metheno-1H-cyclobuta[cd]pentalene, in which 1H indicates that the CH_2 group in the original was in the 1-position. Substitution of the chlorine atoms in the proper positions now gives 1a,2,3,3,3a,exo-5,5b-heptachlorooctahydro-1,2,4-metheno-1H-cyclobuta[cd]pentalene. The compound known as photoheptachlorketone, which is produced by oxidative removal of the chlorine atom at C_5, is therefore 1a,2,3,3,3a,5b-hexachlorooctahydro-1,2,4-metheno-1H-cyclobuta[cd]pentalene-5-one. These molecules are asymmetric.

In mirex, the two five-membered rings are fused together in a slightly different manner to give a fully symmetrical molecule; the basic structure differs from those previously discussed only in that the C_6-metheno-group is linked now to the 1,3 and 4-positions giving (since the molecule is fully chlorinated) 1,1a,2,2,3,3a,4,5,5,5a,5b,6-dodecachlorooctahydro-1,3,4-metheno-1H-cyclobuta[cd]-pentalene, or simply dodecachlorooctahydro-1,3,4-metheno-1H-cyclobuta[cd]pentalene. Since Kepone has the two chlorines at C_2 (choosing the lowest numbered carbon rather than C_5) replaced by carbonyl, it is decachlorooctahydro-1,3,4-metheno-2H-cyclobuta[cd]pentalen-2-one.

There are two examples of heterocyclic nomenclature in the cyclodiene group which may cause some difficulty. The compound "Telodrin" (Figure 17, structure 5) is also called isobenzan, a common name derived directly from the systematic name exo-1,exo-3,4,5,6,7,8,8-octachloro-1,3,3a,4,7,7a-hexahydro-4,7-methanoisobenzofuran. The basic structure from which this name derives is isobenzofuran (Figure 8, structure 17), and the name is obtained just as the names of other structures of Figure 8 are derived from naphthalene or indene. "Thiodan," commonly called endosulfan, is regarded as deriving from 2,4,3-benzodioxathiepin-3-oxide (Figure 8, structure 18), and is 6,7,8,9,10,10-hexachloro-1,5,5a,6,9,9a-hexahydro-6,9-methano-2,4,3-benzodioxathiepin-3-oxide. In these names the methano-bridges are, of course, C_8 for isobenzan and C_{10} for endosulfan. Neither name, as normally used,

specifies the stereochemistry of the ring fusions (3a,7a, or 5a,9a, respectively), and strictly speaking, this should be done by saying "endo-4,7-methano" and "endo-6,9-methano," in analogy with the dimethanonaphthalene series. In the case of endosulfan, the pyramidal configuration of the sulfite ester moiety introduces isomeric possibilities and two isomers, designated α and β-, are found.

In the sections which follow, the full systematic names will only be used if this is considered essential; otherwise, simple names will be used which are derived from the nearest well-recognized molecule, as has usually been the custom in organic chemistry. One example is trans-6,7-dihydroxydihydroaldrin (or trans-dihydroaldrin-diol) for the trans-diol from dieldrin and this name is used here, although the molecule is perhaps better named trans-4,5-dihydroxydihydroaldrin, using the polycyclo-system. Another is trans-1,2-dichlorodihydrochlordene, for trans-chlordane. It is hoped that these names will be clearly understandable with the aid of the figures in the text.

In view of the increasing popularity of the "polycyclo" nomenclature, extracts from the relevant IUPAC Rule (A. 32) for deriving names on this system are appended here for convenience:

Rule A.32.12*
A polycyclic system is regarded as containing a number of rings equal to the number of scissions required to convert the system into an open chain compound.

32.13. The word tricyclo, tetracyclo, pentacyclo, etc. is followed by brackets containing, in decreasing order, numbers indicating the number of carbon atoms in:

the two branches of the main ring
the main bridge
the secondary bridges

32.21. The main ring and the main bridge form a bicyclic system.

32.22. The location of the other or so-called secondary bridges is shown by superscripts following the number indicating the number of carbon atoms in these bridges.

32.33. For the purposes of numbering, the

*From IUPAC Nomenclature of Organic Chemistry, 3rd. ed., Butterworth and Co. Ltd., London, 1971. With permission.

secondary bridges are considered in decreasing order.

32.31. When there is a choice, the following criteria are considered in turn until a decision is made: (i.) The main ring shall contain as many carbon atoms as possible, two of which must serve as bridgeheads for the main bridge. (ii.) The main bridge shall be as large as possible. (iii.) The main ring shall be divided as symmetrically as possible by the main bridge. (iv.) The superscripts locating the other bridges shall be as small as possible.

B. SYNTHESIS AND CHEMISTRY

1. Hexachlorocyclopentadiene and its Self-condensation Products: Chlordecone, Despirol®, Mirex, Pentac®

Hexachlorocyclopentadiene is clearly the key compound in the synthesis of cyclodiene insecticides. A number of synthetic routes are available for this compound, which along with the analogous hexabromo-derivative, was first described in 1930 by Straus, Kollek, and Heyn, who prepared it by what might be called the "wet" method.[173] This method starts from cyclopentadiene (Figure 11, structure 2) which is available as a by-product from the manufacture of butadiene. Cyclopentadiene dimerizes readily at ambient temperature by an addition reaction (the Diels-Alder reaction) in which one molecule acts as diene and the other as dienophile (Vol. I, Chap. 3A.1). The dimer (dicyclopentadiene; Figure 11, structure 1) is reconverted into the monomer (2) by thermal depolymerization at 180°C, a process which results in some loss due to the formation of higher polymers from the dimer.

Basically, the preparation of hexachlorocyclopentadiene (Figure 11, structure 3) involves the chlorination of cyclopentadiene with 6 to 11 mol of aqueous 0.25-4.5 molar sodium or potassium hypochlorite at temperatures ranging from −5°C to +50°C. Solvents such as carbon tetrachloride, pentane, hexane, chloroform, isopropyl ether, or benzene may be used in up to 10:1 ratio with the cyclopentadiene, but the reaction may equally well be conducted without solvent, and this avoids the necessity for its removal afterwards. A number of side reactions can occur which lead to in-

efficient utilization of cyclopentadiene and hypochlorite. These can be minimized by the addition of 1 mol %, based on the cyclopentadiene, of sodium sulfamate or related compounds, both the yield and quality of the product being thereby improved.[174] In some preparations, an emulsifying agent such as sodium lauryl sulfate is also added.[175] The reaction is moderately exothermic and is soon completed. In a typical preparation, cyclopentadiene (0.5 mol) is added to a vigorously stirred solution of aqueous sodium hypochlorite (3 mol) containing sodium sulfamate (0.005 mol). The initial temperature of 25°C is maintained by cooling, and 82% of the sodium hypochlorite is utilized in 20 min. The organic layer is separated and fractionated *in vacuo* to give 55% of hexachlorocyclopentadiene (bp 60 to 62°C/1 mm). The main by-product of this synthesis is 1,2,3,4,5-pentachlorocyclopentadiene (Figure 11, structure 4), which gives a dimer, mp 214°C, on standing, and is accompanied by other isomers as subsidiary products. If the amount of sodium hypochlorite and the reaction time are reduced, the proportion of pentachlorocyclopentadiene can be considerably increased. Compounds such as 1,4,5,5-tetrachlorocyclopentadiene can also be isolated from the low boiling material preceding the hexachloro-compound (called "hex") in the fractionation.[176]

The wet method may also be used to prepare pentachloro-methylcyclopentadiene as well as hexabromocyclopentadiene, and has been employed for the commercial preparation of "hex"; the other industrial processes involve the chlorination of mixtures of C_5-hydrocarbons.[177] Thus, pentane, isopentane, or mixtures of the two are chlorinated photochemically in the liquid phase at 80 to 90°C until they have an average composition of approximately $C_5H_5Cl_7$ and a density of 1.63 to 1.70. These polychloropentanes are then vaporized with excess chlorine and passed over a porous surface-active solid such as Fuller's earth maintained at 300 to 430°C, then through a tube at 450 to 525°. Octachlorocyclopentene is formed as an intermediate and can be isolated if the conditions of thermal chlorination and temperature are suitable; at 500°C, its dechlorination to hexachlorocyclopentadiene proceeds to completion, and the latter is obtained in 90% yield. Polychloropentanes or cyclopentanes having about seven chlorine atoms per molecule are most

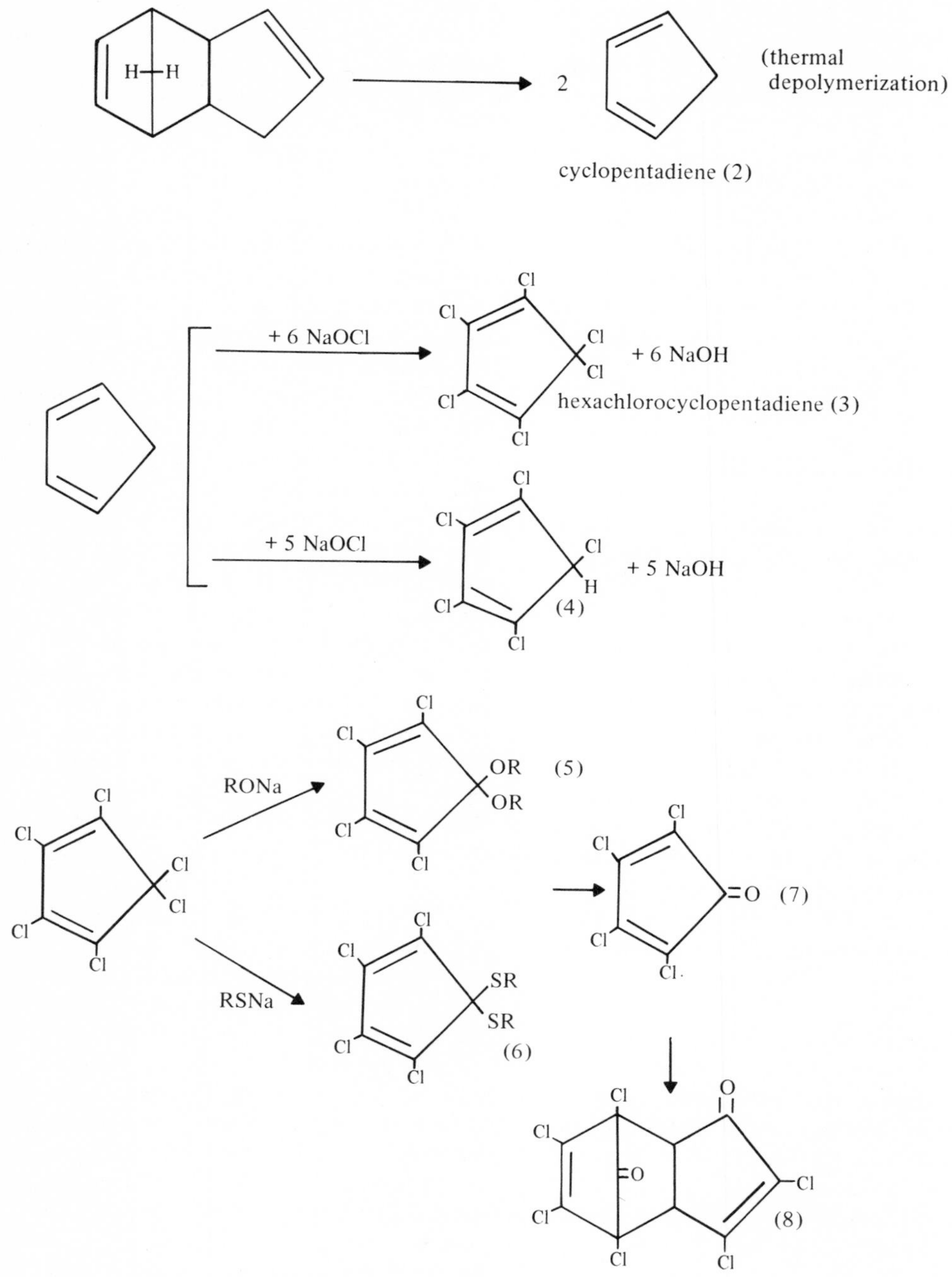

FIGURE 11. Synthesis and reactions of hexachlorocyclopentadiene.

suitable for thermal chlorination; if compounds with lower chlorine content are employed the molar ratio of chlorine to chlorohydrocarbon must be between 6:1 and 9:1 to avoid the risk of explosion, and an optimal ratio of about 6:1 is required to produce good yields of hexachlorocyclopentadiene from the higher chlorinated compounds. Yields of the diene obtained can be from 55 to 75% depending on the composition of the mixture used for thermal chlorination, these yields being improved by the use of a catalyst. The slow, but high yielding dechlorination of octachlorocyclopentene at 500°C is speeded up and made quantitative by the presence of colbalt or nickel surfaces. In another process, cyclopentadiene is chlorinated in the liquid phase at temperatures below 50°C to give tetrachlorocyclopentane, which by catalytic chlorination over arsenious oxide or phosphorus pentachloride at 175 to 250°C gives octachlorocyclopentene nearly quantitatively; the latter is then thermally dechlorinated in the usual way.

Octachlorocyclopentene and hexachlorocyclopentadiene may also be prepared by the Prins reaction, which is of particular interest in view of its utilization in an efficient synthesis of high specific activity ^{14}C-labeled "hex". In the normal synthesis, carbon tetrachloride and trichloroethylene are condensed at 100°C in the presence of aluminum chloride to give first a mixture of mainly C_5 and C_7 chlorinated hydrocarbons, including octachlorocyclopentene (74% yield), a nonachloropent-1-ene, unsaturated ethylcyclopentane derivatives, and dodecachloro-3-ethyl-1-pentene. When pyrolyzed at 500°C, this mixture affords "hex" (75%) together with some hexachlorobenzene, carbon tetrachloride, and tetrachloroethylene. The reaction may be conducted stepwise, and appears to proceed according to the following sequence:[177]

$$CCl_4 + CHCl=CCl_2 \xrightarrow{AlCl_3} CCl_3CHClCCl_3$$

$$CCl_3CHClCCl_3 \xrightarrow{AlCl_3} CCl_3CCl=CCl_2 \text{ (hexachloro-1-propene)}$$

$$CCl_3CCl=CCl_2 + CHCl=CCl_2 \xrightarrow{AlCl_3} CCl_3CHClCCl_2CCl=CCl_2 \text{ (nonachloro-1-pentene)}$$

$$CCl_3CHClCCl_2CCl=CCl_2 \xrightarrow{AlCl_3} CCl_2CCl=CClCCl_2CCl_2 \text{ (octachlorocyclopentene)}$$

Modifications of the Prins synthesis have been used for the laboratory preparation of "hex". For example, hexachloro-1-propene condenses with dichloroethylene in the presence of aluminum chloride to give an octachloropentene (84% yield) which then cyclizes with dehydrochlorination to give "hex" (70% yield):

$$CCl_3CCl=CCl_2 + CHCl=CHCl \xrightarrow{AlCl_3}$$

$$CCl_2=CClCCl_2CHClCHCl_2 \xrightarrow{AlCl_3} \text{ "hex"}$$

octachloropentene

The physical properties of hexachlorocyclopentadiene are well documented;[177] it is a clear, greenish-yellow liquid with a harsh, unpleasant odor, a viscosity of 37.5 dynes/cm and refractive index n_D^{20} 1.5647 to 1.5652. Its melting point has been variously reported as 9.6°C, 9 to 10°C, to 8.2°C, and 10 to 10.8°C and boiling point 60 to 62°C or 78 to 79°C/1 mm. A crystalline form melting at −0.8° to −0.2°C is said to be converted by strong cooling into the form melting at 10 to 10.8°C. The latent heat of fusion has been given as 10.0 cal/g and the freezing point depression as 16.1°C/mol. The compound has an absorption band in the UV at 323 nm (log ε, 3.17) in ethanol, which undergoes a hypsochromic shift (a shift to shorter wavelengths) as the two allylic chlorine atoms are successively replaced by hydrogen, ε being increased by this process. The IR spectra of "hex" and the compounds derived from it are notable for their content of a number of prominent absorption bands; "hex" has two bands at 1,603 and 1,572 cm^{-1} (6.24 and 6.36 μ) due to the stretching of the conjugated chlorinated double bond system, and three bands in the carbon-chlorine bond stretching region at 678, 704, and 803 cm^{-1} (14.75, 14.21, and 12.45 μ), all of which are strong, except that at 6.36 μ. There are two bands in the double bond region at 1,572 and 1,606 cm^{-1} in the Raman spectrum.

"Hex" is a remarkably useful synthetic intermediate; it undergoes many chemical transformations which have been reviewed in detail elsewhere,[177] and this account will discuss only those reactions that are relevant in one way or another to insecticide (or acaricide) chemistry. The two allylic chlorines in "hex" are labile and can be replaced by alkyl groups by treating the diene with sodium alkoxide or alcoholic potassium hydroxide at 20 to 60°C in such a way that the base is always present in low excess during the reaction, so that extensive decomposition is avoided. In this way, dialkoxy ketals are formed (Figure 11, structure 5), and analogous thioketals (Figure 11, structure 6) may be produced by reacting the diene with the sodium salts of mercaptans in alcohols or benzene. The ketal bridge in these compounds is readily hydrolyzed under acidic conditions (for example, with concentrated sulfuric acid at 0.5°C), to give the corresponding tetrachlorocyclopentadienone (Figure 11, structure 7) which tends to dimerize by undergoing a Diels-Alder reaction with itself. An interesting feature of the dimer (Figure 11, structure 8) is that the bridging carbonyl group is readily lost as carbon monoxide when the molecule is heated. This elimination reaction, which results in aromatization of the ring which contained the keto-group, is a general reaction of molecules of this type and will be referred to again later (Vol. II, Chap. 3B.2e). Such structures contain rings which are rather sterically strained and this is reflected in the position (1,809 cm^{-1}) of the IR absorption band due to the bridge keto-group (an absorption band due to the conjugated keto-group is found at 1,721 cm^{-1}). Another feature of chlorinated compounds with strained, carbonyl substituted rings is that they are frequently isolated from aqueous solutions as the corresponding hydrates.[177]

"Hex" reacts with itself in various ways to form molecules which appear complicated at first sight, but are actually formed by quite simple reactions. Some of the products have useful toxicant activity against insects or mites. Thus, when "hex" is treated with sulfur trioxide, fuming sulfuric acid, chlorosulfonic acid, or sulfuryl chloride at 35 to 80°C, it reacts with itself by addition to give a cage structure in which the allylic chlorine atoms of one molecule are substituted by hydrolyzable sulfur containing groups. This intermediate (or intermediates) loses sulfur on hydrolysis to give the corresponding cage structure (70% yield) containing a carbonyl bridge, which has been marketed as an insecticide under the trademark Kepone®, also known as chlordecone (Figure 12, structure 1).[40] The technical product is more than 90% pure and sublimes, with some decomposition, near 350°C. Kepone has the typical absorption of a saturated monoketone in the UV, exhibits a strained ring carbonyl band at 1,786 cm^{-1} (5.6 μ) in the IR, and is isolated as a hydrate when the intermediate sulfur compounds are hydrolyzed during the synthesis. Kepone also forms solvates with acids, alcohols, amines, and thiols, and is soluble in strongly alkaline aqueous solutions.[178] It can be recrystallized from 90% aqueous ethanol and is readily soluble in acetone, but less so in light petroleum fractions or benzene. A mono- to trihydrate is said to be formed easily at room temperature and under normal conditions of humidity.

When heated with phosphorus pentachloride at 125 to 150°C, the keto-group is replaced by two chlorine atoms to give a $C_{10}Cl_{12}$ chlorinated hydrocarbon (Figure 12, structure 2) of mp 485°C, identical with the product formed directly from "hex" by heating it with aluminum chloride either without solvent or in methylene chloride, carbon tetrachloride, tetrachloroethylene, or hexachlorobutadiene.[177] The enhanced reactivity of "hex" in the presence of aluminum chloride is said to be due to the formation of $C_5Cl_5^+$, the pentachlorocyclopentadienyl cation. The $C_{10}Cl_{12}$ structure has little contact insecticidal action, but is a stomach poison which is especially effective against ants; it has been marketed for this purpose since 1959 and is referred to by the common name "mirex."[40] It is a white crystalline, free-flowing solid of molecular weight 546 (vapor pressure 6 x 10^{-6} mm Hg/50°C; density 0.020 g/ml at 24.3°C). Although virtually insoluble in water, it is soluble in organic solvents to the extent of 5.6% in methyl ethyl ketone, 7.2% in carbon tetrachloride, 12.2% in benzene, 14.3% in xylene, and 15.3% in dioxan. Mirex is rather inert chemically; it is reported to be unaffected by sulfuric acid, nitric acid, zinc dust, and hydrochloric acid or sulfur trioxide. The chemical reactions, physical properties, and absorption spectra of Kepone and its conversion into mirex have led to its formulation as the cage structure shown (Figure 12, structure 1),[179]

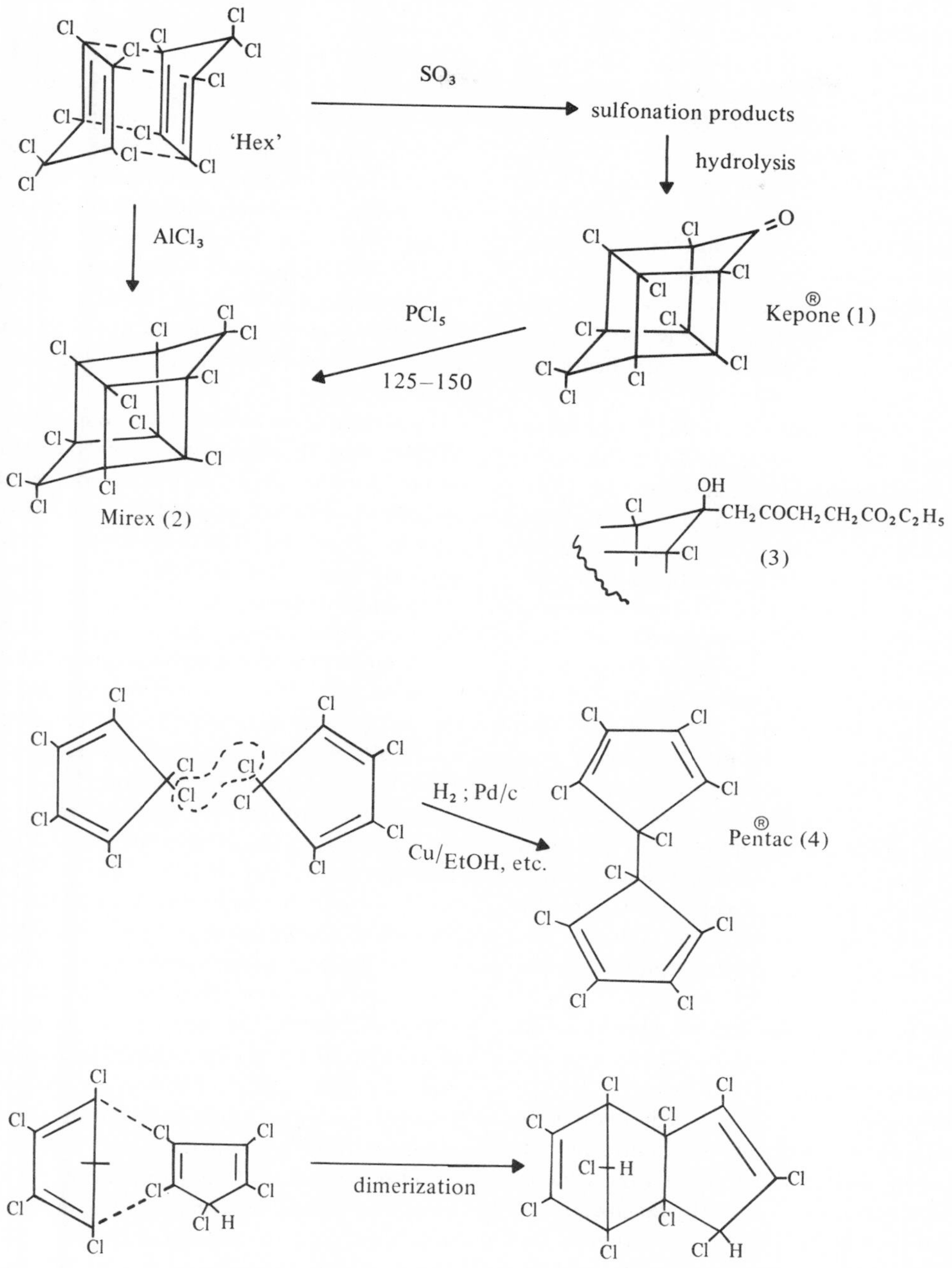

FIGURE 12.   Reactions of hexachlorocyclopentadiene; synthesis of chlordecone, mirex, etc., and Pentac®.

rather than as a dimeric product derived from "hex" via the Diels-Alder reaction, as was first suggested. Other insecticidal compounds have been made from Kepone by derivatizing the keto-group, as, for example, the compound "Despirol" (Figure 12, structure 3), which is active as a stomach poison towards various insects and mites.[180] The technical product is a light tan to brown solid and the pure compound, which is insoluble in water but soluble in acetone, alcohol, chloroform, benzene, toluene, and xylene, melts at 91°C with decomposition. Among others, it has found some small use against the Colorado potato beetle in Europe.

Reductive dechlorination of "hex" results in either removal of chlorine atoms from the molecule or in dechlorination with dimerization. Thus, its reduction with hydrogen at atmospheric pressure using a palladium on carbon catalyst produces 20% of bis(pentachloro-2,4-cyclopentadien-1-yl), mp 123 to 124°C (vapor pressure 10^{-5} mm Hg/25°C), which was introduced as an acaricide by the Hooker Chemical Corporation in 1960 under the trademark Pentac.[40,181] Several other methods give better yields. The compound (Figure 12, structure 4) can be prepared in 73% yield by coupling two molecules of "hex" in 80% ethanol or methanol at ambient temperature with cuprous chloride or powdered copper as catalyst, by refluxing with copper bronze in light petroleum (bp 100°C), or by refluxing with copper powder in toluene. On catalytic reduction the product gives bicyclopentyl.[177]

Stepwise removal of the vinylic chlorine atoms from "hex" may be effected by hydrogenation with a platinum catalyst; the liquid 1,2,3,4,5-pentachlorocyclopentadiene results if hydrogenation is stopped when one molar equivalent has been absorbed, and the solid 1,2,3,4-tetrachlorocyclopentadiene (mp 62 to 63°C) by the absorption of two molecules of hydrogen.[182] Further hydrogenation of this last compound gives cyclopentane. In the 1,2,3,4-tetrachloro-derivative, the presence of the free allylic methylene group can be demonstrated by the formation of highly colored fulvenes when the diene is heated under reflux with an equimolar amount of an aromatic aldehyde in ethanol. A single chlorine may also be removed from the allylic position of "hex" by its reduction with lithium aluminum hydride at −50°C or with stannous chloride in acetone. The above

tetrachloro-derivative is also formed in 66% yield in one step by adding concentrated hydrochloric acid (diluted 1:1 with water) to a stirred slurry of "hex" in light petroleum (bp 40 to 60°C) and zinc dust previously treated with copper sulfate (to form a zinc/copper couple). The petroleum is allowed to reflux gently during the addition, and the product is readily isolated by cooling the organic layer after the zinc dust has been filtered off.[183] The pentachloro- and tetrachloro-dienes each undergo a Diels-Alder dimerization (Figure 12) reaction at room temperature or on warming, but the monomers are recoverable when the dimers are heated.

Cyclopentadiene derivatives containing both chlorine and fluorine atoms have been of some interest for the preparation of fluorine containing cyclodiene insecticides, which are mainly of interest from the structure-activity relationship point of view. Thus, Soloway[184] has mentioned compounds derived from 1,2,3,4,5-pentachloro-5-fluorocyclopentadiene and others derived from 5,5-difluorotetrachlorocyclopentadiene are well documented. When octachlorocyclopentene is fluorinated with antimony tri- and pentafluorides, two of the chlorine atoms are replaced by fluorine, and the product, $C_5Cl_6F_2$, loses two chlorine atoms on treatment with reducing agents such as zinc and hydrochloric acid to give $C_5Cl_4F_2$.[185] This molecule (bp 70°C/20 mm; d_4^{20} 1.7382; n_D^{20} 1.4990) has been shown to contain its two fluorine atoms in the allylic positions by degradation of its Diels-Alder adduct with methyl acetylenedicarboxylate to dimethyl tetrachlorophthalate. When the adduct is pyrolyzed, the methylene bridge is eliminated with aromatization and the absence of fluorine in the aromatic product shows that the fluorines must have been allylic in the original diene.[177]

This liquid difluoro-analogue of "hex" dimerizes readily at ambient temperature and the monomer can be regenerated by pyrolysis at 480°C. Dimerization is the result of the Diels-Alder reaction between two molecules of diene, and the ease of its occurrence parallels the rate at which 1,2,3,4-tetrachloro-dienes with various allylic sutstituents react with dienophiles in general. Thus, for various allylic substituents, dimerization occurs in the order $CF_2 > CH_2 > CHCl$, while the -CCl_2 group tends to inhibit the Diels-Alder reaction with a double bond having vinylic chlorines (as occurs in dimerization; Figure 12), and hence inhibits the dimerization, which does not occur readily with "hex." The 5,5-dialkoxy-tetrachlorocyclopentadienes are also more reactive than "hex" in the Diels-Alder reaction.[177] Both 3,4-dichloro-1,2,5,5-tetrafluorocyclopenta-diene[186] and hexafluorocyclopentadiene[187] have been prepared and shown to give adducts with various dienophiles.

Riemschneider has described the preparation and use in the Diels-Alder reaction of trimethyltri-chlorocyclopentadiene and several dialkyltetra-chloropentadienes. These compounds are prepared by adding "hex" to a sixfold molar excess of methylmagnesium bromide in ether, when a number of the chlorine atoms are replaced by -MgBr groups. The Grignard reagent so prepared is then treated with excess of the appropriate alkyl halide, so that the chlorines initially displaced from "hex" are replaced by alkyl groups:[188,189]

$$C_5Cl_6 + 6CH_3MgBr \longrightarrow C_5Cl_{6-n}(MgBr)_n + (6-n)MgBr + nCH_3Cl$$

$$C_5Cl_{6-n}(MgBr)_n + (6-n)CH_3MgBr + 6RBr \longrightarrow C_5Cl_{6-n}R_n + 6MgBr_2 + (6-n)CH_3R$$

$$(n = 1 - 3)$$

The reaction results mainly in the formation of dialkyltetrachlorocyclopentadienes, but mono- and trialkyl-substituted products are also formed.

2. The Diels-Alder Reaction with Hexachloro-cyclopentadiene and its Relatives

a. Alodan®, Endosulfan, Isobenzan, Heptachlor, Chlordane, and Analogues

The ability of the halogenated cyclopentadienes to participate in the Diels-Alder reaction is clearly the most important aspect of their chemistry and many such addition reactions, especially involving "hex" as the diene, have been reported. A lengthy list of the adducts made from "hex"; 1,2,3,4-tetra-chloro-; 1,2,3,4,5-pentachloro-; 5,5-dimethoxy-1,2,3,4-tetrachloro-, and 5,5-diethoxy-1,2,3,4-tetrachlorocyclopentadienes up to 1958 is given by Ungnade and McBee in their review on the chemistry of "hex" and some related com-pounds,[177] and many more adducts have been made since then. The adducts produced from "hex" and their derivatives have found applica-tions in areas ranging from extreme-pressure additives for lubricating oils to biological activity as fungicides, plant-growth regulators and insecti-cides.

Some aspects of the chemistry of compounds derived from "hex" and related dienes are quite complex, and the discovery in recent years of a whole range of molecular rearrangement products produced from them in various ways, especially by photochemical methods, tends to complicate the picture further. The present section will therefore consider synthetic routes to the various groups of compounds that are of interest, but the details of these compounds are mainly reserved for consider-ation in subsequent sections.

The principles of the Diels-Alder reaction, in which an olefinic compound adds to the terminal carbon atoms of a conjugated 1,3-diene system to give a cyclic structure, were discussed previously (Vol. I, Chap. 3A.1). A simple reaction of this type involving "hex," and one which is also of further synthetic utility, is the addition to it of vinyl chloride.[190] This addition occurs when gaseous vinyl chloride is passed into well-stirred "hex" at 200°C, and the reaction is assumed to be complete when an increase in weight is achieved that corresponds to the uptake of one molecule of vinyl chloride. The addition takes place as shown in the upper sequence of Figure 13, in which the isolated chlorine atom in the product (1,2,3,4,5,7,7-heptachlorobicyclo[2.2.1]-hept-2-ene, also called heptachloronorbornene) is attached by a wavy line, indicating that it may have more than one stereochemical configuration. By analogy with other products of the Diels-Alder reaction with "hex," the product (Figure 13, structure 1) is likely to have the endo-structure (Figure 13, structure 3a) in which this chlorine atom is attached by one of the two equivalent bonds directed away from the dichloromethano-bridge (that is, below the plane of the ring) rather than by one of the two exo-bonds (carrying hydrogens in Figure 13) that are directed towards it (above the plane of the ring). Heptachloro-norbornene is biologically inactive and the related compound hexachloronorbornene (Figure 13, structure 1a) may be regarded as a basic inactive

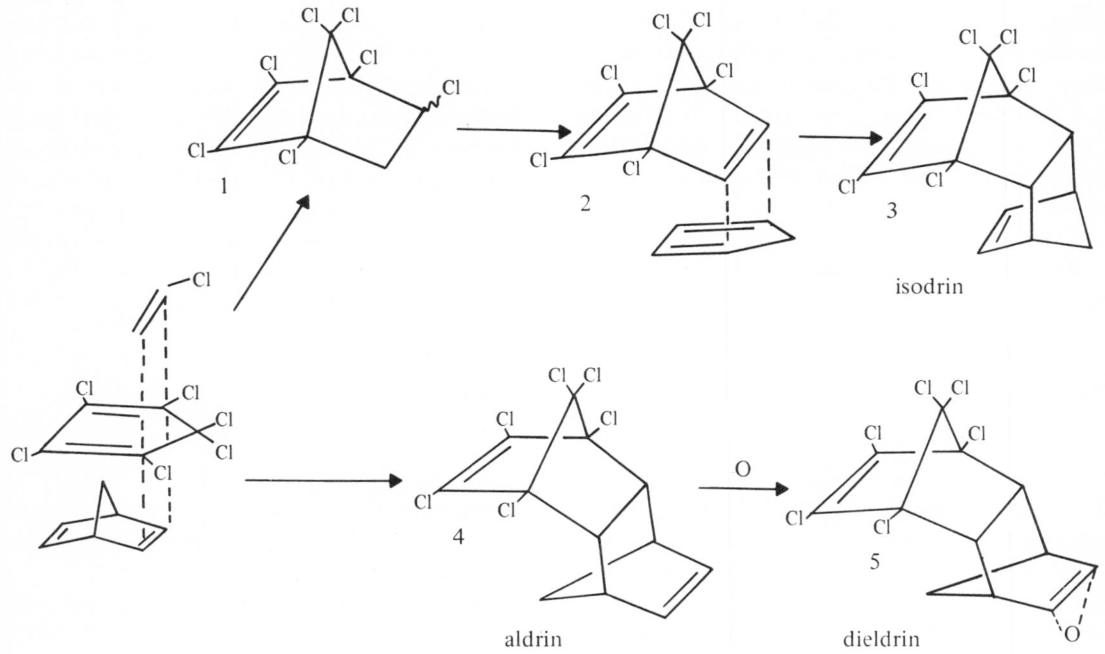

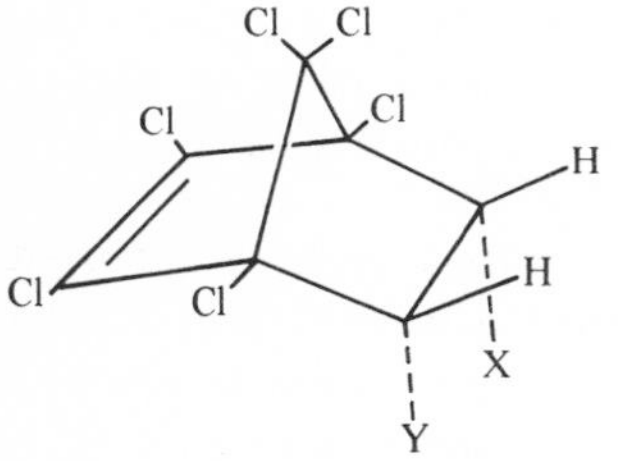

Open chain derivatives: 1a. X=Y=H; 2a. X=Y=Cl; 3a. X=H, Y=Cl; 4a. X=Y=CH_3; 5a. X=Y=CH_2OCH_3; 6a. X=Y=$CH_2OC_2H_5$.

Simple cyclic derivatives: 7. X—Y= —CH=$CHCH_2$— —$CH_2CH_2CH_2$—; —CH=$CHCH_2CH_2$—; —CH_2CH=$CHCH_2$—; —$CH_2CH_2CH_2CH_2$—; —CH_2OCH_2—.

FIGURE 13. Synthesis of isodrin, aldrin, and dieldrin. Structures of some other derivatives of hexachlorobicyclo[2.2.1] hept-2-ene (hexachloronorbornene).

structure from which most of the well-known cyclodiene insecticides are derived.

As prepared by the Diels-Alder reaction, heptachloronorbornene probably contains some of the *exo*-isomer as well, but it is usually dehydrochlorinated anyway, to give the liquid 1,2,3,4,7,7-hexachlorobicyclo[2.2.1]-hepta-2,5-diene (also called hexachloronorbornadiene; Figure 13, structure 2), a key intermediate in the synthesis of isodrin, endrin, and some other compounds.[190] The dehydro-chlorination, which is effected with boiling ethanolic potassium hydroxide, gives besides hexachloronorbornadiene an analogous compound in which one of the vinylic chlorines, rather than one of the methano-bridge chlorines, as was first thought, has been replaced by an ethoxy-group. Some interesting compounds of the isodrin-endrin series have been prepared from this last compound, and their chemistry has been described by Mackenzie.[191]

Other adducts are easily made from "hex" by heating it with dienophile, with or without solvent, in an open or a sealed vessel, depending on the nature of the dienophile.[192] In some cases, the reaction occurs so readily that no heating is required. It is advisable to conduct the reactions in a sealed tube when the dienophile is volatile and especially when radiolabeled adducts are being prepared. The most active dienophiles are those in which the double bond involved in the addition reaction is activated by conjugation with an electron-withdrawing group, as for example, in maleic anhydride or benzoquinone; in each of these cases the adducts are readily formed when the dienophiles are mixed with "hex" and heated to about 100°C for a few minutes. A longer period of heating and a higher temperature are frequently required when the dienophile is a simple olefin. If the adduct contains double bonds, the possibility exists that "hex" may add to them in favorable circumstances to give multiple adducts, in which case it is advisable to employ an excess of dienophile, rather than of "hex" in the reaction. The addition of a little quinol sometimes helps to inhibit oxidation or polymerization reactions when long periods of heating are involved. Adducts are usually easily isolated by direct recrystallization of the crude reaction mixture for solid products following a suitable decolorization process if necessary, or by distillation of liquids following prior distillation of the more volatile "hex". Since normal adducts formed from "hex" (including *bis*-adducts) have a chlorinated double bond, they exhibit a characteristic, sharp IR absorption at about 1,600 cm⁻¹ ("hex" has two absorption bands in this region) due to the stretching vibrations of this group, which is a useful way of distinguishing them from compounds such as rearrangement products in which the chlorinated double bond may no longer be present.[170]

As an example of the above general method, equimolar amounts of "hex" and *cis*-2-butene-1,4-diol, when heated at 140°C for 15 hr in dioxan in a sealed tube, gave 72% of the expected adduct (Figure 13; X=Y=CH₂OH). The related compound Alodan® (X=Y=CH₂Cl, mp 102°C) is similarly prepared from *cis*-1,4-dichlorobutene-2, and the halomethyl-compounds (X=H; Y=CH₂Cl, CH₂Br, or CH₂I, respectively) from the corresponding allyl halides and "hex" by heating the reactants in sealed tubes at 125°C for 12 to 20 hr.[193-195] These compounds have melting points of 55, 78, and 109°C, respectively. Although the simple derivatives of hexachloronorbornene numbered 1a-6a (Figure 13) have no insecticidal activity, such activity appears in alodan and the compounds having a single halomethyl group.[184] The next logical elaboration of the hexachloronorbornene system is the addition of simple ring systems instead of the freely rotating substituents just discussed, and the simplest molecule of this type is

produced by the reaction between "hex" and cyclopentadiene. This reaction occurs readily, and is so exothermic that it can be dangerous if the two reactants are mixed without solvent.[196]

Since in this reaction both components are dienes, the question arises as to which component acts as the dienophile, and it turns out that both are able to fulfill this role; the lower the temperature of the reaction, the greater is the tendency of "hex" to act as the dienophile, so that the resulting adduct mixture (at a reaction temperature of 50 to 80°C) may contain up to 25% of this product (Figure 14, structure 2), which is an undesirable component from the point of view of biological activity.[196] When heated to 160°C, this adduct readily undergoes thermal isomerization to the major product of the reaction which is called chlordene (Figure 14, structure 1; 4,5,6,7,8,8-hexachloro-3a,4,7,7a-tetrahydro-4,7-methanoindene). Thus, the higher the temperature at which the Diels-Alder reaction is conducted, the smaller the amount of impurity due to the reverse addition, and the quality of the adduct is improved if it is heated to remove unreacted "hex", because this accelerates the thermal isomerization of the unwanted component. For both adducts, melting begins at about 155°C and is not sharp. The adduct derived from "hex" as a dienophile may be distinguished from chlordene by its participation in a number of reactions associated with the presence of allylic chlorine atoms and the unchlorinated bicyclo[2.2.1]heptene nucleus. Thus, the two allylic chlorine atoms are removed by silver nitrate in acetic acid, and phenylazide adds to the unchlorinated double bond; neither of these reactions occurs with chlordene. The stretching frequency of the chlorinated double bond is reported to be displaced from 1,600 cm⁻¹ in chlordene to 1,632 cm⁻¹ in the adduct containing allylic chlorine atoms. Under appropriate conditions the unchlorinated double bonds of both adducts react with a further molecule of "hex", acting as a diene, to give products containing 12 chlorine atoms. Chlordene absorbs one molecule of hydrogen when reduced in the presence of Adam's catalyst (platinum oxide) at atmospheric pressure, and the product (2,3-dihydrochlordene, mp about 170°C) is identical with the compound produced directly from "hex" and cyclopentene in a sealed tube at 130°C. Chlordene, although not itself a highly toxic compound, is the precursor of highly toxic

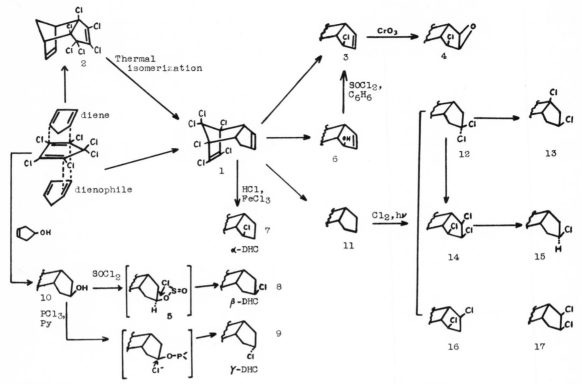

FIGURE 14.   Synthesis of chlordene and related compounds. In this and in subsequent figures, the partial structure implies that the hexachloronorbornene nucleus is retained intact.

molecules such as heptachlor, as will be seen. Another interesting but nontoxic compound is produced by condensing "hex" with 2,5-dihydrofuran in a sealed tube at 130°C. This compound, in which the furan ring is fused symmetrically to the *endo*-bonds of hexachloronorbornene (Figure 13, structure 7; X-Y=-CH$_2$OCH$_2$-), is also the precursor of highly insecticidal molecules.[192]

Compounds in which a cyclohexane or cyclohexene ring is fused to the hexachloronorbornene nucleus are readily prepared via the appropriate Diels-Alder reaction with cyclohexene or a cyclohexadiene, or their derivatives.[192] Thus, "hex" added dropwise to boiling cyclohexadiene gives an adduct of mp 115°C (7, X-Y=-CH=CHCH$_2$CH$_2$-), which is hydrogenated at atmospheric pressure over Adam's catalyst to a dihydro-derivative (7, X-Y=-(CH$_2$)$_4$-) identical with the compound produced directly from "hex" and cyclohexene. A symmetrical cyclohexene derivative (7,X-Y=-CH$_2$CH=CHCH$_2$-) is produced either by the reaction between "hex" and cyclohexa-1,4-diene, or by the reaction between hexachloronorbornadiene (Figure 13, structure 2), acting as a dienophile, and butadiene. As expected, the product, mp 94°C, gives the dihydro-derivative mp 78°C on catalytic reduction. The two unsaturated compounds give bis-adducts by reacting with a second molecule of "hex," when this is present in excess. An unusual system that confers high toxicity on the molecule is obtained by heating the adduct (Figure 13; X,Y=CH$_2$OH) from "hex" and *cis*-2-butene-1,4-diol with excess thionyl chloride in benzene. This results in the elimination of two molecules of hydrogen chloride with the formation of the cyclic sulfite ester of the diol (Figure 17). The product, endosulfan (Thiodan®),[176] has considerable insecticidal activity both by contact and as a stomach poison; it exists in two isomeric forms, alpha and beta, mp 108 to 110°C and 208 to 210°C, respectively, due to the pyramidal sulfite ester group.[197] The diol intermediate (Figure 17, structure 3) also offers an additional route to the furan-derivative (4), which is obtained by the direct reaction between "hex" and 2,5-dihydrofuran and is the precursor of isobenzan (5).[198]

With the exception of the furan ring, the simple ring systems listed in Figure 13 (7) confer measurable, but not practically useful toxicity when they are added to the hexachloronorbornadiene nucleus. Epoxides may be prepared from the unsaturated compounds and are not by themselves very toxic, but are of interest from a theoretical point of view, and will be discussed in a later section (Vol. II, Chap. 3C.1b). They can be made using peracetic acid, but a more convenient route uses perbenzoic acid, or 3-chloroperbenzoic acid in methylene chloride or chloroform.[192]

Although the simple furan derivative depicted in Figure 17 has no insecticidal activity, the successive introduction of chlorine atoms into the α-positions of the furan ring produces bioactive compounds. The chlorination of tetrahydrofuran occurs in the presence of ultraviolet light and a variety of products are produced, depending on the conditions.[199] At room temperature, the 2,3-dichloro-derivative is formed, but the α-positions can be exclusively substituted by irradiation at low temperatures. Thus, 2-chlorotetrahydrofuran is prepared by conducting the chlorination at $-35°C$ with the theoretical amount of chlorine without solvent, and 2,5-dichlorotetrahydrofuran by irradiation of tetrahydrofuran in carbon tetrachloride with two molar equivalents of chlorine at -28 to $-30°C$. The reaction slows distinctly after introduction of the first chlorine atom. In the case of the furan adduct derived from "hex," substitution is only likely to occur in the α-positions because of steric hindrance at the ring junction. Thus, these positions may be successively substituted with chlorine by irradiating the adduct in carbon tetrachloride with the theoretical amount of halogen at about 40°C. In this way, one to four chlorine atoms may be introduced, but maximal insecticidal activity is found in the symmetrical 1,3-dichloro-derivative isobenzan (Telodrin®; Figure 17, structure 5; 1,3,4,5,6,7,8,8-octachloro-1,3,3a,4,7,7a-hexahydro-4,7-methanoisobenzofuran).[184]

When a seventh chlorine atom is substituted in the allylic position of chlordene (Figure 14, structure 1) there is a very considerable increase in general insecticidal activity. The product, called heptachlor (Figure 14, structure 3; 1-exo-4,5,6,7,8,8-heptachloro-3a,4,7,7a-tetrahydro-4,7-methanoindene), melts at 95 to 96°C when pure. The allylic chlorination may be effected either directly, by heating chlordene in carbon tetrachloride with sulfuryl chloride (SO_2Cl_2) and a trace of an initiator such as benzoyl peroxide,[200] or indirectly by first substituting the allylic position by an acetoxy- or hydroxyl-group which can then be replaced by chlorine.[177] The direct method gives a low melting product contaminated with other compounds and the indirect method is better for laboratory preparations. The latter method proceeds via the 1-hydroxy compound (Figure 14, structure 6) in the following sequence:

$$\text{Chlordene} \xrightarrow[\text{CCl}_4]{\text{N-bromosuccinimide}} \text{1-bromochlordene} \xrightarrow[\text{acetic acid}]{\text{CH}_3\text{COONa}} \xrightarrow[\text{HCl}]{\text{CH}_3\text{OH}} \text{1-hydroxy-chlordene}$$

$$\text{1-hydroxychlordene} \xrightarrow{\text{SOCl}_2/\text{C}_6\text{H}_6} \text{1-chlorochlordene (heptachlor).}$$

A number of variants of this procedure are possible; the allylic acetate may be prepared directly from chlordene by heating it with selenium dioxide (a reagent for allylic hydroxylation) in acetic acid and hydrolyzing the product directly with zinc chloride and concentrated hydrochloric acid in dioxan, or the 1-hydroxychlordene may be isolated and chlorinated with dry hydrogen chloride in an inert solvent. The reactive bromine atom in 1-bromochlordene may also be replaced directly by a hydroxy-group by prolonged heating with potassium carbonate in dioxan.[201] In this series of compounds, no insecticidal activity is found in 1-hydroxychlordene, and 1-bromochlordene is only a little more toxic than chlordene, but 1-fluorochlordene is as toxic as heptachlor; fluorination may be effected by boiling 1-bromochlordene in hexane with twice the theoretical amount of anhydrous mercuric fluoride, and the product melts at 180 to 182°C.[177]

Although heptachlor 2,3-exo-epoxide (Figure 14, structure 4) is produced quite readily from heptachlor in living organisms, drastic conditions, such as treatment with chromium trioxide in sulfuric acid, are needed to produce it chemically.

This shows that the presence of the chlorine atom in the allylic position hinders epoxidation by normal methods, since chlordene-2,3-exo-epoxide is formed without difficulty when chlordene is oxidized with perbenzoic acid. The epimeric endo-epoxides are made by an indirect method involving the formation of intermediate chlorohydrins.[202] To prepare heptachlor-2,3-endo-epoxide (Figure 18, structure 10), heptachlor is stirred at 80°C in glacial acetic acid while an equimolar amount of t-butyl hypochlorite is added; after a further period of heating at this temperature the crystalline acetate that results from the addition of acetyl hypochlorite to the double bond can be obtained by cooling and filtration. This procedure results in the addition of acetyl hypochlorite (or addition of hypochlorous acid followed by acetylation of the chlorohydrin) in such a manner that the added chlorine atom occupies the 3-exo-position and the acetoxy-group the 2-endo-position (Figure 18). The corresponding chlorohydrin is generated by ester exchange when the acetate is boiled in methanol containing hydrochloric acid. Finally, the chlorohydrin is warmed briefly with sodium hydroxide in aqueous dioxan, when the alkoxide ion generated in the endo-2-position displaces the 1- (or 3) chlorine atom by nucleophilic attack from the rear and so forms the endo-2,3-epoxide. Endo-epoxides are formed in a similar manner from chlordene and 1-fluorochlordene. If the halogen atoms in the intermediate halohydrin are different, the one of higher atomic weight is eliminated, so that the epoxide from 1-bromochlordene must be prepared using t-butyl hypobromite rather than the hypochlorite.[203]

An interesting series of compounds, the dihydroheptachlors (DHC), are produced by the addition of gaseous HCl at 100 atm pressure to chlordene.[167] The reaction is an equilibrium which favors the formation of β-DHC (Figure 14, structure 8) when the reaction is conducted at 100°C without solvent using ferric chloride as catalyst.[204] Under these conditions, the reaction mixture contains β-, α- and γ-DHC in 91, 2 and 5% yield, respectively. When the reaction is carried out in a solvent under the same conditions, carbon tetrachloride, toluene, chloroform, or hexane being suitable examples, the proportion of isomers in the product depends on the solvent. Thus, 30% of γ-DHC is formed when toluene is the solvent. When α- and γ-DHC are treated with HCl in a solvent under the reaction conditions used to produce them, each gives the other two isomers by HCl elimination to give chlordene, followed by re-addition of HCl; α-DHC appears to form mainly β-DHC, since a mixture rich in γ-DHC, when isomerized in various solvents at 160°C, showed an improvement in β-DHC content at the expense of the γ-isomer, whereas the original content of α-DHC showed little change. The compounds are insecticidal crystalline solids (α-, β- and γ-DHC melt at 122 to 123°C, 135°C, and 112 to 113°C, respectively). The remaining isomer, in which the seventh chlorine atom would occupy the strongly hindered 1-endo-position, has not been isolated.

Although α-DHC could be obtained, in principle, from heptachlor, by reduction of the 2,3-double bond, catalytic reduction results in removal of the allylic chlorine atom and reduction of the double bond to give dihydrochlordene (Figure 17, structure 1a). Nor can this compound be made by a Diels-Alder reaction between 3-chlorocyclopentene and "hex" as might be supposed; HCl is eliminated during the reaction, so that the product is chlordene. However, 4-hydroxycyclopentene-1 and the derived 4-chlorocyclopentene-1 each add to "hex" to give adducts with the substituent in the 2-exo-position. Thus, mainly β-DHC is produced from 4-chlorocyclopentene. With SOCl₂ in dioxan, the 2-exo-alcohol (Figure 14, structure 10) undergoes nucleophilic substitution by chlorine via the intermediate exo-chlorosulfite ester and the configuration is retained in the product, which is mainly β-DHC. With PCl₃ in pyridine, nucleophilic substitution occurs under inversion conditions, and attack of Cl⁻ from the endo side of the cyclopentane ring affords a 74% yield of γ-DHC, with about 22% of β-DHC. α-DHC is prepared by treating 1-endo-hydroxydihydrochlordene (from 1-ketodihydrochlordene by reduction with sodium borohydride) with phosphorus pentachloride; when similarly treated, 1-exo-hydroxydihydrochlordene does not undergo nucleophilic substitution, but rather eliminates water to form chlordene, a good indication that substitution by chlorine in the 1-endo-position is not possible due to steric hindrance.[205]

The proton-NMR (nuclear magnetic resonance spectra) of these isomers support the structural assignments. In β-DHC, the signal due to the endo-protons on C_1 and C_3 is displaced to lower than normal τ values under the influence of the chlorinated bicycloheptene-ring, and a similar shift of the signal of the C_1 and C_3 *exo*-protons due to the influence of the 2-*exo*-chlorine atom produces a composite signal for these four protons; the spectrum therefore shows this group at 7.95 τ together with the signal due to the single 2-*endo*-proton (5.83 τ) and that of the two tertiary protons (6.43 τ). In contrast, γ-DHC shows distinct signals due to the two C_1 and C_3 *exo*-protons (8.75 τ), the C_1 and C_3 *endo*-protons (7.52 τ), the single 2-*exo*-proton (6.08 τ), and the two tertiary protons (6.85 τ). In the absence of the 2-*exo*-chlorine atom, the signal due to the C_1 and C_3 *exo*-protons lies at higher τ-values than with β-DHC, and the displacement of the signal due to the corresponding *endo*-protons in the opposite direction under the combined influence of the chlorinated bicycloheptene ring and the 2-*endo*-chlorine atom results in complete separation of the two groups.[205]

The chlorination of chlordene or dihydro-chlordene under various conditions gives a mixture of compounds having high insecticidal activity. The mixture thereby obtained from chlordene was the first material of the cyclodiene type to be used for insect control; a material produced in this way, called chlordane, is still employed as an insecticide today, but the nature of the product has changed since it was first introduced. Chlordane is a product of some importance for insect control and its history and development are discussed in detail later. Considerable effort has been devoted to synthetic and analytical work designed to elucidate the structures and properties of the components of technical chlordane and related products, and some of the investigations of general

synthetic interest are discussed here; others will be mentioned when the general chemistry of the cyclodienes is considered, and in relation to the development of chlordane.

In an attempt to obtain the simplest possible uniform chlorination product from dihydrochlordene, Buchel and his colleagues conducted the chlorination under a variety of conditions, including the use of sulfuryl chloride and benzoyl peroxide, or chlorine without UV-irradiation.[206] A crystalline product was obtained by UV-irradiation of a mixture of magnesium oxide (60 g) and carbon tetrachloride (500 ml) containing dihydrochlordene (100 g), and held at 40°C while chlorine (42 g) was passed in during a period of 40 min. This procedure (B) gave a more insecticidal product than a similar procedure (A) which omitted the magnesium oxide and gave an oil; the materials produced by each method were analyzed on a comparative basis, using columns of silica gel for separation and gas chromatography for quantitative analysis. By these methods, A and B were each separated into six components (Table 12). The greater insecticidal activity of product B, as compared with A, appears to be due to the presence of more δ-chlordane, which is a highly toxic compound (Vol. II, Chap. 3C.1b) having a steric resemblance to isobenzan. The structures of the various chlordane isomers found in this analysis and shown in Figure 14 were mainly verified by NMR data and other methods, although the properties of the nonachlor (Figure 14, structure 14) differ from those given elsewhere. It is particularly noteworthy that β-chlordane (Figure 14, structure 17) was obtained from α- and

TABLE 12

Composition of the Chlorination Product of Dihydrochlordene Obtained by Chlorination with UV Irradiation Alone (Method A) or in the Presence of Magnesium Oxide (B)

Number in Figure 14	Melting point °C	% Composition	
		Method A	Method B
12 γ-chlordane	131	17	15
13 ϵ-chlordane	165–167)		
	)	12	19
14 $C_{10}H_5Cl_9$	117–122)		
15 β-DHC	136–137	29	22
16 δ-chlordane	72	23	29
17 β-chlordane	107–108	19	15

(Data adapted from Buchel.[206])

β-dihydroheptachlor by their careful chlorination in daylight to a content of eight chlorine atoms.

α-Chlordane was not detected in the chlorination products of dihydrochlordene with any of the chlorination conditions used, but was synthesized in this investigation by the stereospecific *trans*-addition of chlorine to chlordene by the action of trichloromethylsulfenyl chloride in the presence of ferric chloride. The product (1-*exo*, 2-*endo*-4,5, 6,7,8,8-octachloro-3a,4,7,7a-tetrahydro-4,7-methanoindane) was obtained in 95% yield and melted at 103 to 105°C. This product is α- or *trans*-chlordane, called γ-chlordane by the Velsicol Chemical Corporation. β- or *cis*-chlordane (1-*exo*, 2-*exo*-4,5,6,7,8,8-octachloro-3a,4,7,7a-tetrahydro-4,7-methanoindene) is called α-chlordane by Velsicol. It is curious that when giving his melting point for *trans*-chlordane of 103 to 105°C, Buchel aligns this compound with the constituent (IV) isolated by March from technical chlordane, by citing March's melting point of 104.5 to 106°C for this constituent.[206] Constituent (IV) was, in fact, suggested by March (Vol. I, Chap. 3C.1a) to be the *cis*-compound on the basis of its ease of dehydrochlorination, whereas constituent (V), mp 106.5 to 108°C, did not dehydrochlorinate easily and was assumed to be the *trans*-isomer.[207]

A good laboratory synthesis of *cis*-chlordane appears to be that described by Cochrane,[208] in which the chlorohydrin (1-*exo*-chloro-2-*endo*-hydroxydihydrochlordene) obtained by treating chlordene with t-butyl hypochlorite in acetic acid and hydrolyzing the resulting acetate, is treated with thionyl chloride in pyridine. The *endo*-hydroxy-group is displaced by chlorine with inversion to give *cis*-chlordane in 90% yield. In a rather similar manner, the chlorohydrin precursor of heptachlor-2,3-*endo*-epoxide (Figure 18, structure 10), in which the hydroxyl-group is again in the 2-*endo* position, gives a mono-chloro-derivative of *cis*-chlordane (Figure 14, structure 17), in which the cyclopentane ring chlorines are in the 1-, 2-, and 3-*exo*-positions.[208] This compound (mp 214 to 215°C) is distinct from the compound, melting at 117 to 122°C, assigned the same structure by Buchel,[206] and distinct from a constituent (nonachlor, mp. 128 to 130°C) of technical chlordane which has been assigned the 1-*exo*, 2-*endo*, 3-*exo*-trichlorodihydrochlordene structure from several lines of evidence.[208]

When chlordene (as opposed to dihydrochlordene) is chlorinated directly, the product contains *cis*- and *trans*-chlordane, heptachlor, nonachlor, and some other compounds (Vol. I, Chap. 3C.1a). An approximate composition of technical chlordane produced in this way is given in Table 13. Further aspects of the chemistry of compounds of the heptachlor-chlordane group will be discussed in the sections on general chemistry and analysis.

The replacement of the methano-bridge chlorine atoms by hydrogen in the insecticidal compounds obtained from "hex" causes a drastic loss in toxicity, and so the analogues of chlordene, heptachlor, and chlordane made from 1,2,3,4-tetrachlorocyclopentadiene, which is even more reactive than "hex" as a diene, are only of interest from the structure-activity relationship point of view. It is not easy to obtain 1,4,5,5-tetrachlorocyclopentadiene; only a limited number of adducts have been made from it and those derived from cyclopentadiene appear to lack insecticidal activity, in contrast to the situation found in the corresponding aldrin analogues (Vol. II, Chap. 3C.1b,c). Compounds in the chlordene series with five, or less than four chlorine atoms do not appear to have been prepared, but might be obtained by condensation of pentachlorocyclopentadiene with cyclopentadiene for the former, and by dechlorination procedures from the compounds of higher chlorine content for the latter. A number of compounds with a variety of chlorine arrangements have been made in the aldrin series and are of considerable theoretical interest.[184]

Riemschneider prepared a low melting (ca 37°C) chlordene analogue ($C_{10}H_6Cl_4F_2$) from 1,2,3,4-tetrachloro-5,5-difluorocyclopentadiene and chlorinated it to give a material ($C_{10}H_6Cl_6F_2$; bp 135 to 139°C/0.5 mm), analogous to chlordane, which he called M377.[186] Melnikov reported the 8,8-difluoro-analogue of dihydrochlordene (mp 43°C), obtained by reaction of the same diene with cyclopentene, and observed that yields of adducts were only moderate in some cases on account of the facile dimerization of the diene.[185] Riemschneider also obtained the chlordene analogue, $C_{10}H_6Cl_2F_4$, from 1,2,5,5-tetrafluoro-3,4-dichlorocyclopentadiene, and the chlordane analogue $C_{10}H_6Cl_4F_4$ (bp 98 to 104°C/2.5 mm) by its chlorination. Compounds derived from hexabromocyclopentadiene or from 5,5-dialkoxy-1,2,3,4-tetrachlorocyclopentadiene lack insecticidal activity.

TABLE 13

Approximate Composition of Technical Chlordane[a]

Fraction	Percentage present
Diels-Alder adduct of cyclopentadiene and pentachlorocyclopentadiene ($C_{10}H_7Cl_5$)	2 ± 1
Chlordene ($C_{10}H_6Cl_6$); isomer 1	1 ± 1
Chlordene isomers 2	7.5 ± 2
3 and 4 together	13 ± 2
Heptachlor ($C_{10}H_5Cl_7$)	10 ± 3
Cis-chlordane ($C_{10}H_6Cl_8$)	19 ± 3
Trans-chlordane ($C_{10}H_6Cl_8$)	24 ± 2
Nonachlor ($C_{10}H_5Cl_9$)	7 ± 3
Hexachlorocyclopentadiene (C_5Cl_6)	>1
Octachlorocyclopentene (C_5Cl_8)	1 ± 1
$C_{10}H_{7-8}Cl_{6-7}$	8.5 ± 2
Constituents with shorter GC retention time than C_5Cl_8 (includes hexachlorocyclopentadiene)	2 ± 2
Constituents with longer GC retention times than nonachlor	4 ± 3

Adapted from data of Velsicol Chemical Corporation.[271]

[a]These approximations are derived from moderate resolution gas-liquid chromatography and are influenced by conditions of analysis and the relative detector response to the various components. The profiles obtained under standard conditions (see text and Figure 22) are used to compare Technical chlordane samples with Reference technical chlordane.

b. Aldrin, Dieldrin, Isodrin, Endrin, and Related Compounds

Bicyclo[2.2.1]hepta-2,5-diene (norbornadiene) is a key intermediate in the synthesis of cyclodiene compounds of the aldrin-dieldrin series. It can be made in modest yield by the Diels-Alder reaction between vinyl chloride and cyclopentadiene, followed by dehydrochlorination of the resulting 5-chlorobicyclo[2.2.1]hept-2-ene, or by the elimination of chlorine from 5,6-dichlorobicyclo[2.2.1]hept-2-ene, but the usual method involves the Diels-Alder reaction between cyclopentadiene and acetylene. In the first preparations of norbornadiene, as described by Hyman, Freireich, and Lidov, up to 65% yield of the product (based on the weight of cyclopentadiene employed) was obtained by passing acetylene and cyclopentadiene, mixed in a somewhat more than 2:1 ratio and at a pressure of 80 lb/in.[2], through a hot reaction zone at 343°C.[209] In some experiments the reaction was conducted in a continuous flow system using a mixture of diphenyl ether and diphenyl as solvent. Plate and Pryanishnikova conducted the reaction in a steel autoclave with cyclopentadiene (20 g) in isopentane (20 ml), into which acetylene was passed at 9 to 10 atm.[210] The production of norbornadiene from this mixture was investigated for various temperatures and reaction times, the maximum pressure attained being less than 30 atm. With a reaction time of 30 min, no conversion was found below 230°C, and conversion was maximal (40%) at about 300°. At higher temperatures yields fell off because of resinification of the reaction products. A shorter reaction time (5 min) at 320°C gave the same yield. Hyman recorded norbornadiene to have a boiling point of 83 to 84°C/620 to 635 mm; n_D^{20} 1.4685 to 1.4720 and d^{20} 0.8770 to 0.9100, and the Russian authors recorded bp 89.5 to 90.5°C; n_D^{20} 1.4693; d_4^{20} 0.9056.

"Hex" reacts readily with norbornadiene to give an adduct in which the new norbornene nucleus is fused to the concurrently generated hexachloronorbornene nucleus through the endo-positions X and Y (Figure 13) as found in other cases.[211] As indicated previously, the major pro-

113

duct of this particular reaction is 1,2,3,4,5,10-hexachloro-1,4,4a,5,8,8a-hexahydro-1,4-*endo,exo*-5.8 dimethanonaphthalene, in which the *unchlorinated* norbornene nucleus is fused to hexachloronorbornene through the former's *exo*-positions (Figure 13). The pure compound (mp 104 to 104.5°C), HHDN, constitutes at least 95% of the technical material known as "aldrin." A variety of conditions are suitable for the reaction on the laboratory scale. It may be effected simply by reacting "hex" in an excess of boiling norbornadiene for about 16 hr; solvents may be used, but are not necessary. The use of excess norbornadiene is also advantageous from another aspect; if "hex" is present in excess, a second molecule of it may add to the unchlorinated double bond of HHDN to give a bis-adduct, $C_{17}H_8Cl_{12}$, which although readily separable from the mono-adduct on account of its insolubility in methanol, nevertheless serves to reduce the yield. Under favorable conditions, the use of at least a molar excess of norbornadiene reduces bis-adduct formation to less than 5% of the reaction product.

Depending on the quality of the "hex" used, the product may also contain traces of adducts formed from cyclopentadienes substituted by less than six chlorine atoms, and it also contains traces of isodrin, the stereoisomer in which the norbornene nucleus is fused through its *endo*-positions to the *endo*-positions of the hexachloronorbornene nucleus (Figure 13). Some of the earliest preparations of ^{14}C-aldrin (aldrin is used to denote HHDN here, unless otherwise stated) made at the Radiochemical Centre, Amersham, England, were examined by the author using continuous descending reversed phase paper chromatography (on Whatman strip impregnated with liquid paraffin from a 5% solution in ether, with pyridine/water (3:2) as mobile solvent). Isodrin is slightly more polar than aldrin; its R_F value in this system is slightly higher than that of aldrin, and a clear separation is effected by a continuous descending run with the pyridine/water solvent. This method indicated that the ^{14}C-aldrin contained about 2% of ^{14}C-isodrin.

The unchlorinated double bond is reactive and undergoes a variety of transformations, the most important of these from the synthetic point of view being the epoxidation reaction (Figure 13). The pure epoxidation product, mp 175 to 176°C,

is 1,2,3,4,10,10-hexachloro-6,7-*exo*-epoxy-1,4,4a,5,6,7,8,8a-octahydro-1,4-*endo,exo*-5,8-dimethanonaphthalene, abbreviated to HEOD, which is the main constituent of the technical material (not less than 85% HEOD) called dieldrin.[212] This epoxide is very easily produced, for example, by treating aldrin in chloroform with perbenzoic acid at room temperature for several hours, or from aldrin treated with peracetic acid in benzene as solvent. When 70% peracetic acid (from 90% hydrogen peroxide, acetic acid and acetic anhydride) is used for the epoxidation, the reaction is strongly exothermic and cooling is needed to maintain a temperature of 50°C during the gradual addition of the peracid. After the exothermic reaction is complete, the same temperature is maintained for several hours more, after which the product may be freed from solvent by steam distillation, precipitated by the addition of water and recrystallized. In HEOD, the 6,7-epoxide ring may be fused to the norbornene nucleus through its *exo*- or *endo*-positions (see section on nomenclature of cyclodienes); a number of lines of evidence indicate that *exo*-fusion is involved, one of the most recent being the formation of an *exo-cis*-diacetate by the treatment of HEOD with acetic anhydride-sulfuric acid. In the following discussions it is convenient to use the names aldrin and dieldrin to denote the pure compounds.

Catalytic reduction of the 6,7-double bond of aldrin in ethanol at atmospheric pressure and ambient temperature over a platinum oxide catalyst gives 6,7-dihydroaldrin, mp 78°C, and the 6,7-double bond readily adds chlorine to give *trans*-6,7-dichlorodihydroaldrin, mp 146°C, while bromine in CCl_4 gives a mixture of *trans*- and one of the two possible *cis*-6,7-dibromo-derivatives (presumably the *exo-cis*-dibromide on steric grounds).[211] The *trans*-dichloride and the mixed dibromides lose hydrogen halide in boiling ethanolic potassium hydroxide to give 6-halo-derivatives of aldrin, which, like the latter, can be converted into the corresponding epoxides, and the higher melting dibromide (which presumably is the *cis*-derivative on this account) reacts with sodium sulfide in boiling ethanol to give the episulfide analogous to dieldrin.[211,212] This compound, as well as the derived sulfoxide, is insecticidal to houseflies, although less so than dieldrin:[184,213]

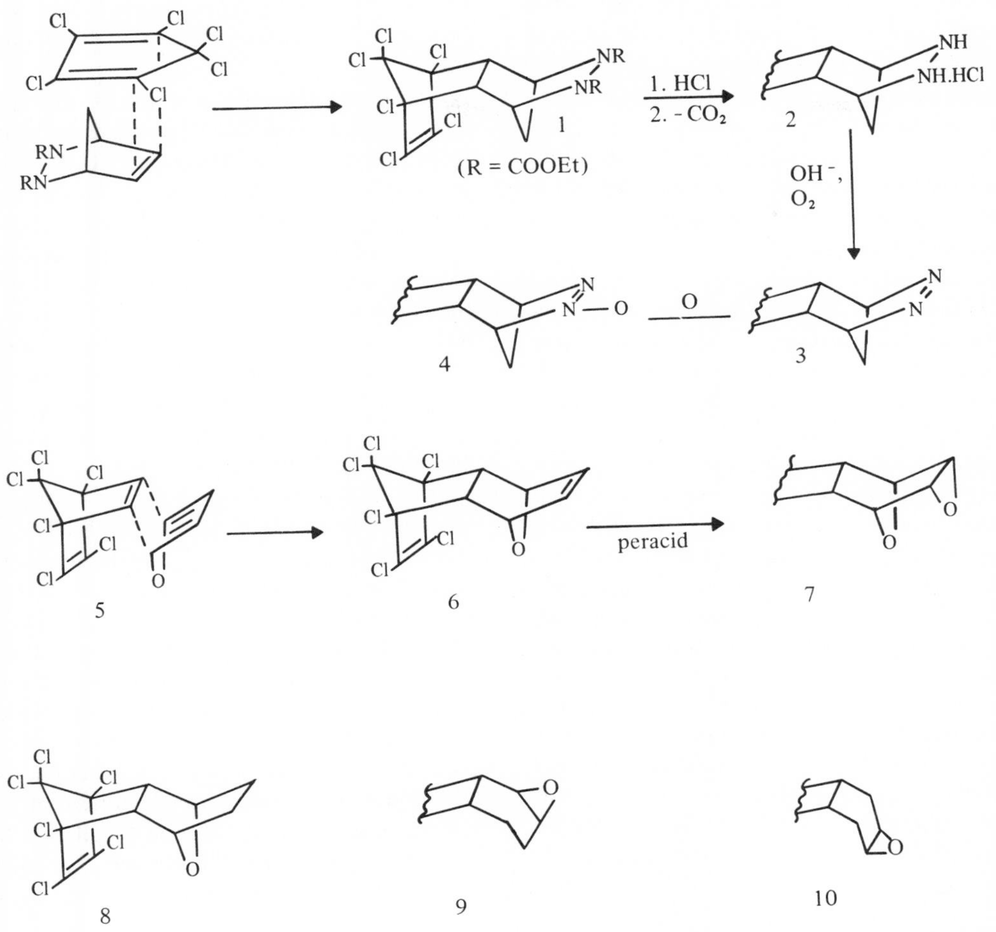

FIGURE 15. Structures of various insecticidal compounds derived directly or indirectly from hexachloro-cyclopentadiene. Use of a partial structure indicates that the hexachloronorbornene nucleus is still present.

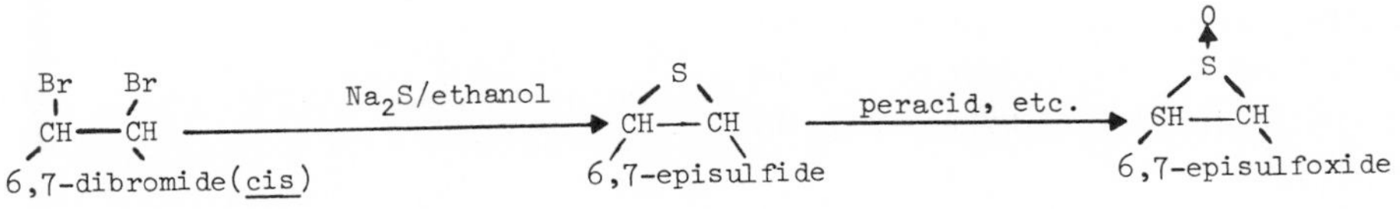

Many other derivatives of the aldrin or dihydro-aldrin structure have been made starting from aldrin or by condensing "hex" with substituted norbornenes. Most of them are less insecticidally active than aldrin or dieldrin, although much depends on the species used for evaluating toxicity. Some of these compounds are of theoretical interest and some will be referred to in connection with structure-activity considerations. An exception to the general observation of reduced toxicity in derivatives occurs in the aldrin analogue in which the 6,7-double bond is replaced by the nearly sterically equivalent azo-grouping.[214] To prepare this analogue, diethyl azodiformate is condensed with cyclopentadiene (Diels-Alder reaction), and this adduct is then condensed with "hex" to give the methano-hexahydrophthalazine derivative (Figure 15, structure 1), which is readily converted by successive hydrolysis, decarboxylation, neutralization, and oxidation into the methano-tetrahydro-phthalazine-derivative (Figure 15, structure 3) via the intermediate (2). The azo-derivative (3) is three to four times more toxic and the derived

N-oxide (4) four to five times more toxic than dieldrin to houseflies.[184] High toxicities are also shown towards some other insects, and even the dimethoxycarbonyl-compound (1,R=COOCH$_3$) is as toxic as dieldrin to houseflies (it may be converted into the azo-compound (3) in vivo). These compounds have never been developed for practical use, since they are highly toxic to mammals, as well as to insects.

Other interesting compounds of the aldrin-dieldrin series have the 5,8-methano-bridge replaced by an oxygen atom.[215] The first of these compounds (Figure 15, structure 6), although analogous to aldrin, is obtained by a route similar to that used in the preparation of isodrin; a Diels-Alder reaction is carried out in which hexachloronorbornadiene is condensed with an unchlorinated diene, in this case furan. The process is therefore virtually the converse of the Diels-Alder reaction used in the preparation of aldrin, the dienophile, instead of the diene, being chlorinated. It may be carried out in the usual way, for example, by passing furan slowly into hexachloronorbornadiene heated to 150°C, or by heating equimolar quantities of the reactants together in a sealed tube. The product is a crystalline solid mp 139°C, which in analogy with the aldrin-dieldrin conversion, can be converted into an epoxide by similar treatment with peracids. Catalytic reduction of the 6,7-double bond of 9-oxa-aldrin (Figure 15, structure 6) proceeds smoothly at atmospheric pressure and ambient temperature, as in the case of aldrin, to give a compound directly analogous to dihydroaldrin. Similar, but less insecticidal compounds are obtained when 2-methyl- or 2,5-dimethylfuran is used, so that the oxygen atom has one or two flanking methyl groups in the adduct.[184]

If hexachloronorbornadiene is condensed with cyclopentadiene, instead of furan, the addition produces isodrin (Figure 13, structure 3), the endo-, endo-isomer of aldrin, which is easily oxidized to the epoxide endrin by the usual methods.[190] The mode of addition contrasts with the steric course of the similar reaction with furan which, in the transition state of the Diels-Alder reaction, is orientated so that the oxygen atom is "beneath" the plane of the hexachloronorbornene nucleus (that is, on the anti-side relative to the dichloromethano-bridge), as shown in Figure 15. In the isodrin synthesis (Figure 13) the conjugated system of the cyclopentadiene ring must lie "beneath" or on the anti-side of the chlorinated ring system.

Analogues of isodrin in which the unsubstituted methano-bridge has been modified are of interest from the structure-activity aspect. If 1,3-cyclohexadiene, rather than cyclopentadiene, is condensed with hexachloronorbornadiene, the result is an analogue of isodrin of apparently lower toxicity which has a 5,8-ethano-bridge, instead of the methano-bridge.[192] Soloway has also described isodrin analogues in which the methano-bridge is replaced by cyclopropane, spirocyclopropane, and spirocyclopentane moieties.[184] The spirocyclopropane derivative is as toxic as aldrin to houseflies and its epoxide is more toxic than endrin. The preparation of simple derivatives by reactions involving the 6,7-double bond of isodrin is limited by the tendency for molecular rearrangements to occur when carbonium ions are generated at these positions; in the presence of acids (or by UV irradiation) the adjacent chlorinated and unchlorinated double bonds interact with the formation of the interesting cage structure (Figure 21, structure 3), which has conveniently been called photodrin.[170] This interesting reaction demonstrates the endo, endo-stereochemistry of isodrin. Endrin reacts similarly to give a half-cage ketone (Figure 21, structure 10), thereby providing an additional confirmation of molecular structure.[216]

As indicated previously, the removal of chlorine atoms from the hexachloronorbornene nucleus results in general in a reduction in insecticidal activity. Two notable exceptions are found, however, in the aldrin and dieldrin analogues lacking the vinylic chlorine atoms on C$_2$ and C$_3$, which are both more toxic than dieldrin (Vol. II, Chap. 3C.1b,c). These are prepared by the Diels-Alder reaction between 1,4,5,5-tetrachlorocyclopentadiene and norbornadiene, respectively, and epoxidation of the resulting adducts with peracids. Investigations on the photochemistry of aldrin and dieldrin during the last few years also provide another, more direct route; the irradiation of aldrin or dieldrin in hexane with UV light of wavelength below about 260 nm results in the successive removal of the two vinylic chlorine atoms.[160] The product of removal of one vinylic chlorine atom from dieldrin can be isolated in moderate yield if the reaction is stopped when the dieldrin has just disappeared, and the doubly dechlorinated product if the reaction is stopped at the point where the mono-dechlorinated one has disappeared. Products of molecular rearrangement are also formed (Vol. I, Chap. 3B.4d).

Aldrin and isodrin analogues prepared from 1,2,3,4,-tetrachlorocyclopentadiene have the chlorines on the dichloromethano-bridge replaced by hydrogen and have low insecticidal activity.[184] The same products may be formed directly by reduction of aldrin or isodrin with zinc dust and acetic acid. The compounds having one chlorine and one hydrogen atom on the methano-bridge are of theoretical interest although they are not of high insecticidal activity to houseflies.[219] Pentachlorocyclopentadiene reacts with norbornadiene to give the aldrin analogue in which the single bridge chlorine atom is directed toward the chlorinated double bond of hexachloronorbornene (called the *syn*-10-chloro-compound), and none of the *anti*-isomer. An isodrin type synthesis in which a mixture of *syn*- and *anti*-pentachloronorbornadienes reacted with cyclopentadiene gave only the *syn*-10-chloro-analogue of isodrin according to Mackenzie.[219] Treatment of isodrin or aldrin with sodium methoxide in methanol-dimethylsulfoxide also effects replacement of the *anti*-10-chlorine by hydrogen, presumably via a carbanion intermediate which is rapidly protonated by the methanol present; in the absence of alcohols, as when the same reaction is conducted with sodium hydride-dimethylsulfoxide, isodrin, but not aldrin, gives a small amount of the *anti*-10-chloroanalogue as well as the *syn*-isomer.[219]

Recent work indicates that reductive dechlorination of the dichloromethano-bridge of cyclodienes of the chlordene and aldrin series with aqueous chromous chloride also gives a mixture of the *syn*- and *anti*-10-monochloro-compounds, with the *syn*-compound predominating as usual.[64] This method involves a period of heating, the rate of reaction depending on the structure of the compound treated. However, aldrin has also been dechlorinated by treating it at room temperature with sodium borohydride in the presence of Vitamin B_{12}, a method which affords a mixture of the *syn*- and *anti*-compounds without reduction of the unchlorinated 6,7-double bond. A similar result is obtained with sodium borohydride in hexamethylphosphoric triamide, which gives also the analogue in which one of the vinylic chlorines is replaced by hydrogen. If cobaltous sulfate or nitrate is used instead of Vitamin B_{12} in the first method a mixture is obtained of the corresponding compounds in which the 6,7-double bond is also reduced. In this method the chlorinated compound (0.05 mol) is dissolved in diethyleneglycoldimethyl ether (40 ml) with cobaltous nitrate (0.1 mol), and the well-stirred mixture treated with sodium borohydride (0.5 mol) at 20°C. After some hours the reaction is ended and the products are isolated by chromatography on silica gel. Other dechlorinated analogues of aldrin that have been made include the 1,4-dichloro-derivative (only two chlorine atoms in the molecule, obtained by reduction of the 1,4,10,10-derivative with zinc and acetic acid) and the 10,10-dichloro-derivative and its epoxide.[184,218]

Soloway reports that the aldrin analogues in which one methano-bridge chlorine is replaced by fluorine (*syn*-10- and *anti*-10-fluoro-isomers possible), presumably prepared from the corresponding fluoropentachlorocyclopentadiene, are only slightly less toxic than aldrin.[184] Melnikov prepared 10,10-difluoro-analogues of aldrin and dihydroaldrin by the reaction of 5,5-difluorotetrachlorocyclopentadiene with norbornadiene and norbornene, respectively, and reported the aldrin analogue to be slightly less active than chlordane; the activity of 10,10-difluorotetrachloro-analogues appears to be generally low.[185] An aldrin analogue and its epoxide derived from 1,2, dichloro-3,4,5,5-tetrafluorocyclopentadiene have also been reported,[221] as well as two stereoisomers derived from the reaction between hexafluorocyclopentadiene and norbornadiene.[187]

3. Synthesis of Radiolabeled Cyclodienes

A number of the problems involved in the synthesis of radio-labeled compounds are well illustrated by the syntheses which have been devised for the preparation of labeled cyclodienes. In the first place, a choice of labeling with either or both of isotopic chlorine or carbon is available for the basic cyclodiene skeleton of structures such as aldrin or dieldrin. Depending on the purpose of the experiments in which the labeled compounds are to be used, certain other imaginative variations are also possible by derivatization of the molecules. For example, Winteringham took advantage of the facile addition of bromine to the double bond of aldrin by using [82]Br-6,7-dibromo-6,7-dihydroaldrin in studies of the rate of excretion of this compound by houseflies.[213] Although the half-life of ^{82}Br is only 35 hr, the compound can be prepared at high specific activity, and it was possible to show by using it that there was no significant difference between dieldrin-susceptible and resistant strains of houseflies in the rate at which they excreted this nontoxic compound. At the same time it was demonstrated that there was no significant metabolic attack on this molecule in vivo, an observation of some importance at the time.

Further advantage was taken of the known reactions of the dibromo-compound by reacting the *cis*-isomer with Na$_2$^{35}S, in the reaction mentioned in the previous section to give the 6,7-^{35}S-episulfido-analogue of dieldrin, which is a toxic molecule and clearly is a much closer relative of dieldrin than the dibromo-compound.[213] In both derivatives the label is peripheral; at least one bromine might be metabolically eliminated from the dibromide by dehydrobromination and the sulfur atom can be oxidized to give the corresponding sulfoxide and sulfone. However, there was no apparent removal of either label, although it is probable that a low degree of oxidation of the sulfur atom occurred in vivo. The 87-day half-life of ^{35}S and the high specific activity attainable adds to the utility of this approach, and again it was possible to show that the molecule did not undergo any major changes in vivo, in either susceptible or dieldrin-resistant houseflies. These examples are particularly good ones; when using such molecular modifications, the possibility must be borne in mind that the behavior of derivatives in vivo may differ significantly from that of the parent molecules.

The previous synthetic section shows that the total synthesis of aldrin is resolvable into that of "hex" and norbornadiene (bicyclo[2.2.1]hepta-2,5-diene), while for isodrin synthesis, hexachloronorbornadiene and cyclopentadiene are required. For each synthesis, there is the possibility of labeling either or both reactants with carbon-14, while the chlorinated reactants can also, in principle, be labeled with ^{36}Cl (or ^{38}Cl). Norbornadiene is made from cyclopentadiene and acetylene, and offers some synthetic attractions because ^{14}C$_2$-acetylene is available from ^{14}C-barium carbonate. However, a number of problems make the norbornadiene route unattractive and attention turns to ^{14}C-hexachlorocyclopentadiene, which is a potential key intermediate in the synthesis of both aldrin and isodrin. However, the reaction of

"hex" with vinyl chloride and the subsequent dehydrochlorination to give hexachloronorbornadiene are not attractive propositions for small scale preparative work, so that for isodrin the preparation of ^{14}C-hexachloronorbornadiene is not favored. Another point is that it is advantageous to use the dienophilic component in excess in the final condensation that leads to aldrin or isodrin, so as to avoid the formation of bis-adducts by further condensation with the reactive dienes involved;[211] if these dienophiles are unlabeled an economy of radio-labeled material is achieved on this account and, hopefully, optimal conversion of the costly labeled reactant (the diene) can be achieved.

These practical considerations indicate that ^{14}C-"hex" and ^{14}C-cyclopentadiene are the most desirable intermediates. The route to isodrin via ^{14}C-cyclopentene and ^{14}C-cyclopentadiene (Figure 16, route I) provides a possible additional route to ^{14}C-"hex" (route IA), and therefore a method of preparing the different isomeric forms of the cage structure from a common labeled intermediate. As shown in the figure, labeling in two different ring systems can be effected by using this route, but for each isomer only one type of ring label is available. As indicated earlier, the Diels-Alder condensation between "hex" and norbornadiene produces a small amount of isodrin, but its recovery is not a practical proposition.

The route to ^{14}C-isodrin (Figure 16, route I) requires six steps from [1,6-^{14}C$_2$]-adipic acid and nine steps from ^{14}C-barium carbonate, and gives an overall 10 to 12% chemical yield.[222] The low yield (36%) of cyclopentadiene from 1,2-dibromocyclopentane may result from its dimerization, which occurs rather readily, and it is this step which confers the low yield on the total synthesis, since the others are high yielding. Later repetitions of this first synthesis have given very similar results.[223] Since adipic acid is labeled in the carboxyl groups, one of which is lost during the formation of cyclopentanone, it would be advantageous to use the methylene labeled acid to avoid the resultant halving of specific activity. Carbonium ions are generated in the conversion of ^{14}C-cyclopentanol into cyclopentadiene, which results in random labeling of the cyclopentadiene ring and hence also the terminal ring of ^{14}C-isodrin. The final condensation stage leading to isodrin in route I is advantageous chemically, since ^{14}C-cyclopentadiene is trapped by an inactive

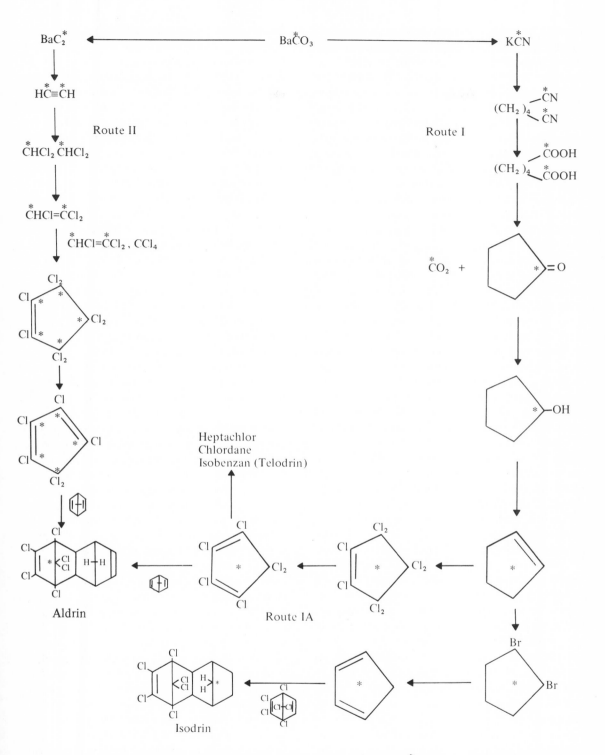

FIGURE 16.   Synthetic routes for ^{14}C-labeled cyclodiene insecticides. Route I, $K\overset{*}{C}N \rightarrow$ cyclopentadiene $\rightarrow$ isodrin; 11% chemical yield. Route IA, $K\overset{*}{C}N \rightarrow$ cyclopentene $\rightarrow$ hexachlorocyclopentadiene $\rightarrow$ aldrin; 37% overall yield from $Ba\overset{*}{C}O_3$. Route II, $Ba\overset{*}{C}O_3 \rightarrow$ octachlorocyclopentene $\rightarrow$ aldrin, via acetylene and trichloroethylene; 28% overall chemical yield.

119

reactant to give a crystalline product having more than five times the original molecular weight. Radiochemically, the ^{14}C-isodrin produced has the same specific activity as the ^{14}C-cyclopentadiene precursor, but the large increase in molecular weight reduces the activity in terms of counts/min/μg of the labeled isodrin at the available counting efficiency; this is an important practical consideration, and must be allowed for in estimations of the final specific activity likely to be acceptable from an analytical point of view. Since the actual mass of compound available can always be increased by the addition of inactive carrier at the end of operations, such syntheses are best conducted at the highest possible specific activity consistent with the avoidance of decomposition problems. Frequently, this means that a large amount of isotope has to be used to make the scale of chemical operations feasible, which may only be possible when labeled compounds are made commercially to meet a wide market.

The synthesis of ^{14}C-aldrin via [14]C-"hex" is more efficient both chemically and radiochemically than the synthesis of isodrin by the above method. From 2 mmol of ^{14}C-barium carbonate at 20 mCi/mmol, Korte and Rechmeier[224] synthesized ^{14}C-aldrin (HHDN) in 37% yield overall, using route IA (Figure 16), in which ^{14}C-cyclopentene is converted via ^{14}C-octachlorocyclopentene into [14]C-"hex" in two high yielding stages; the label is randomized during this process. Since there are no low yield steps in this ^{14}C-aldrin synthesis, the chemical yield is correspondingly better than in the ^{14}C-isodrin synthesis. Furthermore, the high molecular weight [14]C-"hex" is condensed with the low molecular weight unlabeled norbornadiene (only a 30% increase in molecular weight is involved) so that there is only a small reduction in radioactivity per unit mass in this case (the specific activity being unchanged). As mentioned before, the choice of hexachloronorbornadiene and norbornadiene as the unlabeled reactants enables them to be used in excess in the final condensation, thereby ensuring optimal conversion of the labeled reactant without the formation of by-products.

Two syntheses of ^{14}C-aldrin published in 1960 employ the Prins reaction described earlier for the preparation of [14]C-"hex." Both syntheses take advantage of the doubling in specific activity that occurs when [14]C$_2$-acetylene is made from ^{14}C-barium carbonate, and the doubling which again

occurs when one molecule of inactive carbon tetrachloride condenses with two molecules of [14]C$_2$-trichloroethylene (Figure 16, route II). Thus, [14]C-"hex" is effectively built up from four carbon atoms of ^{14}C-BaCO$_3$, which is the most efficient type of labeled synthesis. Although this method is suitable for the preparation of high specific activity "hex," the McKinney and Pearce synthesis[225] was conducted on a fairly large scale with materials of moderate specific activity; a 22% chemical yield of ^{14}C-dieldrin of specific activity 1.32mCi/mmol was obtained starting from 100 mmol of ^{14}C-BaCO$_3$ (specific activity 0.4 mCi/mmol), following a 19% dilution with inactive carrier at the trichloroethylene stage.

The same route was used by Thomas and Kilner[226] with certain differences in detail and is designed to produce high specific activity products. In the first recorded synthesis, [14]C$_2$-trichloroethylene was prepared on a 4 mmol scale and having a specific activity of nearly 17 mCi/mmol with an overall 12% yield from ^{14}C-BaCO$_3$. In this pilot preparation, it was necessary to add inactive carrier acetylene, octachlorocyclopentene, and hexachlorocyclopentadiene at the appropriate intermediate stages in order to maintain the desired chemical scale. The authors pointed out at the time that by using barium carbonate of specific activity 30 mCi/mmol or greater, and increasing the chemical scale of the initial steps, dilutions with inactive carrier could be avoided and a specific activity of 30 to 50 mCi/mmol should be attainable. ^{14}C-aldrin and dieldrin are currently marketed at 40 to 80 mCi/mmol, demonstrating the general improvement in techniques since then.

In the early synthesis, ^{14}C-barium carbide, prepared in 80% radiochemical yield by heating the ^{14}C-carbonate with excess of barium in an argon atmosphere, was hydrolyzed to give [14]C$_2$-acetylene. At this stage, the [14]C$_2$-acetylene was diluted with carrier and converted (5 mmol, 75 mCi) into symmetrical [14]C$_2$-tetrachloroethane by chlorination in a 2-l blackened flask containing some pure sand so that the potentially explosive reaction could be moderated to give a nearly quantitative yield. [14]C$_2$-trichloroethylene (4.15 mmol, 68 mCi) was prepared by dehydrochlorination of the [14]C$_2$-tetrachloroethane with concentrated aqueous potassium hydroxide at 0°C, and the chlorinated olefin converted into octachlorocyclopentene (50 to 60% yield based on trichloroethylene) by heating it in a sealed tube with a 100% excess of inactive carbon tetrachloride (10 ml) and aluminum chloride (350 mg) at 90°C for 3 hr. ^{14}C-octachlorocyclopentene (1.7 g, 38 mCi), together with some ^{14}C-"hex" (0.39 g, 29.6 mCi), was separated from the reaction mixture by gas phase chromatography. Various other compounds totaling 11.5 mCi were obtained and some materials were probably lost on the column. An unexplained loss of about 30% of the expected specific activity is apparent in the ^{14}C-octachlorocyclopentene produced, but this activity was not present in the carbon tetrachloride recovered after the Prins reaction (Vol. I, Chap. 3B.1) as might have been expected; the symmetrical intermediate $\overset{*}{C}Cl_3\overset{*}{C}HClCCl_3$ from $^{14}C_2$-trichloroethylene and inactive CCl_4 can theoretically equilibrate with the excess inactive CCl_4 present, so that a considerable loss of $\overset{*}{C}Cl_4$ into the unlabeled carrier is possible. In fact, very little activity was found in the recovered CCl_4, and furthermore, there appeared to be no loss in specific activity through exchange reactions in the McKinney and Pearce synthesis either. One of the other hazards in the high specific activity synthesis is the possibility of radiation decomposition at the trichloroethylene stage, and this material was stored in the dark at $-80°$.

In the final stage of the synthesis, ^{14}C-octachlorocyclopentene, recovered by gas-chromatography in 85 to 90% yield (based on that in the original crude reaction mixture) and containing a little ^{14}C-"hex," was diluted with carrier (29.6 mCi, total 2.3 mmol) and distilled under reduced pressure over a nickel spiral at 490°, to give ^{14}C-"hex" (1.5 mmol, 18 mCi), which was diluted with carrier (1 mmol) and converted successively into ^{14}C-aldrin and ^{14}C-dieldrin of 95% radiochemical purity by reaction with excess norbornadiene and then epoxidation.

Heptachlor and the chlordane isomers labeled with carbon-14 in the hexachloronorbornene nucleus have been prepared by the Diels-Alder reaction between ^{14}C-"hex" and cyclopentadiene to give ^{14}C-chlordene which is then further converted.[223] For example, the stereospecific synthesis of *trans*-chlordane from chlordene described by Buchel (Vol. I, Chap. 3B.2a) is adaptable for this purpose. In principle, the members of this series may easily be generally labeled with ^{14}C by first condensing ^{14}C-"hex" with ^{14}C-cyclopentadiene to give fully labeled ^{14}C-chlordene. For aldrin and isodrin, general labeling of this kind requires the synthesis of ^{14}C-norbornadiene and ^{14}C-hexachloronorbornadiene. Molecules having the labels in different parts of the carbon skeleton are clearly valuable for indicating the nature of metabolic routes; the fact that ^{14}C-isodrin and ^{14}C-aldrin labeled in different cyclopentene rings (Figure 16) give one major and similar metabolic product in each case (that is, the corresponding epoxide) shows that the carbon skeleton does not suffer extensive disruption involving the loss of the unchlorinated cyclopentene ring.

The normal synthesis of isobenzan (Telodrin; Figure 17) in which "hex" is condensed with 2,5-dihydrofuran and the adduct chlorinated, has been adapted for use with ^{14}C-"hex," so that the molecule can be labeled in the hexachloronorbornene nucleus.[227] A different approach permits the labeling of this molecule in the furan ring. In this synthesis, use is made of the facile condensation of "hex" with the best known of all dienophiles — maleic anhydride. $^{14}C_2$-succinic acid labeled in the carboxyl groups is converted via $^{14}C_2$-fumaric acid into maleic anhydride labeled in both carbonyl groups. The $^{14}C_2$-maleic anhydride reacts (Diels-Alder) very readily when warmed with "hex", and hydrolysis of the resulting acid anhydride gives hexachloronorbornene-5,6-*endo-cis*-dicarboxylic acid (called chlorendic acid; Figure 17, structure 1) labeled in the carboxyl groups. This acid is esterified, the ester reduced to the corresponding diol (5,6,-bishydroxy-^{14}C-methyl hexachloronorbornene; Figure 17, structure 3) with lithium aluminum hydride, and the diol dehydrated with alumina at 140°C to give the isobenzan precursor (Figure 17, structure 3) labeled in the two carbon atoms flanking the oxygen atom. Chlorination then gives 1,3-$^{14}C_2$-isobenzan.[227] Double labeling is clearly possible with a molecule of this sort and can be effected by chlorination with $^{36}Cl_2$; by using a molecule labeled in the α-chlorine atoms as well as the α-carbon atoms, or simply an appropriate mixture of molecules separately labeled in these positions with either ^{14}C or ^{36}Cl, the loss of chlorine can be measured in a metabolic study from the change in $^{36}Cl/^{14}C$ ratio in the product. Several other combinations are possible. For example, a combination of ^{14}C-"hex" with ^{14}C-maleic anhydride leads to the generally labeled molecule, which can be additionally

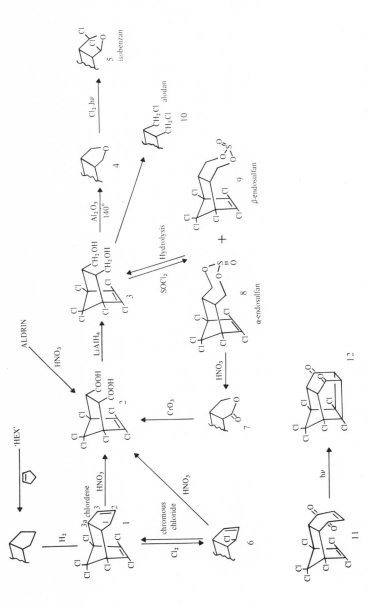

FIGURE 17. The relationship of various cyclodiene insecticides to chlorendic acid and the photochemical formation of a cage molecule, demonstrating the *"endo"* configuration of the original Diels-Alder adducts.

labeled with ^{36}Cl if required, and so on. The choice of isotope for labeling (if a choice is possible) and the pattern to be used when alternatives are available depend on some degree of foresight regarding possible degradation pathways.

Labeling with ^{36}Cl has not found wide application with organochlorine compounds, because of the low specific activity formerly available. With specific activities up to 350 μCi/m-atom more recently available, specific activities up to 2 mCi/mmol should be possible for ^{36}Cl-aldrin and dieldrin (with a theoretical maximum of about 7mCi/mmol). This represents a considerable improvement, bearing in mind that the distribution of ^{36}Cl-dieldrin has been measured in rats using material having specific activity as low as 36 μCi/mmol. As in the case of isobenzan, a combination of the molecules labeled separately with ^{14}Cl and ^{36}Cl might find application for some purposes in experiments with other cyclodienes.

Chlorinated molecules containing ^{38}Cl can be made by thermal neutron irradiation ($^{37}Cl(n,\gamma)^{38}$) of the inactive compounds and are of interest for studies in vivo. Carbon tetrachloride, for example, when irradiated (0.1 ml;1 mmol) in a thermal neutron flux of 10^{12} neutrons/cm^2/sec for 10 min, gives 2.1 mCi of ^{38}Cl activity. Trichloroethylene gives the highest yield of organic material (86%) out of a number of simple chlorinated hydrocarbons that have been irradiated in this manner, 32% of ^{38}Cl-labeled trichloroethylene being produced in this case, although no indication has been given of the level of radioactivity actually attained. ^{38}Cl is a β,γ emitter producing high energy β particles (53% with energy of 4.81 MeV) along with γ-radiation (of energies 1.60 MeV; 31%, and 2.15 MeV; 47%) and having a half-life of 37 min. The short half-life means that high starting specific activities would be essential for total synthesis and it is doubtful whether this approach would be possible for elaborate molecules like the cyclodienes, although it is interesting that the key intermediate trichloroethylene gives the best yield in the irradiation process. The Szilard-Chalmers reaction (neutron irradiation) applied directly to complicated molecules is likely to give many products, but there is an early report that metabolism work with ^{38}Cl-isobenzan in mammals led to inconclusive results, so that the method has apparently been used with a cyclodiene insecticide. Because of the high emission energies, ^{38}Cl is potentially useful in experiments involving direct counting in vivo, although the short half-life necessarily limits the duration of such experiments to a few hours. Recent work on the metabolism of dieldrin and its photoisomer (photodieldrin) shows that there is a difference in the way that these compounds are distributed in male rats in vivo; photodieldrin, but not dieldrin, tends to accumulate in the kidneys.[229] This is discovered by examining the distribution of the ^{14}C-labeled molecules in the tissues after sacrificing the animals (Vol. II, Chap. 3B.2d), and it is clear that for investigating pharmacodynamic problems of this sort it would be useful to be able to follow distribution changes by direct measurement in the intact animal, as might be done using ^{38}Cl as well as by whole body autoradiography.

4. Constitution and Chemical Reactions of Cyclodienes

a. Configuration of the Ring Systems

Complex polycyclic systems of the cyclodiene type can theoretically exist in a number of stereoisomeric forms. First, there is the question of the stereochemistry of the ring junction between the hexachloronorbornene system and the addendum introduced during the Diels-Alder reaction, and secondly, that of the configurations adopted by substituents in the attached ring system or systems.

As far as the stereochemistry of the ring junction is concerned, the proof that the fusion is through the endo-positions of hexachloronorbornene rests on the early structural studies of Riemschneider and Soloway, supported by a number of more recent investigations on the general reactions of cyclodienes which provide more chemical evidence, as well as NMR data, consistent with the endo-ring fusion. Some of this evidence was mentioned previously when the syntheses in the dihydroheptachlor series were discussed (Vol. I, Chap. 3B.2a). Adducts of special interest to Riemschneider were those from "hex" and maleic anhydride (called chlorendic anhydride) and from "hex" and benzoquinone (Figure 17), and the relationship between chlorendic acid and chlordene is crucial from a structural point of view. As mentioned in the synthetic section, "hex" reacts with cyclopentadiene to give two adducts, depending on the reaction conditions (Figure 14).[196] That chlordene (Figure 17, structure 1), the compound of major interest and the main product of this reaction, contains the hexachloronorbornene system is shown by its lack of reaction with phenylazide due to the chlorine substitution in the norbornene double bond; the cyclopentene double bond does not react with this reagent. Furthermore, it is very stable to alkali, does not give reactions expected of a compound containing allylic chlorine atoms, and is catalytically hydrogenated to the adduct (Figure 17, stucture 1a) obtained from the reaction between "hex" and cyclopentene. Since cyclopentene can only act as dienophile, this interrelation is good evidence that the chlordene skeleton is as shown (Figure 17, structure 1).

When oxidized with a mixture of nitric and sulfuric acids or with hot nitric acid, chlordene affords a mixture of products from which chlorendic acid can be isolated, as well as a related acid formed with the retention of atom C_2 of chlordene.[230] Heptachlor also gives chlorendic acid on oxidation. The fact that chlorendic acid can be reduced to the corresponding diol (Figure 17, structure 3), which is the precursor of both endosulfan (Thiodan), alodan and the furan derivative (4), links these compounds structurally to this acid. Since aldrin is also converted into chlorendic acid by oxidation, a relationship is established between most of the well-known primary adducts and their insecticidal derivatives through this oxidation product.[188] The six cyclohexene ring carbon atoms of hexachloronorbornene are pulled into the boat conformation by the bridging seventh carbon, which is symmetrically placed relative to the others with the normal tetrahedral angle of 109.5°C between its chlorine atoms. According to Riemschneider's calculations, the bridgehead chlorine atoms (that is, those at C_1 and C_4 in the dimethanonaphthalene nomenclature) are at an angle of 22° to the plane containing the carbon atoms which carry them. Using the value of 1.90D for the dipole moment of a C-CCl system and 1.44D for that due to C=CCl, together with the above information, a total moment of 3.20D was derived for the hexachloronorbornene nucleus, and values of 4.96D and 3.07D, respectively, for the exo- and endo-isomers of chlorendic anhydride. The measured value of 2.84D therefore led Riemschneider to propose the endo-stereochemistry for the ring fusion in this molecule, and through the relationships outlined above,

for the stereochemistry of all the other adducts.[188]

Further striking evidence for the endo-ring fusion comes from what has become the most fascinating aspect of cyclodiene chemistry — molecular rearrangements. In 1949, Riemschneider showed that the adducts from "hex" and benzoquinone and from difluorotetrachlorocyclopentadiene and benzoquinone form cage structures of the type depicted in Figure 17, when irradiated with ultraviolet light.[188] Some years later, Lidov and Bluestone[231] reported that isodrin undergoes a similar rearrangement (Figure 21) to the chlorinated cage structure which has been called "photodrin" for convenience (see section on nomenclature of cyclodienes). This and similar rearrangements were later investigated in detail by other workers. In these reactions, the stereochemistry of the ring junction has to be endo- in order to place the reacting moieties in the correct position for cyclization to occur. Molecular rearrangements occur especially easily in the isodrin-endrin series of compounds, but it was not until some time later that molecular rearrangements involving the unchlorinated methano-bridge of dieldrin were demonstrated. This reaction clarifies the structure of dieldrin, since it can only occur if the ring junction is endo- and the methano-bridge is placed close to the double bond. Similarly, the rearrangements of isodrin and endrin without involvement of the methano-bridge show that the ring junction is endo-, and that the ethylene- or epoxyethane-bridge, rather than the methano-bridge, is close to the chlorinated double bond. Since the discovery of the dieldrin rearrangements, environmental problems have greatly stimulated interest in the photochemistry of cyclodiene compounds, and cage structures derived from the heptachlor-chlordane series are now well characterized, so that there is a great deal of additional evidence for the endo-ring fusion. With the mode of fusion of the hexachloronorbornene ring system established, the stereochemistry of the rest of the molecule and its influence on chemical reactivity can be considered in terms of the various structural types in the cyclodiene group.

b. Heptachlor, Chlordane, and Related Compounds

For chlordene (Figure 18, structure 6), there are four possible isomeric forms: two because of

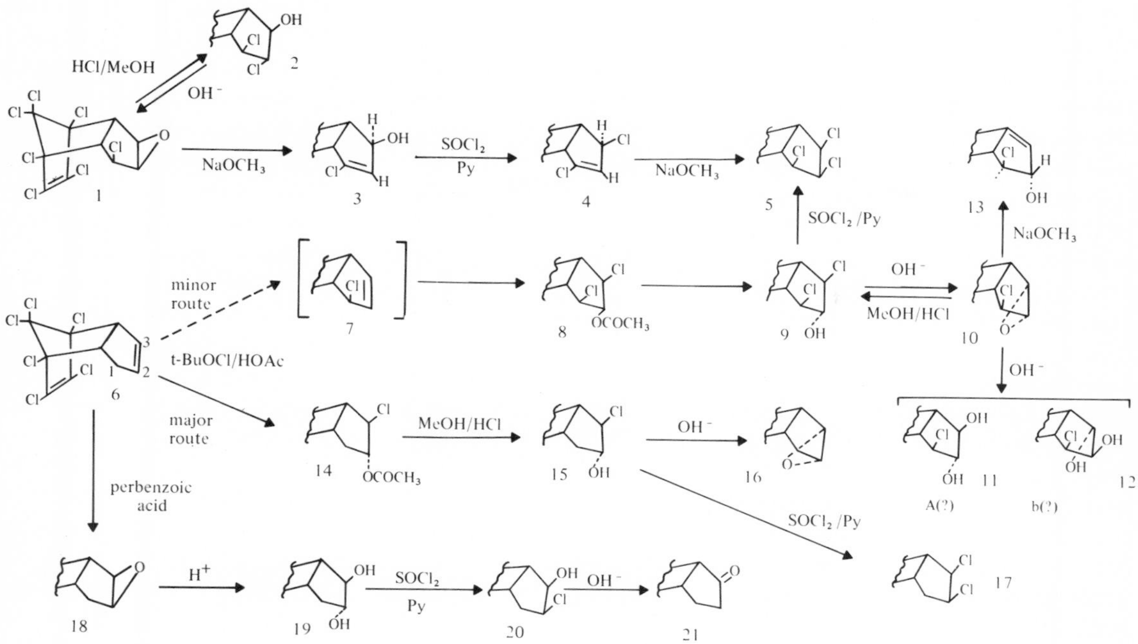

FIGURE 18. Chemical reactions of various compounds derived from chlordene.

the possibility of *exo-* or *endo-* ring fusion, and two more because the molecule is asymmetrical, and there are two possible positions of the double bond in the cyclopentene ring. The last possibility confers optical isomerism, but since the *endo-*ring fusion is established, the actual number of stereoisomers is two, these being the enantiomorphic forms of the *endo-*fused structure shown in Figure 18. Hence, chlordene should be a racemic compound, but so far has not been resolved into enantiomorphic forms. When a single extra chlorine atom is introduced into chlordene, to give heptachlor (Figure 18, structure 7), it might occupy either the 1-*exo* or 1-*endo*-position on the cyclopentene ring (a variety of evidence has now ruled out the alternative, ring junction allylic position (3a) for this chlorine atom). With the *endo-*ring fusion established, this doubles the number of stereoisomers possible in chlordene, two optical isomerides being possible for each position (*exo* or *endo*) of the substituted chlorine atom. If the *exo-*ring fusion is also considered (although such a molecule is unknown), the total number of possible isomerides becomes eight. Fortunately, it turns out that there is only one heptachlor molecule, although being asymmetric it should be capable of being resolved into enantiomorphic

forms; the 1-*endo*-position is strongly sterically hindered since a substituent in this position is in close proximity to the highly chlorinated ring system, and it is now clear that the chlorine is in the 1-*exo*-position, as found in α-dihydroheptachlor (Figure 14, structure 7), discussed earlier in the sections on synthesis and nomenclature. One or two extra chlorine atoms introduced into α-dihydroheptachlor give the various isomeric chlordanes or nonachlors, respectively, and the isomeric possibilities are correspondingly increased at each substitution.

These isomeric possibilities are of considerable interest in relation to the nature of the constituents of technical chlordane, as is seen in the section on the development of this insecticidal product; the need for unequivocal methods for micro-identification of members of this series in relation to metabolic studies and residue analysis has resulted in some interesting chemical work in recent years. The general chemical reactions of heptachlor relate to the reactivity of the allylic chlorine atom and the cyclopentene double bond. Catalytic reduction of the latter does not give α-dihydroheptachlor; the chlorine atom is eliminated instead and the double bond reduced to give dihydrochlordene (Figure 17, structure 1a). The allylic chlorine atom may be replaced by bromine, fluorine, or hydroxyl as indicated in the synthetic section, and its removal as silver chloride in the presence of silver nitrate in boiling acetic acid is the basis of a specific analytical method for heptachlor. The compound 1-hydroxychlordene (Figure 14, structure 6) is the basis of a specific confirmatory test for the presence of heptachlor residues,[64] since it is readily produced when heptachlor is heated with silver carbonate in aqueous ethanol; its observation in gas-liquid chromatographic (GC) analysis is facilitated by conversion to the corresponding trimethylsilyl ether, which gives a much better response when the electron-capture detector is used, although aldrin interferes in some GC systems. The treatment of heptachlor in acetone with aqueous chromous chloride solution at 50 to 60°C for a short time regenerates chlordene by removing the allylic chlorine atom. Some hydrolytic replacement by hydroxyl also occurs to give a little 1-hydroxychlordene and, as is frequently found with reductive processes in the DDT-group, the reductive removal of allylic chlorine between two heptachlor molecules results in dimerization (minor product) by the linking of the allylic positions of their cyclopentene rings. With longer reaction times the chlordene produced reacts further, the dichloromethano-bridge suffering reductive removal of chlorine to give mainly the syn- (that is, remaining chlorine atom directed towards the chlorinated double bond) pentachloro-derivative. Some other reducing agents, mentioned in the synthetic section, appear also to effect more drastic changes including the removal of a vinylic chlorine atom, or double bond reduction with some compounds.[64,232]

The chemical conversion of heptachlor into its 2,3-exo-epoxide (Figure 18, structure 1) requires drastic oxidation, and the corresponding 2,3-endo-epoxide (Figure 18, structure 10) is formed via the intermediate trans-chloroacetate. shown. Like dieldrin, the 2,3-exo-epoxide is chemically rather stable, but it does participate in reactions expected of the epoxide ring under vigorous conditions. Formation of this epoxide from heptachlor was first demonstrated in vivo in dogs by Davidow and Radomski,[233] who converted it into the corresponding chlorohydrin (Figure 18, structure 2) by extended heating with methanolic hydrogen chloride; a bromohydrin was formed from hydrogen bromide in a longer time at room temperature. On treatment of the halohydrins with a base such as methanolic potassium hydroxide, the original epoxide is regenerated, thus providing good evidence for the presence of the epoxide ring. The 2,3-endo-epoxide is cleaved more readily with methanolic hydrogen chloride, to give the same chlorohydrin (Figure 18, structure 9) as is derived via the reaction of heptachlor with t-butyl hypochlorite/acetic acid. It is interesting that a comparable series of reactions occurs with chlordene; in fact, when chlordene is treated with the t-BuOCl/HOAc reagent the main product is the trans-chloroacetate (Figure 18, structure 14), but there is evidently some degree of allylic chlorination also, so that heptachlor is formed as a minor intermediate which leads to the presence of trans-chloroacetate (8) and hence the octachloro-chlorohydrin (9) as impurities in the desired heptachloro-chlorohydrin (15).[234]

The formation of epoxide (16), which is different from the product (18) of peracid oxidation of chlordene, shows that (15) must be a trans-chlorohydrin, and indicates that its hydroxyl-group must be endo-; this has been confirmed by its use to prepare cis-chlordane (17).[208] The hydroxyl group in (15) might also be at C_1, with the chlorine at C_2, but position 2 is strongly indicated for the hydroxyl on account of the steric hindrance in the 1-endo-position, and the fact that when the bromo-analogue of heptachlor is treated with t-BuOCl/HOAc and the intermediate bromo-chloroacetate (analogous to structure 8 in Figure 18) is cyclized with base, the epoxide produced is (10) rather than the bromo-epoxide, showing that the acetate moiety in (8) must be symmetrically disposed with respect to the halogens. In analogy with the conversion of chlorohydrin (15) into cis-chlordane (17), chlorohydrin (9) is converted into a cis-nonachlor (5), with a melting point (214 to 215°C) rather similar to that (209 to 211°C) reported for a nonachlor-compound isolated from technical chlordane by Vogelbach. Since epoxide (16) has the endo-configuration, the peracid oxidation product (18) must have the exo-configuration, which is supported by trans-ring opening, followed by chlorination of the diol (19) under inversion conditions to give the cis-chlorohydrin (20) which, as expected, gives the ketone (21) with base, rather than an epoxide.[234]

Although the trans-chlorohydrins are readily converted into the corresponding epoxides under basic conditions, they can also be preferentially converted into methyl ethers by treatment with methyl *p*-toluenesulfonate in boiling light petroleum in the presence of a little added sodium methoxide; under these conditions, the sodium methoxide is present as a solid and ether formation occurs with minimal cyclization to the epoxide.[235] When heptachlor 2,3-*exo*-epoxide is heated with a strong base such as sodium methoxide, in anhydrous methanol, the 1-*endo*-proton is eliminated in a concerted process that results in opening of the epoxide ring to give 1-hydroxy-3-chlorochlordene (Figure 18, structure 3) which in turn affords 1,3-dichlorochlordene (4) with the thionyl chloride/pyridine reagent. The reaction with base is facilitated by the *trans*-relationship between the 1-*endo*-hydrogen and *exo*-2,3-epoxide ring; a similar reaction with the 2,3-*endo*-epoxide involves the equivalently sited 3a-hydrogen and affords the isomeric allylic alcohol (13).[236-239]

From a metabolic point of view, considerable interest attaches to reactions of these epoxides in which the epoxide ring is hydrolytically opened to give the corresponding *trans*-diols, since such conversion is an obvious prelude to elimination in vivo. The same two *trans*-diols (Figure 18, structures 11 and 12) are expected to be formed by hydrolysis of the epoxide ring of either of the heptachlor epoxides. The *exo*-2,3-epoxide (called HE 160 in this account from its melting point of 160°C) is rather stable chemically; it is not hydrated when heated for 50 hr at 100°C in aqueous tetrahydrofuran containing perchloric acid, but when treated with boron trifluoride in acetic anhydride slowly gives a mixture of diacetates from which the diols can be liberated by ester exchange with methanolic hydrogen chloride. In this way, two diols, A and B, are formed in 9:1 ratio. The corresponding reaction with the 2,3-*endo*-epoxide (called HE 90 from its melting point of 90°C) is fast and gives a 3:2 ratio of A:B.[240] Assuming that the major product is produced by attack on the least sterically hindered 2-*endo*-position of HE 160, it is the 1-*exo*-chloro-2-*endo*, 3-*exo*-dihydroxy-derivative (Figure 18, structure 11) of dihydrochlordene, which is also the diol expected from attack on the 3-*exo*-position (furthest from the 1-*exo*-chlorine and therefore presumably least sterically hindered) of HE 90. The same mixture of diols is produced from HE 90 by some hours of incubation at 37°C with pig or rabbit liver microsomes; with HE 160 the conversion under these conditions is much slower and, as

in the chemical ring cleavage, appears to give mainly the diol (presumed to be Figure 18, structure 11) that preponderates in the mixture obtained from HE 90 (Vol. II, Chap. 3B.2c and 3a).

The photochemical formation of cage structures from molecules of the heptachlor-chlordane group (Figure 19) neatly illustrates the *endo*-fusion of the hexachloronorbornene nucleus in the primary adducts as well as providing additional information regarding other stereochemical problems. Two main types of reaction are evident: (a) photodechlorination, which involves the removal of one or more of the vinylic chlorines from the hexachloronorbornene nucleus without molecular rearrangement and is suggested to be mediated by an excited singlet state involving only the chlorinated double bond, and (b) cage formation, a sensitized triplet state reaction which involves interaction between the double bonds.[241,242]

The irradiation of chlordene in *n*-hexane with wavelengths below 300 nm effects removal of a single vinylic chlorine atom. Since the molecule is asymmetrical, two products are formed, corresponding to removal of a chlorine atom from different ends of the chlorinated double bond. At 300 nm in the presence of a triplet sensitizer such as acetone, the double bonds interact to give a cage molecule retaining all the chlorine atoms. Further irradiation of the mono-dechlorinated compounds under these conditions gives two further cage compounds which are isomers differing in the position which lacks the sixth chlorine atom. When irradiated as a solid film in air with light of wavelength 253.7 nm, chlordene is about 75% converted to other compounds in 16 hr; the products include about 20% of bridged compounds, together with 80% of other derivatives. Since even a 15-min exposure of chlordene in a thin film to UV irradiation in air produces chlordene epoxide, 1-hydroxychlordene and other oxygenated products that probably include 1-*exo*-hydroxy-2,3-*exo*-epoxydihydrochlordene and ketonic material, the major portion of the material arising from long term exposure is likely to consist of complex rearranged and oxygenated products.[243]

The situation with heptachlor is similar; the photolysis of 10^{-4}M solutions in hexane or cyclohexane at 253.7 nm for 3 hr gives only the isomeric monodechlorinated products (2) and (3)

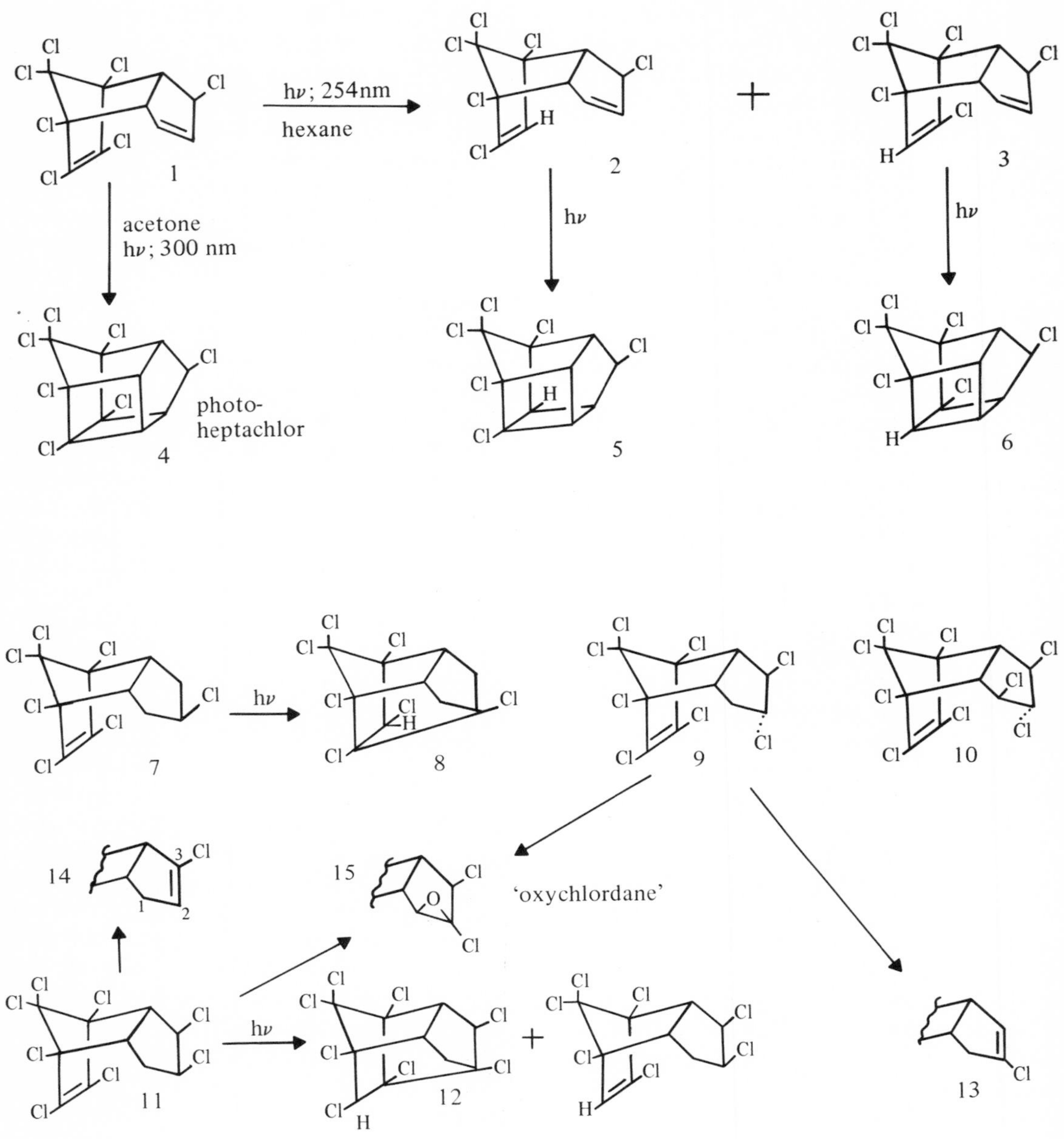

FIGURE 19. Various photochemical reactions and other conversions of heptachlor and relatives.

(Figure 19).[242] If the reaction is carried out in acetone at 300 nm, triplet sensitized cage formation occurs to give exclusively the cage isomer called photoheptachlor (4). Sensitized irradiation of the monodechlorination products likewise gives the corresponding pentachloro-cage structures 5 and 6. Since the systematic names of these cage molecules are somewhat complicated (see the section on nomenclature), simple names will be used here as far as possible. If the photolysis of heptachlor is conducted at 300 nm in cyclohexane containing 10 to 50% of acetone, the formation of photoheptachlor is markedly reduced and the major product is a molecule arising from replacement of the allylic chlorine atom by cyclohexyl-. In this reaction, sensitized by acetone, a chlorine radical abstracted from the allylic position is thought to generate solvent radicals which in turn react with the allylic position. β-Dihydroheptachlor (Figure 19, structure 7), when irradiated in *n*-hexane with wavelengths below 300 nm, gives a little of the corresponding photoisomer (8) and a little of the parent with two vinylic chlorines replaced by hydrogen.[244] The main product is the monodechlorinated derivative with one vinylic chlorine replaced by hydrogen; only one structure is possible in this case, since β-DHC is symmetrical to start with. Further irradiation of these products in acetone with wavelengths greater than 300 nm gives the corresponding pentachloro- and hexachloro-bridged structures. If the irradiations are carried out in aqueous methanol or dioxan at wavelengths below 300 nm, the same products are formed as in hexane, but a longer time is required and the bridged molecules preponderate. With this molecule, photoreactions appear to occur very slowly in the solid phase under short wave irradiation, only traces of the bridged photoisomers being detectable.

The photolytic behavior of the *cis-* and *trans*-chlordane isomers and of the nonachlor of melting point 128 to 130°C (a constituent of technical chlordane, (Vol. I, Chap. 3C.1a) is of considerable interest in relation to their stereochemical features.[244] In *trans*-chlordane (Figure 19, structure 9) and in this nonachlor (which has been assigned a *trans*-configuration (10) for other reasons), the chlorine on carbon-2 of the cyclopentane ring is in the *endo*-position and so is directed towards the chlorinated double bond in a manner expected to prevent bridge formation. Recent work on the photoisomerization of these chlordane isomers and of nonachlor indicates that the *endo*-chlorine does have this effect. Neither *trans*-chlordane or nonachlor gave bridged photoisomers under a variety of reaction conditions although the former is changed to some extent (5 to 10%) when irradiated in oxygen, and nonachlor under these conditions is extensively decomposed (>40%) within 24 hr to compounds not so far identified.* However, as expected, both compounds undergo the usual reactions at the chlorinated double bond; nonsensitized photolysis (wavelength < 300 nm) of *trans*-chlordane in aqueous-organic solvents affords two isomeric monodechlorination products (since the parent is asymmetrical) and eventually the bis-dechlorination product, while nonachlor (symmetrical) gives the monodechlorination product in acetone with short wavelength light. In acetone/water, the double bond of this last compound is reduced.[244,245]

In contrast to the above findings, *cis*-chlordane (Figure 19, structure 11) is converted into a fully chlorinated bridged photoisomer (12) when irradiated in the solid phase in aqueous methanol (sensitized) or aqueous dioxan, and gives also the expected series of dechlorination products, the end product of dechlorination in aqueous media having the fully reduced, dechlorinated double bond.[244] Thus, *cis*-chlordane as a solid film is up to 70% converted after 16 to 20 hr irradiation, nearly half of the products being bridged isomers (the dechlorinated compounds may also give bridged molecules by further isomerization under appropriate conditions).[245] In aqueous dioxan (2:3, v/v), irradiation at wavelengths greater than 300 nm gives 40% conversion to an equal mixture of bridged and other compounds in about 20 hr, while in aqueous methanol (1:1) under these conditions the conversion is slower and gives mostly the bridged rearrangement product (Figure 19, structure 12). The conditions discussed above are obviously artificial, but may serve to indicate the general environmental stability of the compounds concerned; it seems likely that out of *cis-* and *trans*-chlordane, heptachlor, nonachlor, and chlordene, all of which occur in technical chlordane, the compounds likely to be changed most slowly under natural conditions are *trans*-chlordane and nonachlor.

Apart from the photochemical transformations, the most important reactions involving the chlordane isomers and nonachlors are reductive dechlorinations, as effected, for example, with chromous chloride, and the dehydrochlorinations which have recently been carefully explored in relation to organochlorine residue analysis. There is also a somewhat unexpected oxidation reaction involving the chlordane isomers which occurs in vivo and recently came to light as a result of

*However, a recent investigation by Casida shows that with both *trans*-chlordane and *trans*-nonachlor, similar intramolecular bridging does occur, but in each case involves carbon-1, which has a noninterfering *exo*-chlorine.[244a]

metabolic studies (Vol. II, Chap. 3B.2c). The reductive dechlorination of compounds in the cyclodiene series has been shown by Cochrane to occur in the order -CH=CH CCl (allylic) > CCl$_2$ (geminal) > -CCl-CCl-(vicinal) > CHCl(secondary) > C=CCl(vinylic).[232] Thus, the allylic chlorine of heptachlor is removed readily, attack on the dichloromethano-bridge is always to be expected, vicinal chlorines in saturated ring systems are likely to be attacked, and the vinylic chlorine atoms in the hexachloronorbornene nucleus are unaffected. From the cis- (Figure 19, structure 11) and trans- (Figure 19, structure 9) chlordanes, the major products of chromous chloride reduction are mixtures of the syn- and anti- (relative to the chlorinated double bond) heptachloro- compounds produced by removal of one chlorine atom from the bridge, with small amounts of chlordene formed by vicinal dechlorination and, after long reaction times, bridge monodechlorinated chlordene. Similar reactions occur with the cis- (Figure 18, structure 5) and trans- (Figure 19, structure 10) nonachlors; in this case chlordene is a prominent reaction product, being presumably formed by vicinal dechlorination to give heptachlor, which then readily loses the allylic chlorine atom. Nevertheless, the major products appear to arise by the removal of a single chlorine atom from the chlorinated bridge.

The famous dehydrochlorination reactions of cis- and trans- chlordane are the basis on which their stereochemistry was originally assigned. Thus, March[207] assigned the cis- configuration to the isomer which dehydrochlorinated most readily, and Cristol reported that one isomer dehydrochlorinated readily in 0.04 M ethanolic sodium hydroxide while the other did not, again indicating the cis- configuration for the former (Vol. I, Chap. 3C.1a). The trans-isomer in fact loses hydrogen chloride fairly readily under appropriate conditions and the mechanisms of the reactions have been discussed by Cochrane.[237] Complete dehydrochlorination of cis-chlordane is effected by heating it in methanolic sodium methoxide for 15 min or by shaking at room temperature for 20 min with the more powerful reagent potassium t-butoxide in t-butanol. The trans-isomer reacts incompletely with sodium methoxide in 30 min, but is completely converted after 30 min heating with the second reagent. In the case of the cis-isomer, the 1-endo-, rather than the 2-endo-proton is removed by the base, so that

the adjacent 2-exo-chlorine is removed in the ensuing trans- elimination, the sole product being 3-chlorochlordene (Figure 19, structure 14). With the trans-isomer, the 2-exo-proton is more exposed than the 1-endo-proton and is therefore expected to participate in the difficult cis- elimination which results in elimination of the 1-exo-chlorine, to give 2-chlorochlordene. The change in numbering should be noted; in derivatives of dihydrochlordene such as the chlordanes, numbering is by the substituents (Figure 8), whereas it has proved useful to regard the methylene group in chlordene as C$_1$. Thus, heptachlor is 1-chlorochlordene, and C$_1$ in cis-chlordane becomes C$_3$ in 3-chlorochlordene.

The situation is precisely similar for the cis- and trans- nonachlors, which differ from the chlordanes only in having an additional 1-exo-chlorine atom. Accordingly, cis-nonachlor loses HCl rapidly with sodium methoxide to give 1-exo-3-dichlorochlordene (Figure 18, structure 4), whereas the trans-isomer requires a period of heating and gives 1-exo-2-dichlorochlordene, identified by its chromous chloride reduction to the known 2-chlorochlordene derived from trans-chlordane.[208] Thus, all the available evidence clearly indicates the trans-structure for the nonachlor melting at 128 to 130°C, and the cis-structure for the isomer melting at 214 to 215°C. The compounds 1,2-dichlorochlordene and 1,3-dichlorochlordene react differently with the chromous chloride reagent. Whereas the former suffers allylic dechlorination to give 2-chlorochlordene, the latter reacts more rapidly to give chlordene and some 1-hydroxychlordene, presumably via allylic dechlorination followed by rearrangement to heptachlor and further dechlorination. Allylic hydroxylation can occur during this sequence in the presence of water.

Each of the dichlorochlordenes can be oxidized to a different. chloroheptachlor 2,3-exo-epoxide. The epoxide (Figure 19, structure 15) obtained from 1,2-dichlorochlordene is of particular interest because it has recently been shown to be a metabolite of both the cis- and trans-chlordanes that is produced in vivo in pigs, rats, and dogs and is stored in their fat.[246,247] Its precursor, 1,2-dichlorochlordene, reacts very slowly with m-chloroperbenzoic acid to give about 15% of the epoxide, which can, however, be obtained in 60% yield by heating the precursor at 100°C with chromium trioxide in acetic acid. It is interesting to note that the direct oxidation of *cis*-chlordane with this reagent gives about 1% of the epoxide, whereas *trans*-chlordane gives about 10% of it under the same conditions. Vicinal dichlorochlordene has been detected during the oxidation of *trans*-chlordane, suggesting that a dehydrogenation is effected by the reagent as a prelude to epoxidation. The epoxide is deoxygenated in low yield by treatment with triphenylphosphine, and 1,2-dichlorochlordene is also formed with the CrCl$_2$ reagent, with bridge-monodechlorination predominating in the last case. These reactions indicate the presence of a 1,2-epoxide ring and physical data support the assigned structure. The effect of the allylic chlorine atom in inhibiting oxidative chemical attack on the adjacent 2,3-double bond is well illustrated by these dichlorochlordene isomers and by heptachlor. In this context it is interesting that the epoxidation of 1-*exo*-hydroxychlordene, which is isosteric with heptachlor, also occurs very slowly when conducted with *m*-chloroperbenzoic acid.[235]

c. Isobenzan and Endosulfan

Although isobenzan ("Telodrin") production ceased in 1965, the compound (Figure 17, structure 5) is of considerable interest due to its structural relationship with both the chlordane isomers and heptachlor epoxide. Isobenzan has high mammalian toxicity (Vol. II) and it is notable that the virtually isosteric member of the chlordane series called δ-chlordane (1-*exo*, 3-*exo*-dichlorodihydrochlordene) is also one of the most toxic compounds in this group (Vol. II), although it contains no oxygen atom. As shown in Figure 17, isobenzan has been related to chlorendic acid by synthesis and is converted into this precursor by oxidation with chromium trioxide in acetic acid thus confirming the *endo*-ring fusion for this molecule. The chlorine atoms in the furan ring may be replaced by various other groups; for example, by acetoxy groups with silver acetate in acetic acid and by alkoxy groups using the appropriate sodium alkoxide.[198] When the diacetate or dialkoxy-derivatives are treated with aluminum chloride, isobenzan is recovered, which is good evidence that the chlorine atoms are in the 1,3-*exo*-positions, the corresponding *endo*-positions being "within" the structure and rather sterically hindered. When treated under pressure with liquid hydrogen fluoride, one or both of the furan ring chlorine atoms are replaced to give a

mixture of the 1-fluoro- and 1,3-difluoro-analogues. Various other derivatives may be prepared directly from the isobenzan precursor (Figure 17, structure 4), which can be converted into a mono-bromo-derivative (substitution in the furan α-position) by light catalyzed bromination or with *N*-bromosuccinimide and peroxide. This intermediate undergoes bromine replacement by alkoxy- or mercapto-groups, or by fluorine when treated with mercuric fluoride in acetonitrile. Photochemical chlorination of the monofluoro-compound affords the 1-fluoro-, 3-chloro-analogue of isobenzan. A corresponding series of compounds can be made starting from the adduct of 2,5-dihydrofuran and 5,5-difluoro-1,2,3,4-tetrachlorocyclopentadiene; for example, this adduct can be photochemically chlorinated to give the isobenzan analogue which is then treated with liquid hydrogen fluoride as indicated above to replace one or both of the chlorine atoms at C$_1$ and C$_3$, the products being the 1,8,8-trifluoro- and 1,3,8,8-tetra-fluoro-analogues, respectively, of isobenzan.

In endosulfan ("Thiodan"), the two isomeric forms have been shown to arise on account of the pyramidal structure of the sulfite moiety which can take up two configurations relative to the ring system containing it. That each of the isomers derives from the same diol (Figure 17, structure 3) is shown by the synthesis (Vol. I, Chap. 3B.2a) and by the fact that when oxidized with nitric acid, the same lactone is produced from each; it is identical with the one obtained directly by oxidation of one hydroxymethyl group of the diol, followed by cyclization of the resulting γ-hydroxy acid. The intermediate lactone may be further oxidized to chlorendic acid with chromium trioxide, showing that the ring fusion is *endo*- in endosulfan (Figure 17, structures 8 and 9). Since the diol is common to both isomers, the isomerism must arise from the mode of closure of the sulfite ring. In the seven-membered ring which includes the heterocyclic moiety of endosulfan, the four carbon atoms are in the same plane, and so this ring is comparable to a cyclohexane ring; it can exist in boat or chair forms and in each case the -O.SO.O-group has a pyramidal structure and the S=O bond may lie in an equatorial or an axial position. The boat form can be fused to the *endo*-positions of the hexachloronorbornene nucleus in two different ways, and for each mode, the S=O bond may be either axial or equatorial so that four isomers are theoretically possible. Four more possibilities arise in which the chair form of the seven-membered ring is similarly fused to the hexachloronorbornene nucleus with the same conformational possibilities for the S=O bond. In some of these molecules, called the *syn*-isomers, the S=O bond is directed toward the dichloromethano-bridge, while in others, called *anti*-isomers, it is directed away from this bridge. Two of the isomers involving the boat form of the cyclic sulfite ring appear so sterically strained as to be unlikely to exist. Riemschneider calculated the dipole moments of the most likely isomers and compared them with the observed values, the conclusion being that in the two known isomers, the chair form of the large ring is involved and the S=O bond is axial in each isomer.[188] An interesting consequence of this result is that one isomer (that melting at 108 to 110°C according to Riemschneider and called the α-isomer) is "fixed" in the same configuration as is found in the isodrin-endrin series of cyclodienes, while the β-isomer is "fixed" in the configuration seen in the aldrin-dieldrin series; in each case, the sulfur atom corresponds approximately in position to the epoxide oxygen found in endrin or dieldrin. However, the stereochemistry of the endosulfan isomers has been reexamined since the earlier investigation, and a 1965 report concludes that the dipole moment measurements do not allow a reliable distinction to be made between conformational isomers of cyclic sulfites of this type; rather different values were measured for the two dipole moments, and this did not agree with any of the previously calculated values for the isomers having the established *endo*-ring fusion. In this investigation, it was concluded on the basis of NMR measurements and other data that the basic conformations are as suggested by Riemschneider, but the actual assignments are opposite to those first suggested; in other words, the α-isomer is "fixed" in the dieldrin type configuration and the β-isomer (mp 208 to 210°C) in the endrin type configuration (Figure 17, structures 8 and 9, respectively).[197]

Endosulfan is subject to slow hydrolysis to the corresponding diol and sulfur dioxide and, when heated under reflux with methanolic sodium hydroxide, gives sodium sulfite which can be titrated iodometrically using starch as indicator, thus providing a valuable and specific analytical method for this compound when a distinction between the isomers is not required.[248] The sulfur dioxide evolved when endosulfan is heated in an alkaline medium may also be detected colorimetrically. Mild oxidation of the endosulfan isomers converts them into endosulfan sulfate (mp 180 to 181°C) in which isomeric forms are no longer possible; this conversion is effected in moderate yield when the α-isomer is oxidized in chloroform/acetic acid with aqueous calcium permanganate, or similarly from the β-isomer with barium permanganate.[197] Some of the diacetate of the alcohol is formed at the same time through hydrolysis and esterification. This conversion, which again shows that the isomerism involves the pyramidal sulfite moiety, also occurs in vivo, and the sulfate is toxic, so that this conversion is a toxication reaction which parallels the conversion of aldrin into dieldrin.

The easy conversion of endosulfan into the parent diol, which can be detected by GC-methods as the diacetate or bis-trimethylsilyl ether, has been used for derivatization analysis of the compound in residues,[64] and the chromous chloride reagent, leading to hydrolysis and monodechlorination of the methano-bridge, shows promise for a similar purpose.[232] The possible photochemical conversions of endosulfan are of interest from an environmental standpoint, as in other cases, and irradiation experiments have been conducted using a medium pressure UV-lamp with a pyrex filter to transmit only wavelengths above 300 nm, in some analogy with sunlight. The α-isomer has been reported to lose one or both of the vinylic chlorine atoms under these conditions, while the β-isomer is said to undergo methano-bridge dechlorination and isomerization to the α-isomer.[111]

d. Aldrin and Dieldrin

The simpler aspects of aldrin (used here to denote HHDN) chemistry center on the reactivity of the double bond in the unchlorinated norbornene nucleus. Epoxidation is readily effected by peracids and has already been discussed, along with certain halogen addition reactions of this double bond, in the section on synthesis (Vol. I, Chap. 3B.2b). A reaction of aldrin (Figure 20) that specifically demonstrates the presence of the reactive double bond is the addition of phenylazide to give the phenyltriazole derivative (2) of 6,7-dihydroaldrin.[116] This adduct is obtained by treating aldrin with a solution of phenylazide in hexane, evaporating the solvent, and heating the

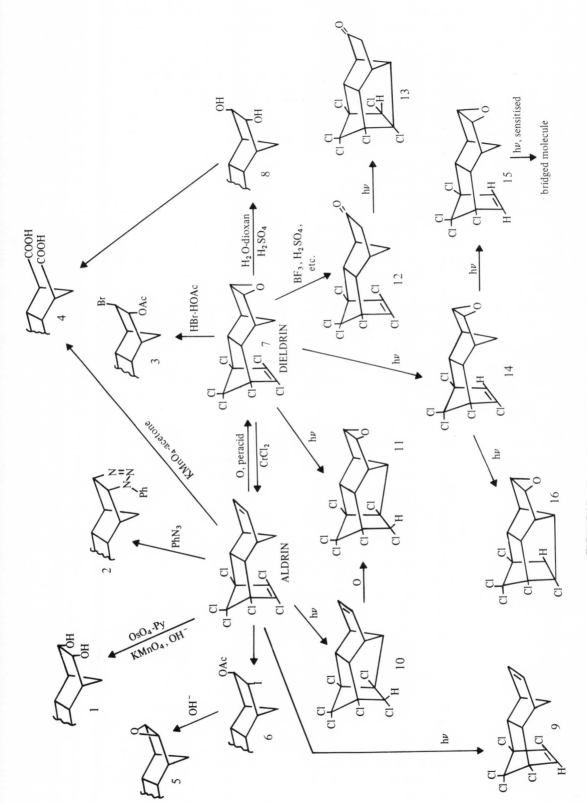

FIGURE 20. Some chemical conversions of aldrin and dieldrin.

residue at 85°C for about 30 min. A similar derivative is also prepared from dieldrin (used here to denote HEOD) by a three stage process involving cleavage of the epoxide ring with anhydrous hydrobromic acid in acetic acid to give the corresponding bromoacetate (3), from which the 6,7-substituents are then removed by reduction with zinc in acid solution to give aldrin, which may have lost bridge chlorine under these conditions; derivatization then proceeds as before. This sequence of reactions provides evidence of the relationship between aldrin and dieldrin (Figure 20, structure 7), but the main value of the phenylazide derivative lies in its use for the colorimetric analysis of aldrin and dieldrin (see analytical section on these compounds).

The epoxide ring in dieldrin is well known to be chemically rather stable, like that in the *exo*-2,3-epoxide of heptachlor (HE 160), but as in the case of HE 160, it does undergo the normal reactions expected of epoxides under appropriate conditions. Thus, ring opening to give the corresponding *trans*-bromohydrin is effected by treatment with anhydrous hydrogen bromide in dioxan at room temperature for about 2 hr. A similar product is obtained from aldrin directly by the addition of acetyl hypoiodite (iodine and silver acetate) in ether at 17 to 26°C. In this reaction the positive species (iodine) is presumed to add to the *exo*-6-position of aldrin, so that the acetoxy-group is in the *endo*-position (Figure 20, structure 6); prolonged hydrolysis of this iodo-acetate with aqueous potassium hydroxide accordingly gives an epoxide different from dieldrin, and which is derived more rapidly by similar treatment of the corresponding free iodohydrin.[168] With *t*-butyl hypochlorite in excess acetic acid, aldrin gives a chloroacetate,[237] which presumably has the same configuration as the iodoacetate, assuming analogous addition of the elements of acetyl hypochlorite to the double bond.

Prolonged boiling with aqueous dioxan (1:1) containing concentrated sulfuric acid (about 7% by volume) converts dieldrin into 6,7-dihydroxy-dihydroaldrin (8) by hydrolytic *trans*-opening of the epoxide ring.[158] The corresponding *exo*-6,7-cis-diol (1) may be prepared by the oxidation of aldrin with alkaline potassium permanganate or osmium tetroxide,[158] and from dieldrin through the corresponding *cis*-diacetate formed when the epoxide is heated in acetic anhydride containing sulfuric acid.[249] In this diol, the coupling seen in

the NMR spectrum between the protons at C_6 and C_7 and the methano-bridge proton directed toward the acetoxy-groups suggests that the latter are *exo*-cis, which is another piece of evidence for the *exo*-configuration of the original epoxide ring. The osmium tetroxide procedure for the *cis*-diol may also be used for the preparation of the corresponding *exo*-6,7-*cis*-diol from isodrin. These *cis*-diols are further characterized by their conversion into acetonides when heated in acetone with ferric chloride.[158] Besides affording the *cis*-diol, permanganate oxidation of aldrin also gives a dicarboxylic acid (4), which is also obtained by oxidation of the *trans*-diol (8).[168]

Lewis acids such as boron trifluoride, and mineral acids such as sulfuric and perchloric rearrange dieldrin to a keto-dihydroaldrin indicated to have structure 12 by McKinney et al.[158] This ketone forms a colored 2,4-dinitrophenyl-hydrazone which may be used for the colorimetric determination of dieldrin. Aldrin adds acetic acid in the presence of sulfuric acid to give 6-*exo*-acetoxydihydroaldrin,[170] which by hydrolysis to the alcohol, followed by oxidation (CrO_3) gives the above ketone. Stereoselective reduction of the latter gives the 6-*endo*-hydroxydihydroaldrin which is formed along with the *exo*-isomer as a metabolic product of dihydroaldrin in some biological systems. When heated with antimony trichloride in boiling CCl_4, dieldrin gives a high yield of a crystalline complex melting at 123 to 124°C, from which dieldrin can be recovered by washing a solution in ether with hydrochloric acid to remove the antimony.[171]

Aldrin and dieldrin participate in a number of reactions which lead to their reductive dechlorination, and a number of these were discussed in the synthetic section; the reactions with sodium methoxide in methanol-dimethyl sulfoxide, or with sodium borohydride and cobaltous salts, which lead predominantly to molecules lacking the anti-chlorine atom of the dichloro-methano-bridge (Vol. I, Chap. 3B.2b), are of particular interest. Some variants of the last method appear to remove a vinylic chlorine atom from the molecules, as well as effecting reduction of the unchlorinated double bond in compounds such as aldrin.[111] Catalytic hydrogenation of dieldrin in ethanol containing triethylamine, using palladized charcoal as catalyst, results in bridge-monodechlorination followed by reductive removal of the vinylic chlorines and, finally, reduction of the dechlorinated double bond.[250] The reactions of chromous chloride with cyclodiene molecules of the heptachlor-chlordane series were discussed earlier (Vol. I, Chap. 3B.4b) and similar conversions occur in the aldrin-dieldrin series. When heated at 60°C for about 6 hr with this reagent, dieldrin is mainly deoxygenated to aldrin, as well as suffering bridge monodechlorination to give the expected mixture of isomers.[232] Two further compounds are formed by monodechlorination of the aldrin produced in the reaction and after 24 hr, only these products remain; when epoxidized they give a pair of compounds whose behavior on GC corresponds with that of the isomeric compounds derived from dieldrin. These conversions parallel those found with chlordene epoxide, which affords chlordene as a major product, together with monodechlorination products from both the epoxide and the derived chlordene. The deoxygenation reaction of cyclodiene epoxides by chromous chloride appears to be quite general, although it does not apparently extend to heptachlor epoxide, presumably because approach to the epoxide ring is hindered by the 1-exo-chlorine atom. As might be expected, it does not occur with endrin, which rearranges to the half-cage ketone under the acidic conditions of the reaction, the final product being the latter molecule minus one chlorine atom.

The reactions of the double bond of aldrin or of the epoxide ring of dieldrin show that these molecules behave quite normally in addition reactions, in marked contrast to the situation with the isodrin-endrin series of compounds, which participate quite readily in molecular rearrangements under conditions normally used to effect such additions. However, molecular rearrangements do occur readily in the aldrin-dieldrin series, when these molecules are exposed to sunlight or short wavelength UV-radiation. Early indications of the environmental photochemical rearrangement of dieldrin were obtained by Roburn, who reported the presence of a different, unidentified compound on grass which had been treated with dieldrin and exposed to sunlight, and found that the same compound is produced when dieldrin spread on glass is irradiated with UV-light. The compound was subsequently prepared for identification by Robinson and colleagues,[251] who irradiated a dieldrin film on filter paper with UV-light of wavelength 253.7 nm. This report, and a simulaneous one by Rosen and colleagues, identified the product as the photo-isomer (Figure 20, structure 11) formed by linking of the methano-bridge of dieldrin with the chlorinated double bond. The same compound had also been prepared at about the same time by Parsons and Moore,[171] who photolyzed dieldrin in ethyl acetate and noted the disappearance of the absorption bends in the IR attributed to the dichloro-ethylene and methylene groups in dieldrin, and that photodieldrin gave no antimony trichloride complex. They considered the 1,2-epoxide (Figure 20, structure 11) to be a less likely product than a different bridged molecule containing a four-membered epoxide ring. This could arise by Wagner-Meerwein rearrangement and would have its norbornane rings in the endo, endo-configuration. Other investigators favored a four-membered epoxide ring with the endo-, exo-configuration retained.

Several subsequent studies have examined the photochemical conversions of aldrin and dieldrin as solids and in solution. The results are generally similar to those observed in the heptachlor-chlordane series. Irradiation in hexane with wavelengths below 300 nm in a quartz vessel results in the replacement of a vinylic chlorine atom of aldrin or dieldrin by hydrogen, to give (9) and (14), respectively (Figure 20). The UV-spectrum of dieldrin has a broad maximum at 215 nm which extends to 260 nm, so that with wavelengths greater than 260 nm, there is little absorption of light energy in the absence of a sensitizer.[160] The mono-dechlorination product has a very similar UV-spectrum, and further irradiation results in the replacement of the second vinylic chlorine atom to give (15), although this second reaction was not observed in the earlier work.

The end point of these reactions, especially if some higher wavelengths are included or if a triplet sensitizer such as acetone is added, is photo-dieldrin, or the corresponding compounds produced by bridging of the dechlorinated molecules.[160] Thus, when dieldrin (1 g) is irradiated in hexane (300 ml) using a quartz vessel so that the solution is exposed to the total emission spectrum, the mono-dechlorination product (14, called pentachlorodieldrin for simplicity) can be isolated in about 38% yield if the reaction is followed by GC-analysis and stopped when dieldrin just disappears. If the irradiation is continued until (14) is just used up, the vinylic bis-dechlorination product (15) can be recovered (27%) by recrystallization from ethanol-water, and the mother liquors contain photodieldrin and its dechloro-analogue (16). GC-analysis at the point when (14) is just exhausted indicates that the mixture contains 20% of photodieldrin, 25% of (16), 45% of the vinylic bis-dechlorination product (15), and 10% of the latter's bridged-derivative. If the molecule (14) is irradiated directly in acetone using a pyrex filter to exclude wavelengths below 300 nm, dechlorination is prevented, and the corresponding bridged molecule (16) is obtained in 78% yield. The vinylic bis-dechlorination product (15) requires short wave radiation to effect the bridging reaction, even in the presence of a sensitizer; the bridged molecule is obtained in 46% yield by irradiating a 0.1% solution of (15) in hexane containing acetone (0.5%) for about 1 hr. Benson observed the formation of photodieldrin and pentachlorodieldrin, with the latter predominating, when dieldrin is irradiated as a 2% solution in ethyl acetate.[252] The photolysis proceeds more rapidly in aqueous methanol to give the same series of products as observed with the reaction in hexane.

Since the atmospheric conversion of dieldrin is of considerable interest, its irradiation in the gas phase has been studied at the Institute for Ecological Chemistry in Bonn. Experiments were conducted at normal pressure with either the total emission spectrum (quartz vessel) or with wavelengths above 310 nm; in nitrogen, oxygen, or nitrogen-oxygen mixtures; in dry or moisture-saturated air; and at temperatures from 55 to 85°C. After irradiation times up to 35 min, the only product which could be detected was photodieldrin.[160]

The compounds formed during the solid state photoconversion of dieldrin have not yet been completely characterized. After irradiation of dieldrin spread on plates at 253.7 nm for 1 hr, photodieldrin (50%), dieldrin (25%), and two unidentified products were detected.[252] Photodieldrin (mp 194 to 195°C) and pentachlorodieldrin (mp 163 to 164°C), when separately irradiated, were found to be relatively stable for up to 5 hr, after which the amounts unchanged decreased progressively. Two unidentified compounds were eventually formed from photodieldrin, while the only product from pentachlorodieldrin after 24 hr appeared to be a rather polar component which remained at $R_f 0$ in the normal analysis by thin-layer chromatography (TLC). No evidence was obtained for the formation of either

of the possible bridged molecules from pentachlorodieldrin in the solid state. The reactions discussed here are quite general and have been used to synthesize dieldrin metabolites that have been found in vivo. For example, the ketone (Figure 20, structure 12), which is produced from dieldrin by microorganisms, rearranges when irradiated in acetone to give equal amounts of the bridged molecule (13) and the isomeric compound involving bridging to the other end of the dichloroethylene bridge.[111] The ketone (13) is also a product of dieldrin metabolism by microorganisms (Vol. II, Chap. 3B.2d).

The intramolecular rearrangements discussed above complete the evidence available for the *endo*, *exo*-configuration of the aldrin-dieldrin series. A molecule raised to an excited state by the absorption of light may, besides undergoing such reactions, lose its energy by collision, exhibit fluorescence or react with other molecules present in its immediate environment. As noted above, photodechlorinations involving an excited singlet state may occur at below 300 nm, but dieldrin does not absorb at the wavelengths provided by sunlight (above 300 nm), and so a sensitizer is required for molecular excitation in these regions. In the presence of light and sensitizers, dieldrin may be expected to interact with atmospheric components such as oxygen, ozone, or nitric oxide, and so its reactions with these compounds have been investigated both in solution, in carbon tetrachloride or perfluorodimethylcyclobutane, and in the gas-phase. The compounds isolated so far from irradiations under these conditions are mainly derivatives of the intact dieldrin molecule in which a single Cl, NO_2, ONO_2, or OH has been inserted at the junction between the norbornane rings.[111] One interesting oxidation product is a chlorohydrin derived from Klein's ketone (Vol. II, Figure 37, structure 4) which should give the ketone by cyclization with base; the yields are not high, however, and no satisfactory chemical synthesis appears to have been devised for this molecule so far, although it is formed quite readily from photodieldrin in vivo.

e. Isodrin and Endrin

The spontaneous rearrangement of endrin to a half cage ketone (Figure 21, structure 10) when stored at ambient temperature or heated briefly to higher temperatures was discussed in relation to methods for endrin stabilization by Bellin in

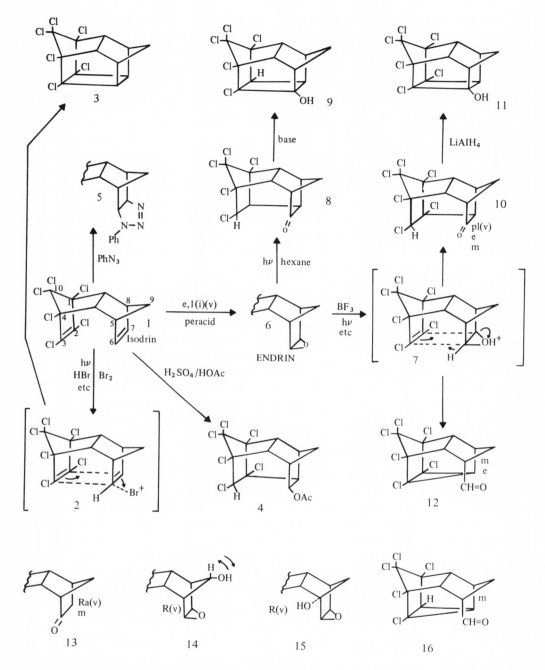

FIGURE 21. Some chemical, metabolic and environmental conversions of isodrin and endrin. e = environment; l = living organisms; m = microorganisms; pl = plants; Ra = rabbit, R = rat; i = in vitro; v = in vivo.

1956,[216] and the same rearrangement effected by boron trifluoride in benzene was examined by Cookson and Crundwell in 1959.[253] These authors also described the conversion of isodrin into a photoisomer (Figure 21, structure 3), which they called "photodrin." A full account of structural investigations in the isodrin-endrin series was given by Soloway shortly afterwards[254] and was followed by the papers of Bird, on the isodrin and aldrin series, and of Parsons and Moore on the reactions of dieldrin.[170,171] An examination of these papers, together with the series by Mackenzie,[157] published since about 1960, gives some idea of the fascinating versatility of the reactions shown by these molecules. If, in addition to these investigations, which are of considerable interest from the standpoint of theoretical chemistry, one considers the work done in recent years in relation to environmental questions, it is evident that the chemistry of the derivatives of hexachlorocyclopentadiene and their relatives is expanding rather than contracting. Although a detailed discussion of many of the reactions is beyond the scope of this book, there is little doubt that some of the products are likely to be of great interest from the point of view of structure-activity relationships in the cyclodiene series, a point well illustrated by the recent report of Weil and colleagues which describes the preparation of certain ketones and derived cage structures related to isobenzan, and mentions their remarkably high toxicity to insects and mammals.[255]

Isodrin (Figure 21, structure 1) undergoes a limited number of the normal reactions expected of the unchlorinated double bond, such as the addition of phenyl azide or mercuric acetate and oxidation with alkaline potassium permanganate to give a *cis*-diol, but reactions conducted under acidic conditions lead to skeletal rearrangement.[254] Thus, isodrin, when treated with HBr, gives photodrin (Figure 21, structure 3) mp 294 to 295°C, which does not add hydrogen and is inert to treatment with other double bond reagents such as chlorine, phenyl azide, peracids, or permanganate. Only about 12% of the expected product of addition of HBr is obtained. The possibility that this hydrobromide was a product of addition with concurrent Wagner-Meerwein rearrangement of the unchlorinated norbornene nucleus was ruled out because the product is not, as would be expected, the same as that obtained from the addition of HBr to aldrin. Treatment of isodrin with bromine

gives photodrin plus a dibromide which also lacks the dichloroethylene moiety and was assigned the structure of a 3,7-dibromide having a new 2,6-carbon, carbon bond (for numbering see Figure 21, structure 1). One bromination experiment also gave a molecule considered to be 6-monobromo-photodrin. These products arise through the competition between various rearrangements that can result from attack on C_6 by Br^+, which generates a carbonium ion at C_7; photodrin then results by a concerted process (shown in Figure 21, structure 2) in which two electrons from the 2,3-double bond form a new bond between C_2 and C_7, thereby neutralizing the carbonium ion at that position and forming another at C_3. This in turn is neutralized by trans-annular linking between C_3 and C_6, with concurrent expulsion of Br^+. Loss of the proton from C_6 instead of Br^+ would give 6-monobromo-photodrin. The structure of the dibromide most compatible with stereochemical requirements involves neutralization of the C_7-carbonium ion by formation of a 3,7-bond, followed by neutralization of the carbonium ion thereby generated at C_2 by addition of Br^-. An alternative structure involves the less likely formation of a new 2,7-bond with adjacent bromine atoms at C_3 and C_6.[254]

In contrast to aldrin, treatment of isodrin with acetic acid containing sulfuric acid gives photodrin, the half-cage *exo*-acetate (Figure 21, structure 4), a little of the epimeric *endo*-acetate, and a trace of the *exo*-acetate formed by normal addition to the double bond. The rearrangements are explained by the mechanism involved in photodrin formation, with the interesting difference that the final carbonium ion generated at C_3 during rearrangement is neutralized by transfer of a hydride ion (H^-) from C_6; neutralization does not occur by addition of CH_3COO^- to C_3, because hydrolysis of such a product would result in ketone generation at C_3 with loss of chlorine from the molecule. The half-cage *endo*-alcohol obtained by hydrolysis of the corresponding acetate mentioned above gives photodrin when dehydrated with phosphorus pentoxide, and with alumina at 185°C the *endo*-hydroxy group attacks C_3 from the rear, to form a new six-membered cyclic ether ring, thereby demonstrating the *endo*-position of the hydroxyl-group and suggesting the *exo*-position for the C_3-chlorine, which would favor such attack. Oxidation of the corresponding alcohol obtained from the half-cage acetate (4) gives the half-cage ketone (10) also obtained from endrin with boron trifluoride in benzene or by UV-irradiation, and this gives the cage alcohol (6-hydroxyphotodrin; Figure 21, structure 11) by reduction with one equivalent of lithium aluminum hydride.[170,254]

Treatment of isodrin with zinc in acetic acid removes both methano-bridge chlorine atoms; chromous chloride effects rearrangement of endrin to the ketone (10) which is then bridge-monodechlorinated as is usual with this reagent. The bridge dechlorinated compound from isodrin is completely dechlorinated by sodium in boiling *n*-amyl alcohol, and the basic diene (*endo, endo*-bisnorbornene) can be converted into the corresponding full-cage-dechlorinated molecule that is also obtained by the similarly affected complete dechlorination of photodrin. These reactions leave no doubt that the isodrin-endrin series of compounds arises from two norbornene molecules fused through their *endo*-positions.[254] The alkali metal total-dechlorination technique has also been used to prepare the basic hydrocarbon, *endo, exo*-bisnorbornene,[256] from aldrin, and this method is the basis of the total chlorine estimation method for cyclodiene compounds (Vol. I, Chap. 2C.1).

Following the reports by Bellin[216] that endrin is converted into insecticidally inactive compounds by traces of strong acid, by traces of iron in certain forms, and by heat, Phillips and collaborators demonstrated the thermal isomerization of endrin to the half-cage ketone (10) and a pentacyclic aldehyde (12); this conversion frequently occurs in the GC-analysis of endrin, for example.[257] In 1963, Roburn reported that endrin was converted into new compounds when exposed to UV-light (253.7 nm) in the solid phase, and in 1966 these studies were extended by Rosen and colleagues[258] to the isolation of a ketone and aldehyde which proved to be identical with the compounds characterized by Phillips. The thermal rearrangement which leads to these compounds may involve either a hydride shift or the abstraction of an epoxy-hydrogen. The same products are formed when endrin is treated with a Lewis acid such as boron trifluoride, or with sulfuric or perchloric acids. In this case the epoxide oxygen is protonated (as in Figure 21, structure 7) and the ring opens to give a hydroxyl group at C_6 and a carbonium ion at C_7; this is then neutralized by transfer of electrons from the chlorinated double bond to form a new bond between C_2 and C_7, the carbonium ion concurrently appearing at C_3. This carbonium ion may be neutralized by transfer of a hydride ion from C_6, which gives the ketone, or by shift of the 6,7-bond to link C_3 and C_7, the hydrogen atom being retained at C_6 to give the aldehyde (12) in this case.

When endrin is irradiated in hexane or cyclohexane with light of wavelength 253.7 nm or 300 nm, or in sunlight, a ketone is formed in up to 80% yield which is different from the half-cage photoisomer already discussed.[259] The product lacks the dichloroethylene moiety, has no methylene group adjacent to the carbonyl function, and contains only five chlorine atoms, as indicated by IR- and mass spectral data. Of the two possible structures consistent with the spectral information, namely, those derived from the half-cage ketone (Figure 21, structure 10) by the replacement by hydrogen of a chlorine at either C_2 or C_3, only compound (8), retaining the chlorine at C_3, is compatible with the NMR spectrum, which indicates only one methylene group in the molecule. The carbonyl group in this ketone proved to be remarkably unreactive, the only significant reaction of the molecule being the *trans*-annular enolization (homoenolization) in the presence of bases to give the pentachloro-cage alcohol (9), analogous to the hexachloro-compound (11). The pentachloro-ketone (8) is evidently not a conversion product of (10) formed as an intermediate, since irradiation of the latter gave none of it. The IR-spectra of these two ketones are very similar beyond 1,350 cm^{-1}, and the pentachloro-ketone behaves similarly to endrin on several gas-chromatographic columns; it was found to constitute about 5% of the total endrin residues found in a muck soil that had been treated with endrin for 5 years.

As in the case of the ketonic rearrangement product obtained from dieldrin in a similar way, the hexachloro-ketone from endrin (often called delta-keto-endrin, which is an abbreviated form of one of its chemical names) has been investigated as the basis of a colorimetric analytical method involving formation of its colored 2,4-dinitrophenylhydrazone. Several of the other reactions mentioned above can be utilized to assist recognition and quantitation of these compounds by GC-analysis, especially if hydroxyl groups are present to facilitate further derivatization.

Like aldrin and dieldrin, isodrin and endrin may

be reductively dechlorinated with sodium methoxide in methanol-dimethyl sulfoxide to give analogues in which mainly the methano-bridge chlorine *anti*- to the chlorinated double bond is replaced by hydrogen, and this conversion occurs without skeletal rearrangement.[250] Chromous chloride cannot be used if the epoxide rings are to be retained, since it deoxygenates dieldrin and rearranges endrin to the ketone (Figure 21, structure 10) before effecting bridge dechlorination.[232] Bridge dechlorination of these epoxides may also be effected without loss of the epoxide function by use of some of the various combinations of sodium borohydride with cobaltous salts mentioned when synthesis was discussed. The reductive bridge-monodechlorinations effected when vitamin B_{12} is used as a catalyst (a mixture of the *syn*- and *anti*- monodechloro-compounds is produced)[220] is interesting in relation to changes which might be expected to occur in vivo, and could lead ultimately to elimination of the bridge with formation of derivatives of 1,2,3,4-tetrachlorocyclohexadiene or 1,2,3,4-tetrachlorobenzene[260] (Vol. II, Chap. 3B.2e). Tetrachlorocyclohexadiene is formed photolytically from hexachloronorbornene, but the more complicated molecules represented by the cyclodiene insecticides are rather stable.[585]

C. HISTORY AND DEVELOPMENT OF THE COMMERCIAL CYCLODIENE INSECTICIDES

1. Chlordane and Heptachlor

Little imagination is required to understand the excitement engendered among those involved in the chemical control of insects by the discovery of DDT. Every chlorinated hydrocarbon might be a potential DDT, and fuel was undoubtedly added to these flames by the observation of Martin and Wain that among the immediate relatives of DDT, highest insecticidal activity resided in those compounds which underwent chemical dehydrochlorination most readily.[261] This observation led to the suggestion that, since the products of dehydrochlorination in this series lack toxicity, either the *act* of dehydrochlorination in vivo is itself in some way involved in toxicity or, more simply still, that the liberated hydrogen chloride produces a toxic effect at the site of action. Clearly, this simple hypothesis suggested many avenues for further synthetic exploration,

although later experiments with DDT isosteres were to show that the picture is by no means so simple.

Although the discovery of both hexachlorocyclohexane (HCH; used here to denote the mixed isomers) and the first cyclodiene insecticides followed hard on the discovery of DDT, much less publicity seems to have attended their appearance, probably because DDT was already available as the war-time insecticide, and the widespread use of these new compounds belongs mainly to the postwar years. Metcalf, writing in 1955,[41] comments that in spite of the wide application of the cyclodienes by that time, little had yet been written about their chemistry or mode of action. It is especially interesting to consider that although the DDT-group, HCH, and the cyclodienes are generally classified together as "chlorinated" or "organochlorine" insecticides, each of the three groups, and the principles governing their synthesis, are quite distinct. With the war-time development of the organophosphorus anti-cholinesterases, we have a situation in which no less than four distinct groups of highly effective insecticidal chemicals were originated within ten years. Indeed, the number of groups increases to five if the carbamate anticholinesterases are added to the list. The latter addition is valid since the development work on these last compounds was begun by the Geigy Company in 1947. One tends to wonder whether a similar situation will ever occur again. Whatever views may be held about the current use of modern insecticides, there is no doubt at all that the compounds which originated between 1940 and 1950 have made a major, but regrettably not widely appreciated, contribution to the well-being of a large section of mankind, and the reasons why these major discoveries occurred together at this particular point in time should make an interesting study for scientific historians.

One must immediately admit that the interest in organophosphorus compounds arose from their military potentialities as anti-personnel toxicants; that the search for other anti-cholinesterases should eventually lead to the carbamates is logical, since water soluble examples of this class were already known to be effective cholinergic drugs. The origin of the organochlorine insecticides has no such basis, however, and although the war-time atmosphere undoubtedly accelerated the development of some of these discoveries, it must be remembered that one of the most significant contributions was made by a neutral nation. The lack of communications due to the war situation and the structural differences between the DDT, cyclodiene, and HCH organochlorines emphasizes the spontaneity of these discoveries in Europe and America.

a. Origins, Physical Properties, Formulations

The cyclodiene group of insecticides depend for their synthesis on the remarkable reactivity of hexachlorocyclopentadiene (called "hex") in the Diels-Alder reaction, which was discussed in the previous sections. A number of reports describing the adducts formed from this diene appeared in the period between 1947 and 1949. Thus, in March 1947, Prill published a short account[262] of the Diels-Alder reaction between "hex" and maleic anhydride, ethyl maleate, p-benzoquinone, acrylonitrile, and methyl vinyl ketone, while a simultaneous account by Riemschneider and Kuhnl described the reactions between cyclopentadiene (acting as the dienophile) and "hex," difluorotetrachlorocyclopentadiene, and dichlorotetrafluorocyclopentadiene.[263] In the publication by Riemschneider, each of the primary adducts was further chlorinated to give the products $C_{10}H_6Cl_8$, $C_{10}H_6F_2Cl_6$ and $C_{10}H_6F_4Cl_4$, respectively. The first of these primary adducts, that formed from "hex" and cyclopentadiene, is the compound called chlordene, and the chlorination product $C_{10}H_6Cl_8$, which Riemschneider called M 410 (according to its molecular weight), is the liquid mixture of compounds called chlordane. Later the same year, Riemschneider published a further account of the preparation of M 410, and its separation into several fractions, including two solids.[264] The reactions between "hex" and maleic anhydride, methylcyclopentadiene, cyclopentene, cyclohexene, 4-methylcyclohexene, 4-ethylcyclohexene, 1-hexene, 1-heptene, 1-octene, naphthalene, and anthracene were also described, as well as certain similar reactions of pentachlorocyclopentadiene. There also appeared in 1947 a circular of the United States Department of Agriculture describing a newly developed organic insecticide called "chlordane." A short account of these developments was given by Riemschneider in 1963.[188]

However, the earliest description of the insecticidal properties of chlordane appears to be that of Kearns, Ingle, and Metcalf,[265] which appeared in the *Journal of Economic Entomology* in 1945, as an account of the evaluation of "A New Chlorinated Hydrocarbon Insecticide" said to have the empirical formula $C_{10}H_6Cl_8$, and therefore called compound 1068. It was found to be "more toxic than DDT and to compare favorably in toxicity to the pure γ-isomer of benzene hexachloride to a number of species of insects." The authors stated that the product was probably a mixture of isomers which had not been resolved at that time.

The material examined in this investigation was a product of the Velsicol Chemical Corporation, and Dr. Julius Hyman recalls its origin as related to the Corporation's interest in cyclopentadiene.[266] A large amount of cyclopentadiene was produced in the United States during World War II as a by-product of the production of butadiene in the synthetic rubber program. The uses of cyclopentadiene appear to have been rather limited, but Velsicol employed it in the production of petroleum resins, which were mixed with drying oils and used to make varnishes and related materials. One of the by-products of resin manufacture was an oily solvent, rich in methylated naphthalenes and a good solvent for DDT, which, at that time, was being introduced for general use in America. Solvents of this type were found by Dr. C. W. Kerns and others to have insecticidal properties by themselves and proved useful as mosquito larvicides and also as extenders for pyrethrum.

The resins mentioned above are products of the Diels-Alder reaction and Hyman had initiated a literature survey on the general reactions of cyclopentadiene as a means to generate ideas among the research staff regarding other possible ways of utilizing it. One of the reactions which appeared in the literature survey, and particularly attracted Hyman's attention because of his special interest in the Diels-Alder reaction, was the preparation of hexachlorocyclopentadiene from cyclopentadiene described by Straus some 13 years earlier.[173] At first sight, the completely chlorinated "hex" molecule seems to be an unlikely participant in the Diels-Alder reaction since bulky substituents on the diene normally tend to inhibit the condensation; 2-chloro-butadiene, for example, reacts only slowly with dienophiles, while 2,3-dichlorobutadiene and higher chlorinated derivatives such as hexachlorobutadiene do not react. Accordingly, it was of considerable interest to determine whether, in fact, "hex" would undergo the Diels-Alder addition, either with itself or with cyclopentadiene, and a batch of "hex" was made and reacted with cyclopentadiene under the direction of Dr. Rex E. Lidov. Besides the main product of the reaction, a secondary product containing twelve chlorine atoms was obtained, and these were quickly shown to be the mono-adduct of "hex" and cyclopentadiene (chlordene; called compound 237 by Velsicol) and a *bis*-adduct produced by the addition of a second molecule of "hex" to the cyclopentene ring double bond of chlordene. Unlike more simple dienes, such as cyclopentadiene, "hex" does not react with itself to form dimers or polymers, and this fact, together with its thermal stability, provides opportunities for adduct formation which would not be available with the simpler compounds.

"Hex," compound 237, and the *bis*-adduct were tested against houseflies by Kearns at the University of Illinois, and only compound 237 was found to have significant insecticidal activity, which was about four to five times lower than that of DDT. This discovery caused great excitement among its authors, since the current price of DDT was 2 dollars per pound and there was the prospect that 237 (chlordene) could be sold at less than 50 cents per pound. Unfortunately, it was soon discovered that besides being less toxic than DDT, chlordene has considerably lower persistence due to its higher volatility. Consequently, the product was chlorinated in an attempt to reduce its volatility and hence to improve the persistence, the first batches of chlordane being made in the latter half of 1944. The oily product is the material called Velsicol 1068[®] which was reported on by Kearns, Ingle, and Metcalf in 1945. Although the vapor pressure of chlordane is higher than that of DDT, the chlorination process reduced the volatility of chlordene sufficiently to give the product practical value as a residual insecticide.[266] This material was found to be several times more effective than DDT against houseflies and some other insects, and so history appeared to have repeated itself within a remarkably short time.

The first patent describing the synthesis of compounds of the cyclodiene group is dated December 1948 and was taken out by Velsicol.[267] It describes the reaction between "hex" and cyclopentadiene, styrene, indene, dicyclopentadiene, isoprene, butadiene, phenylacetylene, and maleic anhydride. In the case of dicyclopentadiene and butadiene, *bis*-adducts were described which resulted from the addition of one more molecule of "hex" to the primary adducts. Adducts were also prepared from the reaction of cyclopentadiene (as dienophile) with hexabromo- and pentabromochlorocyclopentadienes. A second patent by Hyman quickly followed, in which he described the preparation of halogenated derivatives of the primary adducts of "hex" and hexabromocyclopentadiene. This patent, dated February 1949, details the synthesis of chlordane, which was called octachlorodicyclopentadiene-dihydride (abbreviated to Octa-Klor[®]), and a corresponding product, $C_{10}H_6Br_6Cl_2$, obtained by chlorinating the primary adduct from hexabromocyclopentadiene and cyclopentadiene.[268]

The variety of reactions described well illustrates the scope for derivatization of such compounds. A difluoro-analogue of chlordane ($C_{10}H_6Cl_6F_2$) was prepared from chlordene and also a dibromo-analogue ($C_{10}H_6Cl_6Br_2$), the latter being dehydrobrominated to one or both of the corresponding mono-bromides. Advantage was also taken of the reactivity of the allylic position in chlordene to effect the peroxide catalyzed substitution of bromine or chlorine (the product in this case being heptachlor) at this position. As mentioned in the section on general synthesis, the monobromo-derivative so produced is particularly useful for other syntheses, since the reactive bromine is readily replaceable by other halogens or groups (Vol. I, Chap. 3B.2a). Thus, chlorination of this monobromide gave a 1:2 mixture of $C_{10}H_5Cl_9$ and $C_{10}H_5Cl_8Br$, produced by addition of chlorine to the cyclopentene double bond, as well as partial replacement of the allylic bromine atom. In relation to the formation of the allylic halogen derivatives, it will be noted that two allylic positions are available in chlordene (Figure 17, structure 1), one of them being at the ring junction (position 3a); however, substitution is likely to occur at the least sterically hindered methylenic-position (C_1) and much evidence has since accumulated that this is the case. Also described at this time were chlorinated products obtained from the adducts of "hex" with methyl cyclopentadiene or isoprene, and it became evident that in many cases the introduction of additional halogen atoms into the primary adducts effected a considerable increase in insecticidal activity. So it was that the discovery that hexachlorocyclopentadiene participates in the Diels-Alder reaction offered prospects for the development of a whole new range of insecticides.

Shortly after the first preparations of Velsicol 1068 in Chicago, Hyman founded the Julius Hyman Company, based in Denver, Colorado, and there continued the work on the Diels-Alder reactions of "hex." For this reason, two chlorinated products obtained from chlordene were commercially available in the United States, namely, the Velsicol product (Velsicol 1068) and Octa-Klor. These products were both derived by the addition of two atoms of chlorine to chlordene to give $C_{10}H_6Cl_8$, but because the product of this reaction is a mixture of compounds, there were some differences between the materials so produced. There are, in fact, two phases in the history of the production of chlordane: the period between 1945 and 1950 when Velsicol 1068 and Octa-Klor were produced and marketed at the same time, and the period since the production of Octa-Klor ceased in 1950, since when chlordane has been exclusively produced by Velsicol. Because of its nature and the historical events surrounding its invention, the development of chlordane has been particularly influenced by toxicological problems that relate to the early period of its manufacture before 1950, but are not relevant to the product that has been produced for many years since. It is appropriate to examine first the information given in various places about the nature of the chlorination product, beginning first with the early work and coming finally to the present day situation. The toxicological and pharmacological properties are detailed in a monograph on chlordane by Ingel,[269] and such aspects as relate to the history and development of the product will be mentioned in this section for completeness.

In the orginal Hyman patent describing the preparation of chlordane,[268] chlorine was passed into chlordene heated under reflux in carbon tetrachloride until there was no further gain in weight, this situation corresponding to a chlorine uptake of 68 to 69% and an empirical formula $C_{10}H_6Cl_8$. The product distilled as a colorless or very pale yellow, almost odorless oil bp 155 to 160°C/0.5 to 1.0 mm, and having a honey-like consistency at room temperature, refractive index at 20°C 1.575 to 1.585, and density about 1.8.

Riemschneider, citing an American article written in 1948, stated that the American product contained 60 to 75% of chlordane and 25 to 40% of subsidiary products.[188] Metcalf, writing in 1955, which is well into the second phase of chlordane manufacture, states that technical chlordane is a dark brown, viscous liquid with a cedar-like odor, d^{25} 1.61, bp 175°C/2 mm and that the vapor pressure of the refined material at 25°C is 1×10^{-5} mm Hg.[41] The evidence presented to the Food and Drug Administration's Advisory Committee appointed to review the residue tolerances set for chlordane, which reported in 1965, included a specification for technical chlordane of chlorine content 64 to 67%, viscosity 75 to 120 centistokes at 130°F, specific gravity 1.63 to 1.67 at 60°F and a content of unchanged hexachlorocyclopentadiene of not more than 1%. In the *Chlordane Formulation Guide* dated January 1970,[270] the product is described as a viscous, amber-colored liquid, non-crystalline at all temperatures, which is insoluble in water and has low solubility in glycols, but is miscible in aliphatic and aromatic solvents. The chlorine content (64 to 67%), specific gravity and other physical properties are as given in 1965 except that the viscosity range (75 to 135 centistokes at 130°F) is slightly greater. A minimum flash point of 81°F (27.2°C) is given for an 80% solution in kerosene of minimum flash point 60°C. The same specification, with full analytical details is further given in *Standard for Technical Chlordane*, dated August 1971.[271] Both documents refer also to bioassay analysis, and to product control using a comparison between sample and reference standards to technical chlordane based on the statistical matching of a minimum of 16 out of 20 features of the "fingerprint" region of the infrared spectra of the materials compared. Another document, relating to the composition, analytical considerations, and terminal residues of chlordane, and dated November 1966, presents gas-liquid chromatograms of batches of technical chlordane made between 1960 and 1963, along with comparisons of infrared spectrograms of four batches produced between 1953 and 1954, and 24 between 1960 and 1964.[272] A perusal of these documents leaves no doubt that the quality and composition of chlordane has been carefully controlled for many years.

As far as the identity of the individual components of technical chlordene is concerned, the chemical investigations have been greatly facilitated by the application of gas-liquid chromatography with electron capture detection since about 1961. Before that time, several analyses were reported using classical methods such as adsorption chromatography on columns. By chlorinating chlordene in carbon tetrachloride, Riemschneider obtained a product, bp 160 to 185°/1 mm, of which a portion of bp 165 to 175°C was further separated by fractional distillation into a fraction bp 165 to 167°C and one bp 172 to 175°C. By adsorption chromatography, small amounts of two solids, mp 102 to 104°C and 93 to 100°C, were obtained from these fractions.[186]

The number of compounds which might be contained in technical chlordane is potentially large. Neglecting optical isomerism due to the asymmetry of 1,2-dichlorodihydrochlordene (a simple name for chlordane based on chlorine substitution of the reduction product, 2,3-dihydrochlordene, of chlordene), and assuming the endo-ring fusion, there are four possible stereoisomers, two cis and two trans, of this structure. In addition, other position isomers of the formula $C_{10}H_6Cl_8$ may be present, as well as lower chlorinated products (analogues of heptachlor), higher chlorinated products containing nine (called nonachlors) or more chlorine atoms, unchanged "hex" that has carried right through the manufacturing process, unreacted chlordene, adducts formed from traces of pentachlorocyclopentadiene present in the "hex" used, and so on.

Various chromatographic separations of the components of technical chlordane have been reported, and Büchel's investigation of the chlorination products of dihydrochlordene[206] (Vol. I, Chap. 3 B.2a) should be compared with these, in view of the possible similarity of some of the products. Riemschneider has indicated the presence in technical M 410 of four isomers of the formula $C_{10}H_6Cl_8$, two of the formula $C_{10}H_5Cl_7$, and one each of the formulae $C_{10}H_5Cl_9$ and $C_{10}H_4Cl_{10}$, respectively, in addition to unreacted chlordene.[188] Two of the octachloro-compounds, of melting point 103 to 105°C and 104 to 105°C, respectively, were described as isomers of 1,2-dichlorodihydrochlordene. At this point, we enter the deep waters which surround the nomenclature of the isomers of 1,2-dichlorodihydrochlordene (the chlordane isomers).

It is now fully accepted that out of the isomeric possibilities mentioned above for 1,2-dichlorodihydrochlordene, two compounds are known, and their melting points are too close together to permit a reliable distinction on this basis. At an early stage in the history of chlordane, these two compounds, which together comprise a considerable proportion of technical chlordane (Table 13), were designated by two different systems using Greek letters, and the meaning of these in relation to the stereochemistry of the compounds, one of which is accepted to be 1-exo, 2-endo-dichlorodihydrochlordene (therefore trans), and the other 1-exo, 2-exo- (and therefore cis) has been confused in the literature ever since.

According to Riemschneider, his chlordane isomer mp 104 to 105°C has a dipole moment of 1.87D, in good agreement with the value calculated for the 1-exo, 2-exo-chlordane isomer with endo-ring fusion. However, the infrared spectrum given for this compound strongly resembles that of Velsicol γ-chlordane, which has the trans-configuration (1-exo, 2-endo-). The same dipole moment (2.48D) was calculated for either of the trans-isomers with endo-ring fusion, which were therefore not distinguished by this method; it does, however, distinguish between the exo- and endo-ring fusions in these pairs of isomers. In 1951, Vogelbach isolated from a technical sample of chlordane compounds described as α-$C_{10}H_5Cl_7$, mp 143 to 144°C; β-$C_{10}H_5Cl_7$, mp 86°C; γ-$C_{10}H_5Cl_7$, mp 102°C; three octachlors, α-$C_{10}H_6Cl_8$, mp 105.5 to 106.5°C; β-$C_{10}H_6Cl_8$, mp 102 to 103.5°C; α-$C_{10}H_6Cl_8$, mp 141 to 141.5°C; and a nonachlor, $C_{10}H_5Cl_9$, mp 209 to 211°C.[274]

Vogelbach's separation was closely followed by that of March,[207] who used adsorption chromatography on alumina to separate various components from technical chlordane. The fractions obtained in this separation corresponded to small amounts of unreacted "hex" and chlordene (compound 237), together with other constituents described as II ($C_{10}H_5Cl_7$, mp 92 to 93°C), III, IV ($C_{10}H_6Cl_8$, mp 104.5 to 106°C), and V ($C_{10}H_6Cl_8$, mp 106.5 to 108°C); the common feature of all these separations is the pair of octachloro-compounds with closely similar melting points. March's constituent II was identified with heptachlor (mp 95 to 96°C), and Vogelbach's β-$C_{10}H_5Cl_7$, mp 86°C, was probably this same compound; constituent III appeared to be a mixture, and this has been substantiated by more recent gas-chromatographic investigations. Toxicities were measured of the various constituents to female milkweed bugs, (*Oncopeltus fasciatus*), and it was found that constituent IV underwent alkaline dehydrochlorination more readily than constituent V and was also less toxic to the milkweed bug (LD50 for IV, 459 μg/g; for V, 47 μg/g). On the basis of its more rapid dehydrochlorination, constituent IV was judged to be *cis*-chlordane (1-*exo*, 2-*exo*-), and V to be the *trans*- (1-*exo*, 2-*endo*-) isomer.

In this work, March did not use the α,β-nomenclature, but there occurs in the paper the statement "compounds identified with constituents II, IV, and V have been commercially synthesized in pure form."* The asterisk relates to a footnote, "Samples of Heptachlor, Alpha-chlordane and Beta-chlordane, respectively, furnished by Julius Hyman and Co., Denver, Colorado." This footnote clearly links constituents IV and V with the α- and β-chlordanes, respectively, as they were designated by the Hyman Company, and also indicates that Hyman's α-chlordane is *cis* (on the basis of March's dehydrochlorination data). On this basis, the Hyman nomenclature is partly in agreement with the Velsicol system (in which α-chlordane is *cis*-chlordane and *trans*-chlordane is called γ-chlordane). It is not clear to the author at what point the change occurred in the widely used system in which *trans*-chlordane is called α-chlordane and *cis*-chlordane is called β-chlordane. At any rate, Metcalf, writing of March's work in 1955, describes constituent IV as β-C$_{10}$H$_6$Cl$_8$ and V as α-C$_{10}$H$_6$Cl$_8$, which is precisely opposite to the designation implied in March's paper. In the same discussion, Cristol is cited as having shown that in 0.04 *M* ethanolic sodium hydroxide, "β-chlordane readily dehydrochlorinates, while α-chlordane is inert," so that the information is consistent insofar as that isomer which dehydrochlorinates readily is by this time called the β-isomer. Unfortunately, March's constituent IV, called by Metcalf β-C$_{10}$H$_6$Cl$_8$ (page 235 in reference 41), is actually called α- again in the associated Table of Toxicities on the opposite page. The confusion due to this situation is very evident in the literature, and since the melting points are not helpful, some other unequivocal evidence of structure should really be given when these isomers are discussed. According to recent investigations reported by Cochrane, [208] the

nonachloro-compound, mp 209 to 211°C, isolated from technical chlordane by Vogelbach is probably *cis*-nonachlor (1-*exo*, 2-*exo*, 3-*exo*-trichlorodihydrochlordene), while various lines of evidence indicate that another isomer, mp 128 to 130°C, also present in technical chlordane, is the corresponding 1-*exo*, 2-*endo*, 3-*exo*-compound.

Although technical chlordane is a mixture, it has been thoroughly investigated by this time, and as the documentation cited earlier suggests, the constitution of the product has been kept constant by carefully controlled manufacturing processes since about 1950. The product is defined as a mixture of chlorinated hydrocarbons consisting of isomers of 1,2,4,5,6,7,8,8-octachloro-3a,4,7,7a-tetrahydro-4,7-methanoindane (that is, the *cis* and *trans*-chlordane isomers) and closely related compounds, together with small amounts of propylene oxide, added to stabilize the product by reacting with any hydrogen chloride which may be liberated during storage. An important point to note is that while the composition of the product can be just as rigidly controlled during manufacture as can the composition of other mixtures that are successfully used as agricultural chemicals, the *estimates* of the amounts of the two principal chlordane isomers present can vary from 20 to 45% of the product depending on the resolving power of the analytical method used for determination, so that the actual content of these two isomers in technical chlordane is subject to debate; Harris has indicated a combined α- and γ-chlordane content of around 19% in several samples of the technical material.[275] No other individual component represents more than about 10% of the total product, and the amount of unreacted "hex" is not more than 1%;[271] a typical composition, based on gas-chromatographic analysis, is given in Table 13, and a typical gas-chromatogram in Figure 22.

The typical gas-chromatogram contains some 14 peaks, and the identity of the individual components has been reported on by Polen,[272] using technical chlordane, and by Saha and Lee using the material extracted from a 25% chlordane granular formulation.[276] According to Polen, peak A is the Diels-Alder adduct of pentachlorocyclopentadiene and cyclopentadiene. Peak B is described as an isomer of the adduct of "hex" and cyclopentadiene (that is, an isomer of chlordene). Saha and Lee were unable to examine the compounds of peaks A and B, but comment that a

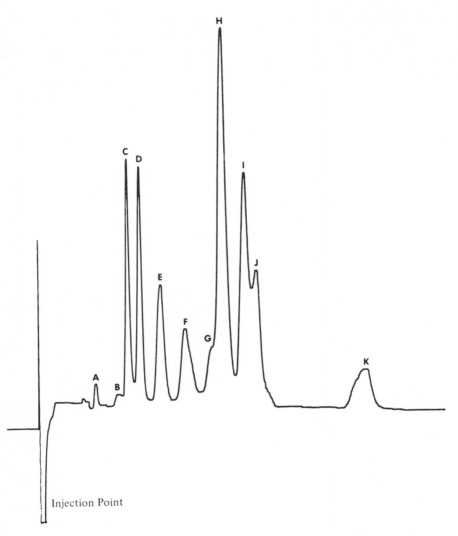

FIGURE 22. A typical gas chromatographic analysis of technical chlordane (by courtesy of the Velsicol Chemical Corporation). See text for discussion.

peak between these two is probably due to chlordene. Polen has indicated peak C to contain a chlorinated derivative of the compound of peak A; Saha and Lee present mass spectral evidence that this compound is indeed a heptachlor analogue lacking one of the methano-bridge chlorine atoms. The prominent peak D is assigned to heptachlor plus an isomer of chlordene by Polen; Saha and Lee claim that it corresponds to heptachlor, but also contains an isomer of chlordane, $C_{10}H_6Cl_8$, mp 168 to 170°C. Peak E (Saha's G) is also assigned by Polen to isomers of chlordene, but to an isomer, $C_{10}H_5Cl_7$, of heptachlor with a chlordane isomer, $C_{10}H_6Cl_8$, by Saha. A melting point

of 135 to 137°C is recorded for the latter, which had a strong IR absorbance at 35 cm^{-1} corresponding to a CH_2 group. Since the IR spectrum of this compound lacked an absorption maximum due to the chlorinated double bond and the mass spectrum provided no evidence of a retro-Diels-Alder reaction, it was suggested that it is chlordene in which two chlorine atoms have added to the already chlorinated double bond, rather than the unchlorinated one.

Peak F (Saha's H) is assigned to a hexachloro-compound that is probably the dichloro-derivative of the Diels-Alder adduct of tetrachlorocyclopentadiene and cyclopentadiene; it was, however, designated as an octachloro-compound $C_{10}H_6Cl_8$ by Saha on the basis of mass spectral evidence. Peak G is assigned the probable structure of a dichloride of the Diels-Alder adduct of pentachlorocyclopentadiene and cyclopentadiene (that is, a chlordane, $C_{10}H_6Cl_8$, minus one methano-bridge chlorine). This peak was not examined by Saha. Peak H represents the chlordane isomer, $C_{10}H_6Cl_8$, called γ-chlordane by Velsicol and assigned the *trans*-structure; it is the individual component present in highest concentration in technical chlordane (Table 13). Referring to this peak (their peak J), Saha and Lee add something to the nomenclature confusion previously discussed, by identifying γ-chlordane as the *cis*-isomer; the source of their authentic material is not revealed. Peak I represents Velsicol α-chlordane, $C_{10}H_6Cl_8$, assigned the *cis*-structure and therefore β-chlordane (according to Metcalf, but *not* according to March, who, in fact, aligned the *cis*-compound only with Hyman's α-chlordane, as indicated earlier). Saha and Lee call this peak α-chlordane and identify it with the *trans*-isomer. Peak J is indicated to be a nonachlor by Polen, and an octachlor, $C_{10}H_6Cl_8$, by Saha, who suggests that this molecule, which according to the IR data has the chlorinated double bond but no CH_2 group, is a nonachlor *minus* one of the methano-bridge chlorine atoms. Peak K is thought to be another chlordane isomer (Polen) or a nonachlor, mp 164 to 165°C (Saha). Saha and Lee also examined another peak (their peak M), which occurs between peaks J and K of Polen, and according to the former source consists mainly of a compound $C_{10}H_6Cl_8$, together with an isomer of nonachlor, $C_{10}H_5Cl_9$. Summarizing their results, Saha and Lee state that current commercial chlordane formulations probably contain, using dicylopentadiene to denote the basic carbon skeleton to these compounds, two hexachlorodicyclopentadienes (chlordene and an isomer of heptachlor lacking one methano-bridge chlorine) and two heptachloro-, eight octachloro-, and two nonachlorodicyclopentadienes. It will be observed that the structural assignments made to some of the peaks differ between the two sources, and the synthesis of some of the additional structures mentioned by Saha would provide valuable information on this point; Velsicol offers reference analytical standards for most of the compounds characterized in their laboratories. Some interesting structural chemistry has been conducted on various compounds in this series in relation to both residue analysis and metabolic studies (Vol. I, Chap. 3B.4b) and the synthetic work by Büchel in this area is also relevant.[204-206]

A Velsicol Chemical Corporation Development Bulletin, *Belt plus Insecticide*, dated April 1971 describes an experimental chlordane (common name AG-chlordane, code HCS-3260) having the trademark BELT plus® (the various formulations of technical chlordane are sometimes marketed under the trademark "BELT").[277] This product is a high purity chlordane which contains, typically, 95% of the mixed α- and γ-chlordanes; in it, the α:γ ratio is approximately 3:1, and the heptachlor content is maximally 1%, but typically between 0.25 and 0.5%. The material is a waxy solid having approximate solubilities (g/100 g solvent at 30°C) of 544, 290, 142, 43, 36, and 0.4 in cyclohexanone, acetone, xylene, pentane, methanol, and water, respectively. BELT plus is currently undergoing test under field conditions for efficacy and residue pattern and it is not clear at present to what extent, if at all, it can replace some or all of the current uses of technical chlordane as used at present.

The foregoing account has related mainly to the chemical aspects of chlordane development. In the early period of chlordane manufacture, before 1950, problems arose with the performance of the product in toxicological evaluations on mammals, and this was later found to be due primarily to variations in the amount of unreacted hexachlorocyclopentadiene which carried through the whole manufacturing process and remained in the final product.[269] From chronic oral toxicity studies conducted on dogs after 1946, the U.S. Food and Drug Administration (FDA) concluded that in those trials chlordane appeared to be five to ten times more toxic than DDT. In some trials, Octa-Klor was used exclusively, in some, Velsicol 1068 was used exclusively, and in others the two products were used interchangeably; the mammalian toxicity of Octa-Klor appears to have been somewhat higher than that of Velsicol 1068. In another study, the results of which were communicated to Dr. Hyman late in 1946,[278] two batches of Velsicol 1068 were compared with DDT in a dietary study on growing rats. The results showed that the toxicities of DDT and Velsicol 1068 were of the same order when administered in this manner. However, it was noticed that the first batch of chemical solidified completely on standing, whereas the second remained liquid, and was apparently less toxic than the first; it caused fewer deaths than the corresponding concentrations of DDT during the test period.

A number of incidents in the period before 1950 pointed to a vapor toxicity effect of chlordane and also indicated that early samples were irritating to mammalian skin. Later experiments showed that dermal irritation and vapor toxicity are particular attributes of hexachlorocyclopentadiene, rabbits being especially susceptible to the vapors, with increasing tolerance in mice, rats, and guinea pigs, the latter being least susceptible. The effect of "hex" on chlordane toxicity was later substantiated when it was shown that mice exposed to air saturated with the vapor from post-1950 chlordane suffered no toxic effects until "hex" was added to the chlordane. These problems involved the use of pre-1950 chlordane in confined spaces, and although the presence of unchanged "hex" is clearly undesirable, this material would hardly present a residue hazard following field applications because of its volatility.[269]

In 1950, public hearings were held on the subject of tolerances for poisonous or deleterious substances on or in fresh fruits and vegetables, and chlordane was one of the subjects discussed. On the basis of their 2-year chronic feeding study of chlordane in the rat, the FDA claimed that minimal "liver damage" occurred at 2.5 ppm in the diet, the lowest level fed. They further asserted that their experiences resulted from a lack of uniformity in the product, which meant that one batch could be more toxic than another. Ingle, on the other hand,[269] concluded that the "no-effect" level was 5 ppm on the basis of trials conducted with material produced later than that used by the FDA, by which time the manufacturing process had improved. As a result of the evidence presented at these hearings, a tolerance of 0.3 ppm was established in 1955 for residues of chlordane in or on 46 raw agricultural commodities, and a little later this was extended to include sweet potatoes.

In 1963, the FDA re-examined the toxicological data available for chlordane up to that time in the light of the then accepted criteria for safety. Since a "no effect" level for the dog was still not established and the 1950 dispute about this level in the rat had not been resolved, it was concluded that the available information was inadequate to

provide any conclusions about the safety of the existing tolerances for chlordane, and notice was given that zero tolerances were to be established for all the listed crops. In their answer to this proposal, Velsicol submitted the results of a further 2-year feeding trial conducted by Ingle, but the FDA did not alter its conclusions, and as a result of a petition by Velsicol to the National Research Council of the National Academy of Sciences, an advisory committee was appointed to review the proposed change in tolerances.

Data submitted to the committee, which included Dr. R. B. March, author of some of the original separative work on chlordane, included the analytical data mentioned previously which supported the uniformity of technical chlordane as manufactured between 1954 and 1963, as well as toxicological data providing evidence for a significant improvement in the product over the pre-1950 one upon which the original tolerance of 0.3 ppm was based. Regarding the toxicological evidence, the committee concluded that while the acute oral LD50 of pre-1950 chlordane to the rat appeared to be approximately 200 mg/kg, the LD50 for the chemical used after 1950 is nearer 575 mg/kg, and that the no-effect level in chronic feeding studies is approximately 25 ppm, which is five times higher than the value proposed by Ingle in 1950. In the interim, chronic feeding trials carried out on rats by the FDA had also shown no liver cell changes at 2 ppm, and only doubtful evidence of changes at 40 ppm; no effects on reproductive capacity were found at this level, nor any evidence of teratogenic effects. One of the points of contention had been the question of the no-effect level in dogs, and early stages of a 2-year dog feeding trial in progress during these hearings had indicated no adverse changes at levels up to 30 ppm in the diet.[269]

On the basis of this and much more evidence detailed by Ingle, the committee concluded that technical chlordane is a uniform product, subject to satisfactory methods of quality control, and therefore adequate for satisfactory toxicological evaluations. Particular note was taken of the relevance of "hex" content to the toxicology of pre-1950 chlordane and the Velsicol internal specification limiting this content to 1% in the subsequent product (based on its ultraviolet absorption), and it was concluded that the tolerance should remain at 0.3 ppm unless further evidence of potential hazard appeared at a later date. The committee reported its findings in February 1965 and an order reinstating the 0.3 ppm tolerance appeared in the Federal Register in April 1965, in which the insecticide chlordane was now defined as containing not more than 1% of hexachlorocyclopentadiene. The tolerance of 0.3 ppm for chlordane is in existence at the present time, which leads to the interesting situation that while chlordane is the least toxic of all organo-chlorines on the basis of acute oral LD50's, this tolerance is much lower than those quite recently in operation for some other compounds (for example, 7 ppm for DDT, 5 to 7 ppm for toxaphene, 10 ppm for γ-HCH). In relation to these figures, Ingle points out that even if toxicity data for pre-1950 chlordane are used as a basis for assessment, a tolerance of 7 ppm for DDT would justify a tolerance of 2.8 ppm for chlordane of that period, the implication being that for present-day chlordane it could safely be considerably higher.[269]

An interesting account of researches conducted in the laboratories of Farbwerke Hoechst AG by Dr. Heinz Frensch details work conducted in the chlordane-heptachlor series around 1949.[176] This work included a detailed investigation of the Straus method of preparing hexachlorocyclopenta-diene, in which it was found that by-products consisting mostly of lesser chlorinated cyclopenta-dienes were present and themselves gave adducts with cyclopentadiene. These adducts were isolated and shown to have insecticidal properties. Specifically, the report mentions chlordene analogues lacking one of the vinylic chlorine atoms that were subsequently chlorinated to give the corresponding chlordane analogues. These were said to correspond to chlordane in activity toward various insects. On the basis of those results a product consisting of chlordane analogues derived from tetra-, penta-, and hexachlorocyclopenta-diene was developed which was said to have some advantages over chlordane and was introduced under the name "Hostatox" in Germany. Corresponding analogues of heptachlor were also prepared about that time. Attempts to simulate the insecticidal properties of chlordane led to the preparation of various "ring cleavage" analogues, one of which was Alodan® (5,6-bis(chloro-methyl)-hexachloronorborn-2-ene). Alodan is a crystalline solid of melting point 105 to 107°C, stable towards acids and dilute alkalis, insoluble in water, but soluble in ethanol (3 to 5%), olive oil (10%), ether (10%), chloroform (40%), and other organic solvents. It has a remarkably low mammalian toxicity and was introduced as a 5% powder or 50% solution for the control of animal ectoparasites. It also has acaricidal properties, but appears to be seldom used nowadays.

As the chemistry shows, the history of hepta-chlor is very much associated with that of chlordane; the original Hyman patent which describes the preparation of chlordane also describes the allylic bromination of chlordene in carbon tetrachloride in the presence of lauryl peroxide, and an analogous preparation of hepta-chlor using benzoyl peroxide and sulfuryl chloride for the chlorination.[268] The heptachlor obtained in this way is a viscous liquid contaminated with other chlorination products. Subsequently, several investigators isolated heptachlor as a solid, mp 95 to 96°C, during their analyses of technical chlordane, as discussed earlier. The analogous 1-bromochlordene is the key intermediate for the preparation of purer samples of heptachlor, since the peroxide catalyzed bromination can be made to give yields of over 90% when carried out in chloroform, carbon disulfide, tetrachloroethylene, or carbon tetrachloride. Carbon tetrachloride is preferred and the reaction is rapidly completed at 55°C. This process, and the subsequent hydrolysis of the bromide to 1-hydroxychlordene, are described in patents applied for in 1947—48 and completed in 1950—51.[201,279] As indicated in the synthetic section (Vol. I, Chap. 3B.2a), the hydro-lysis is effected by heating with potassium hydroxide or potassium carbonate in aqueous dioxan, in boiling aqueous alcoholic solutions containing a little acid, and so on, and the product, being insoluble, is easily recovered by filtration when the reaction mixture is diluted with more water. The alternative method involves replacement of bromine by acetoxy, using an acetate salt in boiling acetic acid, followed by ester hydrolysis. Treatment of the allylic alcohol with thionyl chloride in benzene, or in dioxan with HCl/ZnCl$_2$, then gives heptachlor. The current manufacturing process probably involves direct chlorination of chlordene in the presence of a catalyst such as Fuller's earth, as indicated in a patent of 1970.[280]

Heptachlor was introduced for agricultural use by the Velsicol Corporation about 1948 under the designations "E3314" and Velsicol 104, and quickly gained acceptance as a soil insecticide.[40]

The pure compound is a white crystalline solid melting at 95 to 96°C, and having a camphor-like odor (vapor pressure 3×10^{-4} mm Hg at 25°C). Its solubility in water (0.056 ppm at 25 to 29°C) is very low but it is soluble in a range of organic solvents. At 27°C, typical solubilities (g/100 ml of solvent) are ethanol (4.5), acetone (75), benzene (106), cyclohexanone (119), deodorized kerosene (19), Xylene (102), and methylated naphthalenes (83).[41] The manufacturers describe the technical product as a "soft waxy solid," similar to tallow in appearance and of melting point 46 to 74°C, density 1.65 to 1.67 g/ml at 25°C; 1.59 to 1.61 g/ml at 71°C, and viscosity 46 to 66 centistokes at 71°C. The vapor pressure is stated to be 4×10^{-4} mm at 25°C, 3.75×10^{-3} mm at 50°C, 1.5×10^{-2} mm at 65°C, and 2.8×10^{-2} mm at 71°C. A typical analysis of the product is 73% of heptachlor, 22% of *trans*-chlordane and 5% of nonachlor. It is not readily dehydrochlorinated and is stable on exposure to air, light, and moisture. There is no decomposition at up to 71°C in contact with several metals, glass, and other inert materials. Solutions in aromatic solvents may, however, exhibit decomposition when kept at temperatures over 71°C in mild steel or black iron containers; dust and dirt containing iron powder or iron oxides are efficient promoters of decomposition.[281] Heptachlor is susceptible to epoxidation under environmental conditions and it has become clear during recent years that, among other reactions, a certain amount of hydrolytic replacement of the allylic chlorine atom occurs, to give 1-hydroxychlordene (Vol. II, Chap. 3B.2c).

One of the most attractive features of chlordane, as compared with DDT, is its excellent solubility in most organic solvents. This means that aromatic, aliphatic and chlorinated hydrocarbons, ethers, esters, and ketones, are all available as vehicles for the toxicant, which is also completely miscible with deodorized kerosene, so often used for formulating insecticides. The manufacturers describe it as being stable to 160°F in stainless steel, nickel, passive iron, glass, and many other materials, but subject to decomposition at very high temperatures, and to catalyzed decomposition at elevated temperatures in the presence of iron oxide or iron powder.[270] Like HCH or DDT, it is susceptible to dehydrochlorination by alkaline agents to give nontoxic products and so should not be formulated with other compounds having such action.

As in the case of DDT, chlordane may be formulated as oil solutions, emulsifiable concentrates, wettable powders, dusts, and granules. In the first report of trials with it, Kearns and his colleagues indicated suitable solution carriers to be xylene, methylated naphthalenes, kerosene, deobase (a commercial hydrocarbon based oil), and fuel oil.[265] According to the original Hyman patent, housefly toxicity tests were conducted with solutions of chlordane in deobase, with 1% of the total volume of added "Lethane 384 Special". This compound (2-butoxy-2'-thiocyanodiethylether) does not effect the final kill, but is well known to provide a rapid paralyzing or knockdown effect, and has been frequently included in spray formulations for this purpose. To kill flying insects a space spray containing 0.2% w/v of chlordane and 0.04% of pyrethrum in kerosene was recommended about this time, while crawling insects like bedbugs and cockroaches were said to be quickly killed and reinfestation prevented for several months by use of a spray containing 2% of chlordane in kerosene.

The Illinois group noted that the addition of a small amount of Triton X-100 or other oil soluble emulsifiers gave an emulsion concentrate to which could be added any desired amount of water, or chlordane could be emulsified directly in water by the addition of a soluble emulsifier, although a higher concentration of the emulsifying agent was required in this case. For the preparation of dusts or other dry formulations, absorptive carriers such as diatomaceous earth may be used, or nonabsorptive ones such as pyrophyllite or talc. Since technical chlordane is a liquid, dry formulations are produced by impregnation from a suitable solvent; caking of the dry formulations does not occur since chlordane does not crystallize, even at low temperatures. Many of the carriers used to prepare dry formulations possess catalytic activity for the decomposition of chlordane and other chlorinated insecticides. This activity is frequently related to the presence of acid sites on the surface of the carrier, and problems of decomposition may be avoided either by using a nonactive carrier such as calcium carbonate or by adding a deactivator either prior to or during impregnation with the toxicant. The chlordane formulation guide recommends the addition of varying amounts of deactivator when the carriers are attapulgites, diatomaceous earths, vermiculites, kaolinites, synthetic silicas, montmorillonite dusts, or granules.[270] The nature of the deactivator is not stated, but urea and hexamethylenetetramine have been used as deactivators in other cyclodiene insecticide formulations, and this method is generally applicable to chlorinated insecticides.[116]

In a typical operation, a mixture of technical chlordane, solvent, and deactivator is sprayed on to the carrier in a blender, and the thoroughly blended material is ground so that 90% passes a 325 mesh (for dusts, dust concentrates, and wettable powders). For wettable powders, wetting and dispersion agents (listed in the formulation guide) are added after the initial blend, and the mix is reblended before grinding. A very wide range of liquid and dry formulations may be prepared, among which the most widely used today are 2 and 20% solutions in deodorized kerosene; 45 and 75% emulsifiable concentrates (e.c.); 5, 6 and 10% dusts; 25 and 40% wettable powders; 5, 10, 20, 25, and 33.3% granules. The percentage of chlordane in the emulsifiable concentrates may vary slightly from these figures, depending on the density of the solvent used; a typical 45% e.c. contains 45% technical chlordane, 50% solvent and 5% of emulsifier by weight, while a 75% e.c. (8 lb/gal) might contain 72% technical chlordane, 21% solvent and 7% of emulsifier. A special 8 lb/gal e.c. is also prepared for simultaneous application with liquid fertilizers. In the case of oil solutions, odor and color are important and masking agents or perfumes are frequently added to give an odor more acceptable in homes or commercial buildings. Granules are deactivated and impregnated with the insecticide in the same way as for other dry formulations; typical 20% granular formulations contain (by weight) 20% technical chlordane, 4.7% solvent, 4.3% deactivator, and 71% attapulgite, or 20% technical chlordane, 5.3% solvent, 0.7% deactivator, and 74.0% vermiculite. A 40% dust concentrate (40.0% technical chlordane, 2.0% solvent, 2.5% deactivator, 41.0% attapulgite, and 14.5% celite 209) is made for further dilution as required (with talc, for example) to give the field strength dusts of lower concentration.[270] In addition to these formulations containing technical chlordane, there are various others containing the experimental insecticide BELT plus (a-plus γ-chlordane) both alone and with methyl parathion or Phosvel® (2,5-dichloro-4-bromophenyl methyl phenylphosphonothionate) that have already undergone trial.[277]

Technical heptachlor is formulated in a variety of ways for insect control; the general principles are those applied to the formulation of other organochlorines and are fully detailed in the manufacturer's formulation guide.[281] Typical formulations are emulsifiable concentrates containing 2 lb and 3 lb of technical heptachlor/gal (corresponding to 23.29% and 32.1%, respectively of active ingredient), oil solutions containing 2.0, 3.0 and 3.3 lb/gal for dilution with appropriate solvents, 25% wettable powders, 25% dust concentrates for dilution with appropriate carriers (to make a 2.5% dust, for example), and various granular formulations with heptachlor concentrations ranging from 2.5 to 25%. As in the case of chlordane and other organochlorine insecticides, appropriate deactivators are used with certain carriers to prevent decomposition of the toxicant.

b. Applications in Agriculture and Public Health

The value of chlordane as a broad-spectrum insecticide was soon recognized. In their extensive tests conducted with Velsicol 1068 and reported in 1945 (Vol. I, Chap. 3C.3b), the Illinois group compared Velsicol 1068 with DDT against a wide range of insect pests.[265] The housefly is a favored insect for testing new insecticides because of the relative ease of rearing and manipulation in toxicity measurements. Tests were conducted by the Peet-Grady procedure which involves spraying large groups of the insects with a range of concentrations of the toxicant in a solvent (deodorized kerosene containing 1% of Lethane 384 Special) in a specially designed chamber. Lethane 384 is added to produce rapid paralysis of the flies but does not affect the final mortality observed. These tests indicated LC50's of 0.02% and 0.06%, respectively, for chlordane and DDT, with corresponding LC90's of 0.05% and 0.2%. This was probably the earliest indication that members of the cyclodiene group of insecticides are markedly more toxic than DDT to a number of insect species.

In this context some approximate potencies of newer insecticides relative to DDT were collected together by Busvine in 1952, and are shown in Table 14, which relates to insects of public health concern.[282] Needless to say, much interest centered around the relative effectiveness of the newer compounds for this kind of application. The table shows that chlordane is somewhat less toxic than DDT to adults and larvae of certain culicine

TABLE 14

Insecticidal Potencies Relative to DDT (=1.0) of Several Organochlorine Compounds Toward Pests of Medical Importance

Insect[a]	Methoxychlor	DDD	Chlordane	Aldrin	Dieldrin	γ-HCH	Toxaphene
Housefly (*Musca domestica*)	0.35–0.92	0.25–0.5	1.2–3.0	7–11	10–17	3.3–17.0	0.15–0.44
Mosquito (*Aedes aegypti*)			0.34, 0.66	2.6	4.0	2.0, 2.3	0.05, 0.14
Bedbug (*Cimex lectularius*)	0.97	0.44	0.60	1.9	3.7	13	0.40
Argasid tick (*Ornithodorus moubata*)			2.5		25	25	2.4
Body louse (*Pediculus humanus*)	0.33	0.33	3.0			20	
Aedes spp.	0.09	1.0, 3.3	0.40		2.0	0.5	0.25
Anopheles quadrimaculatus	0.10	1.0	0.75, 1.0	0.53	2.6	0.16, 0.35	0.78
Anopheles albimanus	0.08	1.5					
Culex fatigans	1.05	1.8					

(Data adapted from Busvine;[282] Metcalf and Fukuto.[67])

[a]First five insects, nymphs, and adults; last four mosquito larvae.

mosquitoes (*Aedes* spp.). The Illinois group conducted comparisons of DDT, HCH and chlordane toxicities towards adult anopheline mosquitoes (*Anopheles quadrimaculatus*) by applying emulsified 5% solutions of the toxicants in xylene to squares of wall board upon which the insects were then confined under glass dishes. In these circumstances, the DDT-treated surface gave the most persistent toxic effect when tested at various time intervals after deposition, chlordane being intermediate and the HCH deposit less persistent in its effect. The three compounds were compared as larvicides for this mosquito when applied as acetone suspensions, or xylene emulsions and dusts, and there appeared to be little difference in effectiveness between the three toxicants in any of these applications, although other work indicated HCH to be rather less effective than the other compounds as a larvicide for *A. quadrimaculatus*. Experiments conducted in the Orlando laboratory of the United States Bureau of Plant Quarantine also equated chlordane with DDT in effectiveness as a mosquito larvicide when both were used at a concentration of 0.01 ppm. These results with chlordane are reflected in the relative potencies derived by Busvine in Table 14, which also shows that chlordane is rather less effective than DDT against bedbugs (*Cimex lec-*

tularius), but several times more effective against lice (*Pediculus humanus*). Other sources also reported the outstanding toxicity of chlordane to lice, but indicated a lower persistence of the effect than had been observed with DDT or toxaphene. Chlordane soon acquired a reputation as an agent for the control of cockroaches; the early work of the Illinois group showed that when the two compounds were applied as emulsions, the LD50 for DDT (38 μg/g) was about three times greater than that of chlordane (14 μg/g) after 120 hr, while the corresponding LD90's were 70 μg/g and 25 μg/g, respectively.

In the agricultural insect area, the cyclodiene was soon found to be more effective than DDT at equivalent concentrations against aphids such as *Aphis spiraecola* and *Macrosiphum pisi*. In tests against Colorado beetle larvae (*Leptinotarsa decemlineata*), these were placed on potato leaves dried after previously being sprayed with dispersions of the insecticides in aqueous acetone, and chlordane was found to be a better feeding deterrent than DDT, but was more toxic than the latter in any case. It was also more effective than DDT against squash bugs (*Anasa tristis*), applied as either a dust (2 to 5%) or an emulsion, especially if the formulations were applied to both plant and insect. Adult grasshoppers (*Melanoplus differen-*

tialis) were killed somewhat more effectively by chlordane given as a stomach poison than by DDT, but the cyclodiene appears to have been rather less effective than HCH used in this way. Field tests also showed the material to be an effective control agent for grasshoppers. Other promising applications included the control of certain insect pests of cotton and lygus bugs on alfalfa, but work at the Orlando laboratory indicated a lower effectiveness as compared with DDT against certain lepidopterous larvae such as armyworms.

Chlordane appears to be nonphytotoxic when applied to plants or their foliage at levels necessary for insecticidal action, although there were some reports of deleterious effects on crops in the early years of its development. Thus, one report of 1948 indicated that a 0.1% suspension caused damage to plum trees associated with defoliation, bud injury and slight fruit malformation, and another that damage was caused to spring foliage of cherry and plum, but not of apple. Transient phytotoxic effects were seen with three species of squash following application of 3% chlordane dust, and some injury to greenhouse plants was reported following applications of chlordane. The value of chlordane as a soil insecticide was soon discovered and so the effect of soil applications on germinating or standing crops has long been of interest. There were early reports of deleterious effects of chlordane on the germination of some plant species, but in general, the effects seem to be negligible when applications are made at recommended levels. Thus, chlordane at 10 lb per acre was said to have no effect on annual flowers, grasses, vegetables (including soybeans) or cereals, and plants grown in soil treated at 25 lb per acre showed no injury, except lima beans, which showed some stunting. The germination of tomato, lettuce, and cabbage appeared to be unaffected at 50 lb per acre, although tomato foliage appeared to be susceptible to chlordane sprays; tobacco germination was prevented at this level of application. A report on the effect of chlordane on turf indicated no effect below 100 lb per acre, but greater levels caused destruction of bent grass and clover. These are reports of work conducted prior to or during 1948–49.[24]

In 1952, writing about the use of chlordane against pests of medical importance, Busvine comments that chlordane, as the first of the cyclodienes to be introduced, had by that time been much more extensively tested and used than the others — "it lacks the persistence of DDT and is not quite so insecticidal as γ-BHC; and I do not know of any medically important pests against which it is more efficient than either one or other of these two insecticides."[282] This seems to be fair comment on the properties of chlordane, which has, however, been widely used in the control of insects of public health and veterinary importance, as can be seen by a perusal of the history of events relating to insecticide resistance. In many instances it was found to be quite effective initially even though less persistent than DDT, but in very many cases its use had to be abandoned due to development of the classical cyclodiene resistance; this was especially the case with houseflies and mosquitoes.

The major uses of chlordane today are for termite control, in which treatment of premises with a 1.0% chlordane emulsion is said to give residual control for at least 5 years, for lawn and household insect control, general insect control by pest control operators, and for agricultural usage, especially as a soil insecticide. The turf or lawn usage is for the control of various kinds of ants, several kinds of beetle and moth larvae, chiggers, earwigs, European chafer larvae, ticks, wireworms, and earthworms, chlordane being applied as an aqueous emulsion or 5% dust. Household uses include the control of houseflies, mosquitoes, wasps, ants, cockroaches, fleas and ticks on cats and dogs (used as a 2 to 3% aerosol or oil solution (kerosene) spray or 5 to 6% dust), and stored product pests such as carpet beetles (same formulations used) and clothes moths (2% spray). A 5% dust may also be used for the control of springtails and symphilids on house plants. A 0.1% aqueous emulsion has been prescribed for the control of a variety of insect pests on shade trees, ornamentals, perennials, and annuals. Various leaf beetles, leafminers, caterpillars, thrips, and bugs are controlled by this emulsion, applied to the foliage until run-off occurs.[283]

A large number of soil pests attacking fruit and vegetables are controlled by the application of chlordane to the soil at a rate of 1 to 10 lb of active ingredient (a.i.) per acre, as is evident from the list of Chlordane Federal Label Acceptances, and similar rates are used in soil applications for the protection of cereals and other field crops. There are certain groups of insects, such as Curculionids (weevils), some cutworms (such as the armyworms), Cercopids, and leaf-mining

Agromyzids for which chlordane is a better control agent than DDT, and others for which chlordane is used because DDT is ineffective. The last group includes the Orthoptera (crickets) and larger Hemiptera, which are heavily sclerotized forms upon which DDT fails to exhibit its usual rapid contact action.[24] A certain fumigant effect is valuable in a soil insecticide, and in this respect, the higher vapor pressure of chlordane gives it marked advantages over DDT, which has no fumigant action; such action enables the toxicant to penetrate cracks and crevices which are quite inaccessible to DDT.

Weevils have always presented a challenge to chemical control agents because their larvae normally feed inside seeds or on roots, but many of these problems were solved with the advent of DDT; HCH and chlordane are even better for some species, undoubtedly because of the fumigant action referred to. The boll weevil (*Anthonomus grandis*) is an important pest which falls into this category because the larvae remain within the bolls and are thereby protected from toxicants. Calcium arsenate dust applied neat at 7 to 10 lb per acre was the traditional control agent for this pest from about 1920, and repeated applications resulted in soil poisoning by this material, which was later replaced by copper arsenate. Chlordane, HCH, and toxaphene are effective adulticides, while only the first two are able to reach the larvae within the bolls. Nevertheless, toxaphene is generally superior to chlordane on cotton, since it controls several other pest species more effectively.[24]

DDT and HCH have been used in baits for the control of the fall armyworm (*Laphygma frugiperda*) but chlordane proved to be superior to either of these toxicants. The same pest may be controlled on cotton by applications of chlordane at 0.5 to 1.0 lb per acre (a.i.). Orthopterans, except locusts, were widely controlled by baits containing arsenicals or other inorganic poisons before the advent of DDT. Dinitro-*o*-cresol (DNOC) dusts proved to be effective against the brown locust (*Locusta pardalina*) and the desert locust (*Schistocerca gregaria*) and was widely used against Orthoptera at the end of World War II. DDT proved ineffective as a contact poison although it was effective against *Locusta migratoria* when applied in baits. When chlordane became available, it proved to be superior to DNOC in contact effect, besides having activity as a stomach poison and residual toxicity that was

not found with DNOC. The larger Hemiptera include various pentatomid bugs, some of which are useful pest predators and others crop pests. Chemical control of these insects is difficult; they are not susceptible to stomach poisons and require powerful contact insecticides for control. DDT has a varying effect against these insects and in some cases chlordane is effective where DDT is not.

The experimental high purity chlordane BELT plus has proved to be as effective as technical chlordane for the control of the Mexican bean beetle (*Epilachna varivestis*) and the southern armyworm (*Spodoptera eridania*). It contains a higher proportion of the usually more insecticidal *cis*-chlordane (*cis:trans* ratio is 3:1) and has proved, when tested as a residual deposit against cyclodiene resistant German cockroaches of the Fort Rucker and Virginia Polytechnic strains, to be markedly more active than technical chlordane. Field trials to test its efficacy for control of foliage and soil pests on a wide variety of crops indicate that it is equal to or better than technical chlordane for control of the bollworm (*Heliothis zea*), the tobacco budworm (*Heliothis virescens*), and the alfalfa weevil (*Hypera postica*). The sunflower beetle (*Zygogramma exclamationis*) also appears to be well controlled by this preparation. As a soil insecticide, BELT plus appears to give good control of wireworms (*Limonius* spp.) when applied at 8 to 10 lb active ingredient per acre, and is slightly more effective for this purpose than technical chlordane applied at equivalent rates, as is also found when it is used against leatherjacket (European crane fly; *Tipula paludosa*) infestations in turf.[277]

When considering the applications of chlordane, it must be remembered that heptachlor is one of the constituents of the technical material, and the observed toxicity results from the joint action of the several components present. Harris has recently appraised the performance of technical chlordane in relation to soil type and condition, and in comparison with its individual components (as well as with aldrin and dieldrin), using crickets (*Gryllus pennsylvanicus*), picture winged flies (*Chaetopsis debilis*), and cutworms (*Euxoa messoria*) as test insects.[176] As a contact toxicant, heptachlor was about tenfold more toxic to crickets and flies than either technical chlordane or the *cis*- and *trans*- isomers of chlordane, and the sample of technical chlordane used contained the three components in about equal amounts, totaling 27.58% of the sample. Using this analysis with toxicity data obtained for the individual isomers and for heptachlor, Harris concluded that soil treated with 1.0 ppm of technical chlordane would contain sufficient heptachlor, but insufficient of either of the chlordane isomers acting individually (or even added together, unless more than additive effect occurs), to cause mortality in crickets, so that the toxic effect appeared to be largely due to the heptachlor content of the sample. In laboratory tests on the persistence of toxic action using a Plainfield sand soil, the two isomers and nonachlor ranked with dieldrin as highly residual cricket toxicants showing some toxic effects 48 weeks after application, whereas heptachlor, technical chlordane and chlordene ranked with aldrin as being moderately residual chemicals (activity disappeared within 24 weeks).

Apart from its rather greater volatility, heptachlor has soil insecticide properties rather similar to those of aldrin and also resembles it in being converted into a persistent epoxide in similar circumstances (both conversions occur in soil). A comparison of uses therefore reveals considerable overlap both with aldrin and with technical chlordane, although higher application rates of the latter are generally required for the same purpose. Some better idea of the influence of the heptachlor content on the field performances of technical chlordane will be available when a full evaluation of BELT plus chlordane (typically 95% of chlordane isomers and $< 1\%$ heptachlor) is available. Present information is that the material is about as effective as technical chlordane (better for some purposes) but has to be used at high rates of application and is rather persistent in soil, as expected from the earlier discussion on persistence of the individual isomers.[277]

Like aldrin, heptachlor is used as a seed dressing, a purpose for which chlordane is not employed. In Britain, its use for this purpose ceased in the early 1960's following the bird poisoning incidents, which were linked with high dieldrin and heptachlor epoxide residues in bird tissues, and it has been little used there since.[54] The point of concern about heptachlor relates to this formation of heptachlor epoxide, which was formerly assumed to be the main conversion product. There is now good evidence, however, that more hydrolytic replacement of the allylic chlorine atom occurs to give 1-hydroxychlordene (Figure 14, structure 6) than had been thought,

and this product presents little environmental hazard, being virtually nontoxic compared with heptachlor epoxide (Vol. II, Chap. 3C.2b). Furthermore, recent evidence indicates that heptachlor epoxide can be deoxygenated to heptachlor by soil microorganisms in culture, thus providing another potential source of 1-hydroxychlordene.[284-288]

Current patterns of usage in other countries are obscure. In Canada, where the progress of cyclodiene-resistance has been closely followed for some years by Harris and his colleagues, heptachlor use began to fall off on this account around the middle 1960's, although it was still used extensively for cutworm and corn rootworm control until 1969–70. In 1969, the use of DDT was severely restricted in Ontario Province and that of heptachlor, aldrin, and dieldrin was banned. The Federal Government revised the uses of heptachlor in 1970, and registrations now exist for its use against wireworms, rootworms, and cutworms attacking cereals and tobacco, against pests of lawn and turf, and for control of narcissus bulb fly. Nevertheless, its use is still generally discouraged. The withdrawal of other organochlorines in Ontario has led to an increased use of technical chlordane, since adequate control of soil insects is essential there, as elsewhere. Excluding home usage, which may be considerable, the amount used rose from 4,090 lb in 1969 to 32,552 lb in 1971.[288]

In the public health area, heptachlor would undoubtedly find uses in some of the situations for which aldrin is favored, but it seems to be little mentioned in current accounts of public health pest control. A W.H.O. technical report of 1970 makes only one specific reference to heptachlor (for the control of midge and sandfly larvae) in the section on chemical methods for vector control,[44] but reference is made elsewhere to the use of heptachlor granules, which are said to have a residual effectiveness of 13 weeks in the control of mosquito larvae.[50]

c. Principles of Analysis

As in the case of other chlorinated pesticides, the methods of analysis applied to the determination of chlordane may be divided into those used for the technical material and those used for the detection and quantitation of chlordane residues in various types of produce. The general principles are as discussed in the case of DDT and its analogues (Vol. I, Chap. 2C).

The total organic chlorine content of technical chlordane may be determined directly on a suitable sample (0.7 to 1.0 g recommended) by the Stepanow method using sodium in boiling toluene/isopropanol; this method may also be applied to the formulations, with appropriate modifications, such as solvent extraction of dusts and wettable powders, where necessary. Before such analysis is conducted, an aqueous extract of the material is tested for inorganic chlorine content with silver nitrate solution; if present, inorganic chlorine is then determined separately and subtracted from the total chlorine present. The sodium biphenyl reduction method was adopted as the official final action on this matter by the Association of Official Agricultural Chemists.[271] In this case, the sodium biphenyl reagent is added to the sample in toluene with swirling and the reaction time is 3 min, followed by removal of the aqueous layer and precipitation of the liberated chloride ion with silver nitrate. The *CIPAC Handbook* (Volume I, "Analysis of Technical and Formulated Pesticides") describes a method for the determination of hydrolyzable chlorine in chlordane emulsifiable concentrates by boiling the sample, previously dissolved in acetone, with 0.5 N ethanolic potassium hydroxide for 2 hr, after which the solution is acidified with 0.5 N nitric acid to Congo red, and the chloride precipitated with silver nitrate as usual.

Technical chlordane may also be evaluated by the colorimetric method developed by Davidow; heptachlor appears to be the only other pesticide that will give the same color reaction.[289,290] The Davidow reagent is made by mixing one volume of purified diethanolamine with 2 volumes of 1.0 N potassium hydroxide in methanol (90% v/v). For the analysis, about 0.75 g of samples is dissolved in 500 ml of methanol/benzene (7:3), and a 2 ml aliquot of this solution is heated at 100°C with 2 ml of the Davidow reagent for 45 min. Following its dilution to 10 ml with 90% methanol, an aliquot of the solution is taken for absorbance measurements at 520 and 550 nm. The content of technical chlordane in the sample is then determined by comparing the absorbances at each wavelength with those obtained with reference technical chlordane. This method has been applied for many years to the measurement of chlordane residues in crops and is sensitive to 2.5 to 5 μg of

the technical material, which is equivalent to a sensitivity of 0.025 to 0.05 ppm (based on a 100-g sample of the original crop or other agricultural commodity) or about one sixth to one twelfth of the tolerance (0.3 ppm) established at the present time.

The content of hexachlorocyclopentadiene (specified as $\not> 1\%$) in technical chlordane is determined by measuring its UV absorbance at 324 nm, or by a slightly more complex procedure which involves measuring absorbances at 300, 324, and 350 nm and enables corrections to be made for the presence of other components of chlordane which also absorb at 324 nm.[271]

As outlined in the account of the development of chlordane (Vol. I, Chap. 3C.1a) the quality of technical chlordane is maintained by careful comparison of the infrared spectrum of samples with specially retained reference batches of the material.[271,272] Samples and reference solutions of nearly identical concentration (concentration ratio of reference to sample 0.98 to 1.02) are recorded close together in time and using the same sample cell. Because of the nature of the material, the IR spectrum of technical chlordane contains a considerable number of bands in the "fingerprint region," and certain prominent ones between 7 and 15 μ are lettered A-W. In comparing the patterns of sample and reference, spectra are chosen which have at least 20 prominent features (bands and shoulders) and are examined statistically for the match of a minimum of 16 out of these 20; for this purpose the corresponding maxima and minima in sample and reference must fall in the same 5% transmittance zone. These methods are described in detail, together with those for specific gravity, viscosity, appearance (transparent color), flash point of kerosene solutions and a cockroach bioassay method for the determination of bioactivity, in *Standard for Technical Chlordane*.[271]

The analysis of chlordane, particularly from a residue standpoint, was greatly facilitated by the advent of electron-capture gas-liquid chromatography in the early 1960's. The GC-analysis of technical chlordane has been discussed in relation to the development of this product (Vol. I, Chap. 3C.1a) and a gas chromatogram such as that of Figure 22 is obtained, for example, on a 5 ft x 3/16 in. column containing 5% SE-30 on Chromosorb W[®], at 180°C with injection temperature at 230°C, detector at 250°C and helium as carrier gas. When technical chlordane is applied in the field, successive analysis of crop samples for residues shows that all components disappear quite quickly, the disappearance of the minor components being relatively faster than that of the *cis*- and *trans*- chlordanes. These two compounds give the most prominent peaks, called "signature" peaks, in GC-analysis of technical chlordane or chlordane residues and are therefore used for analysis by comparing their area when the sample is run with the corresponding area given by a known amount of standard technical chlordane. Because of differential loss of the minor components during "weathering", a given area of the signature peaks in the sample corresponds to less technical chlordane than will actually be indicated by the measurement made in this way, so that the residue analysis generally gives a high result for technical chlordane. An examination of the weathering of technical chlordane applied to cabbage revealed that the decrease in the "signature peaks" was accompanied by an increase in two other peaks in the gas chromatogram. These peaks were subsequently identified as the dehydrochlorination products, 3-chlorochlordene and 2-chlorochlordene, respectively, of *cis*- and *trans*-chlordane.[236] The formation of these compounds (Vol. I, Chap. 3B.4b) has been exploited as a derivatization technique for the identification and determination of residues, since the dehydrochlorination products have markedly shorter retention times than the isomeric chlordane precursors. As an example, a mixture of heptahclor, heptachlor epoxide, *trans*-chlordane, and *cis*-chlordane is treated with potassium *t*-butoxide/*t*-butanol, which converts heptachlor into 1-hydroxychlordene, heptachlor epoxide (mp 160°C) into 1-hydroxy-3-chlorochlordene, and the *cis*- and *trans*- chlordanes into the 3- and 2-chlorochlordenes, respectively. These compounds all have quite different retention times. The hydroxy-compounds may be converted into trimethylsilyl derivatives or acetylated for further confirmation, leaving the two derivatives of chlordane unaffected. The level of detectability for the chlordane isomers detected by this method is said to be 0.01 ppm in a 10-g sample of raw crop. Another derivatization technique involves the use of the chromous chloride reagent discussed in the chemical sections to monodechlorinate the methano-bridge; in the case of the chlordane isomers, however, both peaks are shifted to different retention times (R_T), but are not well resolved.[64]

The total organic chlorine content of technical heptachlor may be determined, if required, by methods such as are described above for chlordane. The heptachlor content of the material may be determined selectively, however, by taking advantage of the reactivity of the allylic chlorine atom at C_1. A sample of the technical material is dissolved in xylene and an aliquot of this solution is diluted with acetic acid; an excess of silver nitrate is then added and the mixture is boiled gently for 45 min. The effect of this procedure is to replace the allylic chlorine atom by an acetoxy-group, the liberated chloride ion being precipitated as silver chloride. When the reaction is complete, the mixture is diluted with water and the excess silver nitrate determined by titration with ammonium thiocyanate or potentiometrically with standard sodium chloride solution. Correction for any inorganic chloride originally present is made as for chlordane. An equivalent method involves heating the material with silver acetate in acetic acid, followed by gravimetric determination of the silver chloride produced. Formulations of various types may be analyzed for heptachlor content in the same way, with prior extraction with organic solvent if necessary, as in the case of dusts and powders.[115,291]

The colorimetric method used for chlordane (or the Polen-Silverman modification employing monoethanolamine, butyl cellosolve, and potassium hydroxide) is also applicable to heptachlor, either as technical material or as formulations or residues following suitable extraction. It may also be used for the determination of heptachlor epoxide in the presence of heptachlor, since the color due to the latter (pink-violet) has a single sharp absorption peak at 560 nm, while that due to the epoxide (yellow) shows absorption maxima at 415 nm, and 285 nm. Measurements of the epoxide can therefore be made at 415 nm without interference from heptachlor, but the compounds may also be separated by column chromatography before their determination in this way. In tests of the recovery of heptachlor residues from 500-g samples of 25 different crops, the sensitivity of the method ranged from a maximum of 0.002 ppm for string beans to a low of 0.02 ppm for alfalfa, while recoveries of material ranged from 72% (from alfalfa) to 98% (from sweet potatoes).[113,291]

Gas chromatographic analysis with flame

ionization detection has been used to determine the heptachlor content of formulations; the material is extracted with carbon disulfide (if liquid) or pentane (if solid), and the solutions chromatographed on Gas-Chrom Q[®] carrying 5% Versilube F-50 at concentrations of several $\mu g/\mu l$ using aldrin as internal standard. Heptachlor residues are regularly determined at the picogram level using GC-analysis with electron-capture detection. Estimation may be made directly or by converting heptachlor into derivatives such as 1-hydroxychlordene (Vol I, Chap. 3B.4b), heptachlor epoxide or chlordene. This last conversion is effected by chromous chloride and it must be noted that the dichloromethane-bridge of the resultant chlordene is also slowly attacked, which is useful for qualitative analysis but means that three products have to be accounted for if quantitation is desired. Using techniques of this sort, 0.01 ppm of heptachlor is measurable in a 10 g crop sample.[64]

2. Endosulfan and Isobenzan

a. Origins and Physical Properties

Endosulfan has its origins in the interest of Farbwerke-Hoechst AG in open chain analogues of chlordane, which led to the production of alodan (Figure 17, structure 10).[176] The corresponding bis-hydroxymethylene derivative (Figure 17, structure 3) is a potential precursor of alodan, and investigations on its synthesis showed that the direct condensation of cis-1,4-dihydroxybut-2-ene with "hex" is not easy; the reactants are not miscible in all proportions and undesirable polymers are produced in preference to the desired Diels-Alder adduct. When attempts are made to force the reaction at high temperatures (150 to 200°C), the cis-diol loses water to give 2,5-dihydrofuran, which then itself adds to "hex" to give the tetrahydrofuran derivative (Figure 17, structure 4) which is the precursor of isobenzan. The Hoechst investigators soon found that acetates of alcoholic dienophiles could be easily condensed with "hex" to give acetates of the required Diels-Alder adducts, which on hydrolysis afforded the desired alcohols; bis-hydroxymethylene derivative (Figure 17, structure 3) was therefore prepared by hydrolysis of the condensation product of "hex" and cis-1,4-diacetoxybut-2-ene. Attempts to convert this diol to alodan by methods normally used to chlorinate alcohols were unsuccessful, but thionyl chloride, which is commonly used for such chlorinations, reacted spontaneously with the diol under the normal chlorinating conditions, and the product, formed nearly quantitatively, proved to be the cyclic sulfite ester which is the mixture of isomers known as endosulfan.

Technical endosulfan forms cream to brown colored flakes having a terpene-like odor and consisting of 90 to 95% of an approximately 7:3 mixture of the α- and β-isomers (mp 108 to 110°C and 208 to 210°C, respectively), the remainder containing some of the diol and cyclic ether (Figure 17, structures 3 and 4, respectively). The melting point of this technical product has been reported as 70 to 100°C and 90 to 100°C; Maier-Bode reports 80 to 90°C. There is a tendency for some decomposition to the diol precursor and sulfur dioxide and the product may smell of this chemical; endosulfan is compatible with non-alkaline pesticides, however. The vapor pressure is 9×10^{-3} mm Hg at 80°C and the density is 1.745 at 20°C. It is practically insoluble in water but reasonably soluble in the common organic solvents. Typical solubilities (g/100 g of solvent at 20°C) are chloroform (50), xylene (45), benzene (37), acetone (33), carbon tetrachloride (29), kerosene (20), fuel oil (14), methanol (11), ethanol (5). Endosulfan is stable when stored under normal conditions, but is susceptible to slow hydrolysis by moisture, aqueous alkalis and acids. β-Endosulfan in the solid state exists in one modification with a highly symmetrical crystal form (I) and another with low symmetry (II). The first modification is usually formed during the crystallization of the β-isomer from light petroleum and the second when it is crystallized from methanol. For (I), the SO band in the infrared spectrum is at 1,192 cm^{-1} and for (II) it is at 1,180 cm^{-1}.[248]

The insecticidal properties of endosulfan were first described by Finkenbrink in 1956, and in that year it was introduced by Hoechst for practical use as "Hoe 2671" and carried the trademark Thiodan. Various formulations are available and some of them carry other trademarks such as Thimul[®], Thifor[®], Malix[®], and Cyclodan[®].[40] Thiodan is available as emulsifiable concentrates and wettable powders containing 17.5% and 35% of endosulfan (technical grade is referred to here), and there is also a 50% wettable powder. Thiodan dusts contain 1, 3, 4 and 5% of endosulfan, and 5% granules are also available.[248] As a cyclic sulfite ester which is relatively biodegradable in a number of environmental situations, endosulfan appears to have had a comparatively untroubled development. It is, however, rather toxic to fish and was indicted in connection with the well-publicized Rhine fish kill of 1969, but was later exonerated by an exhaustive official enquiry.[292,293]

Cyclization of the endosulfan precursor (Figure 17, structure 3) affords the isobenzan precursor (4) which was chlorinated about 1954 by scientists at Ruhrchemie AG, who were obviously seeking new chlorinated cyclic structures of the chordane type.[198] The photochemical chlorination described in the synthetic section (Vol. I, Chap 3B.2a) gives a high yield of isobenzan (Figure 17, structure 5) which is highly toxic to a number of insects, and also, unfortunately, to mammals (the acute oral LD50 to rats is 7mg/kg). The alternative route to the isobenzan precursor (Figure 17, structure 4) involves the Diels-Alder reaction of "hex" with 2,5-dihydrofuran, and the problems attending this reaction, especially when the "hex" contains lower chlorinated products (as when made by the Straus hypochlorite method), are fully discussed in a Ruhrchemie patent of 1957.[294] It was found that a number of these problems could be overcome by dissolving the 2,5-dihydrofuran with an excess of "hex" in C_{12}-C_{18} hydrocarbons boiling between 200 and 300°C and passing the mixture through a pipe system heated to 100 to 180°C. Under these conditions, the partial pressure of the cyclic ether is lowered sufficiently to keep it in solution so that the reaction proceeds smoothly and the high melting product (mp 236°C) does not separate until the mixture is cooled. This method permits the use of the cruder Straus "hex" as well as purified "hex." The chlorination step presents no problems once the precursor is available. Technical isobenzan is reported as having melting point 120 to 125°C, specific gravity 1.87 and a vapor pressure of 3.0 x 10^{-6} mm Hg at 20°C and 2.8 x 10^{-4} mm at 55°C. Like other cyclodienes, it is practically insoluble in water but moderately soluble in the common organic solvents. Typical solubilities (g/100 ml of solvent at 25°C) are benzene (38), toluene (34), carbon tetrachloride (34), mixed xylenes (29), acetone (25), and diesel oil (8).[295] Isobenzan showed promise as another broad-spectrum cyclodiene insecticide and was developed by Shell for this purpose and manufactured between 1958 and 1965 alongside dieldrin and endrin at Pernis in Holland. However, a number of difficulties arose, mainly in relation to the high mammalian toxicity, and commercial production ceased in September 1965, although handling and formulation of isobenzan continued for some time afterwards. It showed considerable promise as a soil insecticide and was also used against certain disease vectors; tsetse flies for example. Since it is no longer used commercially, it has been omitted from consideration in the two following sections.

b. Applications of Endosulfan

Endosulfan is a broad-spectrum, nonsystemic contact and stomach insecticide which has found wide application in market gardening, agriculture, and forestry. It is especially effective against butterflies and moths, beetles, and various hoppers (leaf hoppers, tree hoppers, etc.). Grasshoppers and crickets (Orthoptera), termites (Isoptera), thrips (Thysanoptera), various bugs (Heteroptera), ants, wasps (Hymenoptera), etc., but apparently not the honeybee and various flies (Diptera) are also well controlled, as well as some myriapods and mites. In spite of its broad spectrum of activity, it appears to be well tolerated by a number of beneficial insect species when applied at the recommended rates. Thus, a number of reports indicate that, in normal applications, it is not particularly dangerous to honeybees, nor to a large number of pest parasites and predators belonging to the hoverflies (Syrphids), parasitic Hymenoptera (Chalcidids), parasitic beetles (Carabids), ladybirds (Coccinellids), and spiders (Arachnids).[296]

No serious instances of phytotoxicity appear to have been observed at normal rates of application, although discoloration of the blossoms of some decorative plants has been observed with high humidities under glass, and interference with pollen germination occurred when apple blossom was treated with the insecticide, although this apparently did not affect the development of the fruit. An overdosage of endosulfan has also been known to cause bleaching and burns on the foliage of greenhouse cucumbers treated with it. The emulsifiable concentrate and wettable powder formulations containing 35% of endosulfan are suitable for both high and low volume type applications. In the high volume spraying process, the preparations are diluted to give a concentration of 0.15% of the formulation in water with an appropriate increase in concentration for the low volume applications. For agricultural applications 0.9 to 2.5 lb per acre of the 35% formulations are required, depending on the crop treated and the pest to be controlled. Thiodan dust applied to agricultural crops from the ground is generally used at rates of 13 to 22 lb per acre, but up to 27 lb per acre may be required when the treatment is applied from the air. Endosulfan is used for the control of the various pests listed above on cereals (rice, maize), legumes (ground nuts, lucerne), root crops (potatoes, carrots), oil plants (rape-seed, soyabeans, sunflowers, mustard, etc.), fruit and vegetable crops, tobacco, tea, coffee, cocoa, industrial crops (such as cane sugar, cotton, jute and trees providing timber), glasshouse crops, and ornamental plants. In the public health area it has also found application in tsetse fly control. In some situations, for example against pests on cotton, endosulfan is combined with other insecticides. Thus a combination with DDT is used to control pink bollworm, *Pectinophora gossypiella* (0.3 to 0.8 lb per acre of endosulfan plus 0.75 to 1.3 lb per acre of DDT), and when the cotton aphid, *Aphis gossypii*, cotton stainers, *Dysdercus* spp., and mites, *Tetranychus* spp., are present, formulation with methyl parathion is advantageous (0.3 to 0.8 lb per acre of endosulfan plus 0.2 to 0.5 lb per acre of methyl parathion).[296]

Endosulfan is evidently appreciably volatile under field conditions, since it disappears relatively rapidly from plant surfaces. It was found to be less persistent than toxaphene, DDT, or aldrin when applied to grazing grass in meadows. The fact that surfaces coated with it in greenhouses can kill insects seven feet away is testimony to its volatility.[248] Because of its high toxicity to fish, endosulfan has been suggested as a control agent for unwanted species and has been evaluated for this purpose.[297] However, this particular property should be kept in proper perspective since most aquatic invertebrates are less sensitive than fish; in fact, the cyclic sulfite ester structure renders the compound liable to hydrolytic detoxication in water and under environmental conditions so that it is considerably less persistent in toxicant form than some other compounds derived from hexachlorocyclopentadiene.[298,299] The general properties of endosulfan are likely to make it of increasing interest in the future.

c. Analysis of Endosulfan

The methods used for analysis of endosulfan depend on whether or not separate determination of the two isomers is required.[248] For both isomers together, total chloride methods such as the sodium reduction technique and oxygen flask combustion have been used. The mixed isomers may also be determined in total by hydrolysis to the diol precursor (endosulfan-diol) and sodium sulfite in boiling methanolic sodium hydroxide, followed by iodometric titration of the sulfite using starch as indicator. This is a useful routine method which is convenient for the estimation of endosulfan in its various formulations. By its use the extent of endosulfan decomposition can be determined with a reproducibility of ±0.3%. Colorimetric methods are also available for the detection of both isomers. Thus, the sulfur dioxide produced when these chemicals are heated under alkaline conditions gives a red color with *p*-rosaniline and formaldehyde and the absorbance is read at 570 nm. Another colorimetric method involves the treatment of endosulfan with pyridine and methanolic potassium hydroxide and has been applied in residue analysis to the determination of both isomers in hexane extracts without cleanup. Of 45 other pesticides tested, only chlordane, heptachlor, and captan interfered; endosulfan-diol can also be determined in this way.

All of the above methods can be applied to the individual isomers following their separation by other methods. In the method described in the CIPAC Handbook, the technical material extracted by a suitable method from a formulation is separated into its components by column chromatography on acid alumina deactivated with water to activity III, which does not affect the α- and β-isomers. The mixture is applied to the column with carbon tetrachloride, and the α-isomer is eluted with this solvent; a subsequent change to benzene then elutes the β-isomer.[115] In another method, liquid-liquid column partition chromatography with nitromethane on silica gel as the stationary phase has been used to separate the isomers prior to their individual determination by polarography in aqueous acetone. For the separated isomers, the infrared technique is the best method of estimation. The α-isomer is determined in carbon disulfide from the absorbance peaks at 8.4 μ and 10.2 μ (using baseline points at 8.2 and 8.9 μ; 9.6 and 10.8 μ, respectively), while the β-isomer has similar absorbance maxima at 8.4 and 10.4 μ.[115]

Gas-liquid chromatography is a highly convenient method that has been used for the rapid simultaneous separation and determination of endosulfan isomers in the technical material, in formulations and in residue analysis. A suitable method has been described by Zweig and Archer,[300] who used a 1.7-m stainless column packed with 30% silicone grease on 35 to 80 mesh Chromosorb at 250°C for the separation. Preparation of the column packing evidently requires some care if decomposition of the isomers is to be avoided; washing of Chromosorb W with hot hydrochloric acid before coating with the stationary phase has been recommended, and besides the various Dow Corning® greases, the fluorosilicone rubber QF-1 or mixtures have been used for this purpose. Thus, Burke and Holswade used 6 ft columns packed with a mixture (1:1) of 15% QF-1 on 80 to 100 mesh Gas-Chrom Q and 10% DC-200 on the same support, the column temperature being 200°C.[149] This system, which used electron-capture detection, gave retention times, relative to aldrin, of 1.89 and 2.92, respectively (Table 11), for α- and β-endosulfan. With electron-capture detection, the sensitivity limit for detection is 50 to 100 pg of endosulfan. Microcoulometric detection is also used, and, since cells are available which specifically detect chlorine or sulfur in compounds being analyzed, the simultaneous use of both types indicates those endosulfan derivatives (in residues, for example) which retain the sulfur atom. The detection limit for endosulfan with this detector is 5 to 10 ng. Procedures for the extraction and cleanup of residues of endosulfan and its derivatives in agricultural produce, etc. are discussed in the review by Maier-Bode.[248] Degradation products such as endosulfan-diol are best derivatized to facilitate detection by electron-capture gas chromatography, acetylation or conversion into the bis-trimethylsilyl derivative being most suitable for this compound. Lower limits of detection from 0.02 to 0.03 ppm (based on a 10 g crop sample) are possible in residue analysis using one or the other of these derivatives.[64]

3. Aldrin, Dieldrin, Isodrin, and Endrin
a. Origins and Physical Properties

Work on the Diels-Alder reactions of hexachlorocyclopentadiene was continued in Denver following the setting up of the Hyman Company there in 1947. As indicated in the outline of the Diels-Alder reaction given previously (Vol. I, Chap. 3A.1) dienes will add to acetylenic dienophiles to give adducts in which the formerly acetylenic moiety gives rise to an additional double bond in the adduct. Although such Diels-Alder reactions were already known to occur with acyclic dienes to give the corresponding cyclic systems, it seemed unlikely that such an addition could be accomplished with a cyclic diene such as cyclopentadiene and an acetylenic dienophile, since the product must have a highly strained ring system. Thus, the product of addition of cyclopentadiene to acetylene would be cyclohexa-1,4-diene in which C_3 and C_6 are linked by a methano-bridge. The fact that some derivatives of this structure were already known, but appeared to be intrinsically unstable, lent support to the idea that they could only be isolated because the substituents gave stability to the unstable parent molecule. Hence, the discovery that cyclopentadiene does in fact react with acetylene to give bicyclo[2.2.1]hepta-2,5-diene (norbornadiene), which could actually be isolated as a stable product, represented a considerable advance in the chemistry of bridged polycyclic systems. The first synthesis of norbornadiene was actually achieved by Hyman in the experimental engineering laboratory at Denver, due to the special conditions required for the reaction; acetylene is well known to be potentially explosive when subjected simultaneously to high temperatures and pressures and so a continuous flow system was developed to effect the reaction (Vol. I, Chap. 3B.2b). Once the synthesis of norbornadiene had been achieved, its use as a dienophile with hexachlorocyclopentadiene was logical and simple, the product being the insecticide aldrin (HHDN). Aldrin was first synthesized early in 1948, and bioassays conducted by Dr. Y. P. Sun in the Hyman laboratories quickly showed it to be a highly effective insecticide. The carbon skeleton of aldrin differs from that of chlordene, the chlordane precursor, by only two carbon atoms, so that it is not surprising that aldrin has an appreciable vapor pressure (about 100-fold greater than that of DDT). As in the case of chlordene, therefore, ways were sought of making the molecule less volatile, and therefore more persistent, without sacrificing its insecticidal properties. By analogy with the chlordene-chlordane situation, chlorination, or bromination of the unsubstituted double bond in aldrin are obvious, but such products turned out to have little toxicity. However, a consideration of the available reactions of double bonds soon led Dr. S. B. Soloway, who was investigating this particular problem, to the epoxidation reaction (Vol. I, Chap. 3B.2b) and hence to the insecticide dieldrin (HEOD).[266]

Up to this time, all of the chlorinated insecticides devised had the property that under alkaline conditions they lost chlorine by a dehydrochlorination reaction, and in all cases the products had little insecticidal activity. Indeed, it was this marked association between insecticidal activity and ability to dehydrochlorinate quite readily that led Martin and Wain to propose dehydrochlorination as a requisite for insecticidal activity.[261] In the primary adducts derived from "hex," however, the chlorine atoms are completely resistant to elimination by bases. In addition, it was soon found that the epoxide ring in dieldrin is also remarkably stable; in fact, dieldrin and certain other epoxides of the cyclodiene series are among the limited number of epoxides known to possess such high stability, those derived from alicyclic and particularly aromatic systems being usually rather labile in contrast. According to some listings of physical data, the epoxidation reaction has the effect of reducing the vapor pressure of aldrin (2.31×10^{-5} mm Hg at 20°C) more than 100-fold (vapor pressure of dieldrin 1.78×10^{-7} mm Hg at 20°C). However, the vapor pressures measured by the effusion-manometer method are 7.5×10^{-5} and 3.1×10^{-6} mm Hg, respectively, at 20°C, so that according to this method the difference is 24-fold.[301] At any rate, the vapor pressure of aldrin is of the same order as that of chlordane, and like the latter compound, its value as a soil insecticide soon became evident. The lower vapor pressure of dieldrin on the other hand gives it greater persistence, a property which, when combined with considerable chemical stability, provides a good basis for a foliage insecticide.

Several variants of the process leading to the formation of aldrin, as well as the preparation of its 6,7-diethoxycarbonyl derivative, are described in U.S. patent 2,635,977 by Rex. E. Lidov (to the Shell Development Company; patented in April 1953), the original application having been filed in August 1948. Another application by the same author, dated December 1947, appeared at the same time (April 1953) in the form of U.S. patent 2,635,979. This patent dealt with the preparation of 6,7-dihydroaldrin and several of its derivatives,

including a 6-hydroxy-6,7-dihydroaldrin, a 6-chloro-6,7-dihydroaldrin, and 6,6-dichloro-6,7-dihydroaldrin, and was also assigned to Shell. Two other applications filed in August and December 1948 by Velsicol (as assignees of R. E. Lidov and S. B. Soloway) were published in June 1953 (British patents 692,546 and 692,547, respectively).[211],[212] The first one describes the synthesis of aldrin and its 6,7-diethoxycarbonyl derivative, and the reactions of the unchlorinated double bond with halogens, while the second details the synthesis of dieldrin and related epoxides by peracid oxidation of aldrin and its 6-halogen derivatives. Another patent (U.S. 2,676,131) of April 1954 (to Shell Development Co.) also relates to Soloway's 1948 discovery of dieldrin.

The background to these documents is the legal dispute regarding the cyclodiene inventions that surrounded the setting up of the independent Julius Hyman Company in Denver. Judgment was finally given in favor of Velsicol in relation to chlordane, the earliest cyclodiene insecticide, and relatives, but aldrin and dieldrin, developed at Denver, were acquired by Shell Chemical Corporation along with the Hyman Company, which then became its Julius Hyman Division. Aldrin was introduced by the Hyman Company in 1948 as "Compound 118" under the trade mark Octalene® and some of its insecticidal properties and those of dieldrin (introduced in the same year as "Compound 497" or "Octalox") are described in the patents cited above. Using the housefly as a test insect, aldrin is described as being somewhat less toxic than heptachlor and 2.5-fold more toxic than technical chlordane; dieldrin is sixfold more toxic than chlordane, 2.5-fold more toxic than heptachlor and about threefold more toxic than aldrin. The toxicities of dieldrin, aldrin, and chlordane to the German cockroach were indicated to be in the ratio 30:6:1, and the comparisons with other insecticides (using houseflies) clearly show the superiority of the cyclodienes over DDT in regard to their acute toxicities. The first comprehensive account of the evaluation of these compounds as insecticides was actually given by Kearns, Weinman, and Decker in 1949.[302]

Aldrin and dieldrin are still manufactured in Denver and also at Pernis, in Holland, where aldrin production began in November 1954 and dieldrin production in June 1955. The operations at Pernis include the production of intermediates and formulation of the technical materials into commercial products. Facilities for endrin production were added to the Pernis complex in 1957, and isobenzan was made in a pilot plant associated with the aldrin-dieldrin unit from 1958 until 1961, after which a separate plant produced this chemical until its manufacture ceased in 1965. All the products were initially made in batches, but continuous processes were later developed for aldrin and isobenzan. The epoxides, dieldrin and endrin, are still made in batches. An interesting account of insecticide manufacture at Pernis (from a toxicological standpoint) has been given by Jaeger.[303]

Aldrin is the common name approved by ISO (International Organization for Standardization), except in Canada, Denmark, and the USSR, and by BSI (British Standards Institution) for a material containing not less than 95% wt (by weight) of the chemical 1,2,3,4,10,10-hexachloro-1,4,4a,5,8,8a-hexahydro-1,4-*endo*, *exo*-5,8-dimethanonaphthalene,* which is abbreviated "HHDN" for convenience. Technical aldrin, as manufactured, contains not less than 90% wt of aldrin, and is a friable light tan to brown solid with a setting range of 49 to 60° and a mild chemical odor. The specific gravity is given as 1.6 at 20°C and the vapor pressure as 6×10^{-6} mm Hg at 25°C. The chlorine content is 57 to 59% wt, the material insoluble in heptane less than 0.8% wt and the water and free acid contents less than 0.1% wt.[116,295] The composition of a typical production batch of technical aldrin is shown in Table 15. The impurities present are the unchanged diene and dienophile ("hex" and bicycloheptadiene), the product of addition of a second molecule of "hex" to aldrin (HHDN diadduct), isomers of aldrin (including a little isodrin), and intermediates arising from the manufacture of "hex," which nowadays appears to be made via octachlorocyclopentene by the chlorination of pentanes available from straight-run petroleum distillate (Vol. I, Chap. 3B.1). The technical material is virtually insoluble in water and soluble in methanol to the extent of about 0.5 lb/gal; most paraffinic and aromatic solvents will dissolve at least 2.0 lb of technical aldrin/gal (American gallons used here). The approximate solubility of aldrin (95% HHDN) in some representative solvents is given in Table 16.[295] HHDN is a white, crystalline solid melting at 104 to 104.5° C, for which vapor pressures of 2.31×10^{-5}, 4.87×10^{-5}, 7.48×10^{-5}, and 1.03×10^{-4} mm Hg at 20, 30, 40, and 50°C, respectively, have been given. The solubility in water between 25 and 29°C is stated to be 0.27 ppm[304] and in common solvents it is (g/100 ml at 30°C); acetone (159), benzene (350), hexane (98), and methanol (9).[113] When considering the recorded physical constants, it is necessary to note rather carefully the name (and hence the nature) of the material for which the property is given, as well as the method of measurement, as is indicated by the variety of vapor pressures given for aldrin in the literature. The crystal structure of HHDN has been described by Delacy and Kennard.[304a]

Technical aldrin and HHDN are stable in storage, being unaffected by heat, alkali, dilute acids, emulsifiers, wetting agents, and solvents. However, oxidizing agents, concentrated mineral acids, and various acidic catalysts can cause decomposition; the deterioration which sometimes occurs, for example, when aldrin is mixed with certain carriers used in formulation is attributed to acid catalyzed decomposition and can be prevented by treating the carriers with deactivators such as urea, hexamethylenetetramine, and some other alkaline materials. Aldrin is compatible with most common insecticides, acaricides, herbicides, fungicides, soil fumigants, and fertilizers, and can therefore be freely used with them, with the reservation that different formulations cannot always be freely mixed on account of the possibility that there may be undesirable interactions between the nonbioactive components of the mixture.[116]

Dieldrin is the ISO common name, except in Canada, Denmark, and the USSR, and the BSI common name for a material containing not less than 85% wt of the chemical 1,2,3,4,10,10-hexachloro-6,7-epoxy-1,4,4a,5,6,7,8,8a-octahydro-1,4-*endo*, *exo*-5,8-dimethanonaphthalene,"* abbreviated to "HEOD."[116,295] Technical dieldrin, as manufactured, contains not less than

---

*In accordance with the practice adopted in this book, the systematic names given here are based on the American system discussed earlier in this chapter (sections A.1 and 2). British Standards Institution publications list the British name, followed by this American name.

**American name; see earlier footnote regarding systematic names.

TABLE 15

Composition of Typical Production Batches of Technical Aldrin, Technical Dieldrin, and Technical Endrin (for definitions see text)

Technical Aldrin

Component	% by weight
HHDN	85
Other polychlorohexahydro-dimethanonaphthalenes	3
HHDN diadduct	3
Bicycloheptadiene	1
Hexachlorocyclopentadiene	1
Hexachlorobutadiene	3
Hexachloroethane	2
Octachlorocyclopentene	2

Technical Dieldrin

HEOD	83
Other polychloroepoxyoctahydro-dimethanonaphthalenes	2
Hexachloroethane	1
Octachlorocyclopentene	2
Carbonyl compounds	4
Benzene	0.4
Acetic acid	0.3
Balance	7.3

Technical Endrin

Endrin	96.6
HEOD	0.42
HHDN	0.03
Isodrin	0.79
1,2,3,4,7,7-hexachloro-1,4-dihydro-1,4-methanobenzene (heptachloronorbornadiene)	0.03
1,2,3,4,5,7,7-heptachloro-1,4,5,6-tetrahydro-1,4-methanobenzene (heptachloronorbornene)	0.08
1,2,3,4,5-pentachloro-7-oxo-1,4,5,6-tetrahydro-1,4-methanobenzene	0.09
Endrin half-cage ketone (Δ-keto-endrin; $C_{12}H_8Cl_6O$)	1.57
Endrin aldehyde ($C_{12}H_8Cl_6O$)	<0.05
Acidity (calculated as HCl)	0.18
Unidentified components	0.12

Data by courtesy of the Shell International Chemical Company

95% wt of dieldrin; it consists of buff to light tan flakes with setting point not lower than 95°C and a mild chemical odor. The specific gravity is 1.7 at 20°C and the vapor pressure is said to be 1.8×10^{-7} mm Hg at 29°C. The chlorine content is 55 to 56% wt, the material insoluble in xylene less than 0.5% wt, the water content less than 0.1% wt and the free acid (calculated as acetic acid) less than 0.4% wt. Table 15 shows the composition of a typical production batch of technical dieldrin. As expected, some of the impurities carried through from aldrin manufacture are present, together with traces of solvent, isomeric epoxy compounds (a little endrin for example), carbonyl compounds probably produced by rearrangement during the epoxidation of aldrin, and unidentified materials. The technical product is insoluble in water, sparingly soluble in methanol, poorly soluble in aliphatic hydrocarbons (about 0.3 lb/gal), and soluble in aromatics such as xylene to the extent of about 3 lb/gal. Table 16 gives the solubility of dieldrin (85% HEOD) in a selection of solvents. HEOD is a white, crystalline solid, mp 175 to 176°C, specific gravity 1.7 at 20°C, and vapor pressure given above. The solubility in water* at 25 to 29°C is said to be 0.186 ppm,[304] and the chemical is also sparingly soluble in aliphatic hydrocarbons, but more soluble in acetone and quite soluble in aromatic solvents. The stability characteristics of dieldrin (and HEOD) are similar to those of aldrin in many respects, deactivators being again required for some of the solid formulations; like aldrin it is compatible with a wide range of other pesticides and agricultural chemicals.[116]

The synthesis and epoxidation of isodrin, the endrin precursor, were discussed (Vol. I, Chap. 3B.2b) under general synthesis of cyclodienes and are illustrated in Figures 13 and 21. For the formation of isodrin, the synthesis of aldrin is reversed in the sense that the diene (cyclopentadiene) is unchlorinated and the dienophile is the chlorinated version of bicycloheptadiene. This synthesis represents a logical development from that of aldrin, and following the success of the aldrin synthesis and the demonstration of aldrin's insecticidal properties, considerable interest must have accompanied the first attempts to effect this new Diels-Alder reaction in the Hyman labora-

---

*The water solubility of compounds in this series is not easy to define and careful note should be taken of the methods used to arrive at any particular published measurement. For most practical purposes the compounds are insoluble in water.

TABLE 16

Approximate Solubilities of Aldrin (95% HHDN), Dieldrin (85% HEOD) and Endrin in Representative

Solvents at 25°C[a]

Solvent	Aldrin		Dieldrin		Endrin	
	Saturated solution (% by weight)	g/100 ml solvent	Saturated solution (% by weight)	g/100 ml solvent	Saturated solution (% by weight)	g/100 ml solvent
Acetone	58	109	25	26	28	31
Amyl acetate	31	38	28	32	24	28
Benzene	67	183	39	56	37	51
n-Butyl alcohol	11	9	7	5	7	–
Carbon tetrachloride	66	303	24	48	24	51
Cyclohexanone	–	–	–	–	44	–
Deobase	18	18	9	4	–	–
Diesel oil	26	30	17	17	11	10
Dipentene	57	109	21	22	–	–
Ethanol	7	5	5	4	4	3
Ethylene dichloride	72	312	36	70	41	87
Fuel oil	26	30	17	17	11	10
Isopropanol	5	4	4	2	4	3
Kerosene	26	28	6	5	6	5
Methanol	6	5	1	1	3	2
Methyl ethyl ketone	26	28	33	39	33	40
Toluene	75	267	39	54	46	74
Turpentine	60	127	17	17	19	21
Xylene	73	235	37	52	39	55

[a]Adapted from data of Shell Chemical Corporation[116]

tories. Since the reaction involves the addition of cyclopentadiene to the unchlorinated double bond of a heavily chlorinated dienophile, at first sight there exists the possibility that the reaction might be hindered or slowed by adverse steric interactions between the chlorine atoms and the approaching cyclopentadiene molecule, which (from the now known *endo, endo-* stereochemistry of isodrin) must approach in the transition state so that the conjugated system is directed towards the vinylic chlorine atoms (Figure 13). In fact, the condensation is mildly exothermic and proceeds quite smoothly to give isodrin, mp 240 to 242°C. Endrin was prepared by the epoxidation of isodrin with peracetic acid and the total synthesis from "hex" is described in U.S. patent 2,676,132 by Henry Bluestone (to Shell Development Company) of the Julius Hyman Company; the application was made in April 1950, and the invention patented on April 20, 1954.[190] Experimental details are given for the Diels-Alder reaction between "hex" and vinyl chloride and the sub-

sequent dehydrochlorination of the product (Figure 13, structure 1) to the dienophile hexachloronorbornadiene (Figure 13, structure 2) which is then reacted with cyclopentadiene. A detailed discussion of the stereochemical possibilities and nomenclature is given in this document, which describes aldrin and its relatives as the α-series, isodrin and relatives being the β-series. The configurations were not known at that time; in fact, Bluestone suggested that if the α-compounds had the *endo, exo-*configuration (as opposed to the *exo, exo* one) as is now known to be the case, the β-series would be *exo, endo*, which turns out to be incorrect, this series being *endo, endo*. A more recent process for the synthesis of the key intermediate hexachloronorbornadiene involves the direct Diels-Alder reaction between "hex" and acetylene, which can be effected in good yield in a tubular nickel alloy reactor at 155°C and 300 lb/in^2. The Bluestone patent also mentions a β-series of compounds derived from hexabromocyclopentadiene.

Endrin was introduced by the Hyman Company in 1951 as "Experimental Insecticide 269," protected by the patent just discussed, which also describes some of its insecticidal properties. The toxicity to houseflies is stated to be 3.1-fold greater than that of chlordane and a comment is made regarding the much greater toxicity of endrin, as compared with dieldrin, to the Mexican bean beetle (*Epilachna varivestis*) and to various aphids. Also mentioned is its use as a rodenticide in some situations, an application which still continues. One result of the legal settlements in relation to cyclodienes was that the Shell and the Velsicol Chemical Corporations shared the right to manufacture and market endrin, and both do so at the present time, although probably a greater portion of the chemical is distributed by Shell. Endrin is the common name approved by ISO (except in India and South Africa where it is called mendrin) and by BSI for the chemical 1,2,3,4,10,10-hexachloro-6,7-epoxy-1,4,4a,5,6,7,8,8a-octahydro-1,4-*endo, endo*-5,8-dimethanonaphthalene.* Technical endrin, as manufactured, contains not less than 92% wt of endrin, and is an off-white to light tan solid which melts with decomposition (rearrangement) when heated above 200°, and the properties of the purified compound are similar. Its vapor pressure is 2×10^{-7} mm Hg at 25° and the specific gravity at 20° is 1.7. The chlorine content is 55 to 57% wt, residue insoluble in xylene less than 0.5% wt, water content less than 0.1% wt, and free acid (calculated as acetic acid) less than 0.4% wt. It is described as being insoluble in water, sparingly soluble in methanol, of low solubility in aliphatic hydrocarbons (about 0.3 lb/gal), and quite soluble in aromatic solvents (about 3 lb/gal); the solubilities of endrin in some representative solvents are given in Table 16. The technical compound and purified endrin resemble dieldrin in being stable in the presence of basic reagents, alkaline oxidizing agents, wetting agents, and emulsifiers, but the *endo, endo*-configuration is vulnerable to molecular rearrangements catalyzed by heat or acids (Vol. I, Chap. 3B.4c). Thus, endrin is more susceptible than dieldrin to decomposition by some carriers used in formulation and these must be deactivated by addition of hexamethylenetetramine.[116,295]

The half-cage ketone and the aldehyde pro-

duced by rearrangement are inactive as insecticides, and a series of patents by Bellin and colleagues describes the use of inorganic and organic nitrites (0.05 to 15% wt),[216] various simple epoxides such as epichlorohydrin and glycidyl ethers (0.5 to 5% wt) and various alkaline reacting substances such as alkali metal carbonates, urea, etc. (up to 25% wt) to remove traces of acid that are apparently responsible for the sometimes spontaneous exothermic conversion (Vol. I, Chap. 3B.4e). Traces of iron are also highly effective decomposition catalysts, so that chelating agents (0.05 to 15% wt) such as hexamethylenetetramine are good inhibitors of the reaction due to this cause. Endrin is unaffected by admixture with a wide range of pesticides and agricultural chemicals that are chemically neutral or alkaline in reaction. However, when endrin is combined with other insecticides such as organophosphorus compounds, care must be exercised to ensure that any deactivators for endrin decomposition that are present do not decompose the other toxicant, apart from any normal precautions to ensure the compatability of different formulations. Technical endrin is noncorrosive to most metals which are likely to come into contact with it during formulation processes and it does not attack glass or a number of common plastics.[116] The composition of a typical production batch is shown in Table 15, which shows that (as in the case of aldrin synthesis) the Diels-Alder reaction leading to isodrin must be virtually stereospecific, since only traces of dieldrin appear in the endrin finally produced. The other trace components of technical endrin are evidently the rearrangement products formed by its decomposition, together with small amounts of the precursors of isodrin and their derivatives. Isodrin itself has never been used for practical pest control.

b. Formulations and Early Evaluation

The same principles are applicable to the formulation of aldrin, dieldrin, and endrin for practical use as are employed for other organochlorine insecticides, and they are described in detail in the *Handbook of Aldrin, Dieldrin and Endrin Formulations* published by Shell Chemical Corporation.[116] Among the aldrin formulations available are emulsifiable concentrates containing 24, 40, and 48% of aldrin (240, 400, and 480 g/l),

*American name; see earlier footnote regarding systematic names.

wettable powders containing 40, 50, and 70% (400, 500, and 700 g/kg), dust concentrates containing 40 and 75%, 2.5% dust, and 20% granules. Emulsifiable concentrates are generally prepared with aromatic solvents to provide good low temperature storage stability. Kerosene and various oils can be used alone for dilute oil sprays or with at least 50% of aromatic solvent to make concentrates containing smaller than usual amounts of aldrin (for example, 2 lb/gal). Emulsifiable concentrates are normally diluted to such an extent before use that the contained aromatic hydrocarbons are not phytotoxic; ketones such as mesityl oxide and cyclohexanone are less phytotoxic than aromatic hydrocarbons and the solutions are more cold stable, so that these compounds are preferred solvents for some applications such as in liquid seed dressings containing 20 to 30% aldrin. An important consideration is the flash point of the solvent used, which must be as high as possible to minimize fire hazard and meet with transportation regulations. A comprehensive list of emulsifying agents is given in the formulation handbook. Many of them tend to form complexes with mineral acids or with iron salts liberated by these acids; such phenomena cause deterioration of emulsions which therefore normally contain 0.3 to 0.5% of epichlorohydrin or propylene oxide to remove trace acids by reacting with them to form neutral derivatives (for example, chlorohydrins). In household use, as for termite control, 0.1 to 0.5% of odor-masking agent is often added to the formulation. General purpose aldrin emulsifiable concentrates give satisfactory emulsions with dilute fertilizer solutions, but special concentrates are formulated for admixture with undiluted liquid fertilizers, which are nearly saturated solutions of plant nutrients such as ammonia, phosphoric acid, and urea; the emulsifers have to be adjusted to be compatible with this situation, and details of suitable formulations containing 4 lb/gal (ca 45% wt) of aldrin have been given.

Solid formulations of pesticides provide ease of handling, reduced hazard to vegetation, good coverage combined with insecticidal efficiency, and generally reduced toxic hazard to warm-blooded animals. To make wettable powders, dust concentrates, and seed dressings, technical aldrin is mixed and ground with a suitable carrier, deactivated if necessary with 1 to 3% of urea, 0.2 to 5% of wetting and/or dispersing agent being incorporated in the case of wettable powders and seed dressings. Another method is to impregnate sorptive carriers with technical aldrin, using solutions of the toxicant. Carriers may be chosen from the usual inorganic range or from organic materials derived from soybeans, walnut shells, tobacco, and wood. On the basis of liquid sorption, they are divided into high-sorptive materials (for example, diatomaceous earths, synthetic silicates, and certain clays) and low sorptive ones (pyrophyllites, kaolinites, talcs, pumices, sulfur, gypsum, dolomites, etc.), and the usual practice is to prepare a dust concentrate on high-sorptive material for transport and storage, and to dilute it appropriately with a low sorptive carrier nearer to the site of field applications. Wettable powders are normally diluted with water at the time of application, while seed dressings are either applied in the dry condition, as made, or made into an aqueous slurry. Granules are prepared by impregnating bentonites, vermiculites, or attapulgites, usually of 20 to 60 mesh particle size, with low or moderate concentrations of the toxicant. Solid fertilizer mixtures containing aldrin are made by impregnating the solids with solutions of the technical material, or by mixing and blending aldrin granules or dust concentrates with the fertilizer. Aldrin-fertilizer mixtures formulated from a number of commercial fertilizers are stable at ambient temperatures for at least 18 months and for a shorter time at higher temperatures. Examples are activated sewage sludge, ammonium phosphate, ammonium sulfate, superphosphate, treble superphosphate, and diammonium phosphate.

Many formulations of dieldrin are available and it may also be applied to crops as a spray, as a dust or in granule form, the solid carriers being deactivated with urea, when necessary. Typical formulations recommended for use include 18 and 20% emulsifiable concentrates (180 and 200 g/l), 50 and 75% wettable powders (500 and 750 g/kg), a 50%-wt dust concentrate; 1, 2, and 10% field strength dusts, and 2 and 5% granules. Solutions (150 to 200 g/l) are also used and liquid seed dressings containing 30% wt of dieldrin have been described. Since dieldrin is more polar than aldrin, it is generally less soluble in hydrocarbon solvents and those used must have a high aromatic content to take up to 2 lb of dieldrin/gal (ca 20% wt); apart from this consideration, much of what has been said about aldrin formulation applies also to dieldrin. Acid production in dieldrin emulsifiable concentrates occurs either very slowly or is absent, so that the addition of epoxide inhibitors is not normally necessary unless the solvents used contain mineral acids. As in the case of aldrin, special emulsifiable concentrate formulations are required for admixture with undiluted liquid fertilizers and a suitable one containing 1.5 lb/gal of dieldrin has been described. Solutions containing 0.5% technical dieldrin in deodorized kerosene or isoparaffinic solvents have been used as sprays for the control of household insects, either as simple solutions or as aerosol sprays; a typical formulation for the last purpose contains technical dieldrin (0.5%), deodorized kerosene (69.5%) and propellant (30%), often with a knock-down agent and a scented masking agent incorporated.

In liquid seed dressing applications the phytotoxicity factor is critical since solutions of toxicant are applied to seed undiluted. For this purpose, dieldrin and aldrin are ideal, since their solutions in ketonic solvents such as cyclohexanone or mesityl oxide have no effect on seed germination, even after prolonged storage of treated seed. These solutions are compatible with fungicides used for seed treatment, contain sufficient toxicant to maintain efficiency in the presence of a fungicide dressing, adhere well to the seed coat and are stable when stored at low temperatures. As a high melting solid, dieldrin is easy to grind with many carriers of the types mentioned in connection with aldrin, and the same principles apply, including the requirement for deactivation of certain materials with urea.

For many agricultural applications, suspensions of pesticides are made from wettable powders and water and the application equipment incorporates mechanical stirring so that the stability of the suspensions is not very critical. If mechanical stirring is absent, as with knapsack spraying for mosquito control, the stability of the suspension becomes critical, since settling out can result in blocked spray nozzles as well as poor distribution of toxicant over the treated surface. The wettable powders used for this purpose in the tropics have been found to require suspendibility qualities such that 80% of the toxicant remains suspended after 30 min without agitation, if satisfactory performance is to be obtained following their prolonged storage. With dieldrin wettable powders made for this purpose (containing 50 or 75% dieldrin) an average particle size of 1 to 3 μ is necessary and

requires a correspondingly refined milling process. As in the case of aldrin, dieldrin dust concentrates are prepared for dilution with appropriate carriers before field application.

Seed dressing formulations containing up to 75% technical dieldrin have been made using carriers such as deactivated attaclay, and may be used dry or as a slurry for seed treatment, a wetting agent being incorporated when slurries are to be used. Granules are formulated by spraying sorptive granular materials such as attapulgite, bentonite, or expanded vermiculite with solutions of technical dieldrin. Critical factors are the sorptivity of the granules, which limits their dieldrin uptake and the solubility of dieldrin in the solvents used for application, which must have low vapor pressure and high flash point as well as being nonphytotoxic. Vermiculite granules, for example, have high sorptive capacity and will carry up to 20% of dieldrin. Diesel oil is a typical low vapor pressure solvent having low phytotoxicity that is valuable for the formulation of granules for foliage applications (on corn for example), while low vapor pressure aromatic solvents can be used to prepare them for operations in which phytotoxicity is not a problem (for example, in soil treatments in the absence of vegetation, or in some area treatments for disease vector control). As in the case of aldrin, fertilizer mixtures containing up to 1% of dieldrin have been formulated, and, in addition to the solid formulations discussed above, which contain only one toxicant, each of these cyclodienes has been combined with DDT in dust preparations. For example, 2.5% aldrin combined with 5 or 10% DDT, or with 5% DDT and 40% sulfur, and similar dusts containing 1.5 or 2.5% of dieldrin instead of aldrin.

The considerations applied to the formulation of aldrin and dieldrin apply also to endrin, especially those for dieldrin regarding solubility, since both materials are epoxides of roughly similar polarity. Endrin emulsifiable concentrates do not normally require inhibitors since acid formation is slow or absent. Certain solid carriers require deactivation as for the other compounds, and this needs to be particularly thorough because of endrin's susceptibility to detoxicative molecular rearrangements; up to 5% of hexamethylenetetramine (HMT) is incorporated in the carrier for this purpose and appears to cause no reduction in biological activity. Typical emulsifiable concentrates contain 190 to 240 g/l of endrin which may also be formulated with methyl parathion (for example, 1 lb of endrin and 1 lb of methyl parathion per gal of concentrate) for some purposes. Other formulations include 50 and 75% wettable powders, a 25%-wt dust concentrate, 1.5 and 2% field strength dusts, and 1 to 5% granules. The HMT used for endrin stabilization is incompatible with methyl parathion. In this case, field strength endrin dust is prepared using a carrier (such as talc) deactivated with urea, and this is then impregnated with the appropriate amount of technical (80%) methyl parathion, to give dusts containing, typically, 1.5 or 2% endrin with, respectively, 1.5 and 2.5% of methyl parathion. Properly formulated and deactivated solid preparations containing endrin are stable for many months under normal conditions.

In addition to the formulations outlined above, aldrin, dieldrin, and endrin have been incorporated in insecticidal lacquers and used in various other ways, as described for DDT. Pesticide technology is a dynamic, continuously developing entity, so that formulation methods and products are constantly changing as the pest control requirements change and as techniques improve. When discussing the development of DDT, mention was made of the early problems experienced with inadequate wettable powders stored in the tropics during vector control operations, and the subsequent development of the high quality, high toxicant content powders which are so vital for economic public health applications. A bulletin of 1957 describing the uses of aldrin as a soil insecticide states that "wettable powders containing up to 40% of aldrin are available," and another of 1960 on the public health uses of dieldrin mentions that wettable powders contain 50% dieldrin but "the development of 75% powders is in progress." Today, 70 to 75% wettable powders containing these insecticides are commonplace.

An acute problem in pesticide technology is that of adequate communication between those who design toxicants and those who are nearer to the point of practical application. The language used needs to be as simple as possible in order to avoid misunderstanding and it seems desirable that once communication is established, drastic changes should be avoided, even if they might seem to lead to basically more desirable situations. The nomenclature problems of chemists involved with the cyclodienes were discussed in a separate section

and there is another problem related to the difference between the highly pure, crystalline chemicals available in the laboratory and what can be economically prepared in a production plant. This is apt to cause some difficulty for manufacturers for whom the only economically marketable commodity is usually the technical material. As an example, the net weight of technical aldrin in a drum of this material is marked on each drum, and its weight percentage of aldrin (not less than 90%) is also given so that the weight of aldrin present is readily determined. By definition, the weight of HHDN, the active ingredient (a.i.), is at least 95% of that of the aldrin present. The percentage HEOD in technical dieldrin is similarly determined since it is at least 85% of the dieldrin content derived from the figures given with each batch. The names aldrin and dieldrin are very convenient for many purposes and are accepted as common names (with their HHDN or HEOD content, respectively, defined as above) by BSI and ISO, as well as being company trademarks. Moreover, because of the wide use of these compounds, the above procedures for determining active ingredient content for labeling purposes are by now widely understood.

Consequently, a current move by the Food and Agricultural Organization (FAO) to use the abbreviations HHDN and HEOD, instead of aldrin and dieldrin, for labels on the appropriate products seems likely to lead to confusion among users of the FAO specifications, and difficulties in deciding the validity of existing registration and label data. In view of the widely accepted use of these trademarks as common names, even by International Agencies, the normally desirable avoidance of using trade names seems hardly applicable in these cases, especially since no commercial concern produces the pure compounds HHDN or HEOD. Nor does it seem likely that any could do so economically in competition with the current manufacture of the technical materials. According to FAO specifications the aldrin 50% wettable powder (w.p.) would be called "HHDN 47.5% w.p." and the dieldrin 200 g/liter emulsifiable concentrate (e.c.) becomes "HEOD 170 g/liter" e.c., so that in order to bring these nominal active ingredient contents to round figures permitting easy calculation of dilution rates, costly product reformulation and the necessity for reapplication for product registration become necessary.

Since the cyclodienes have now been in use since a little before 1950, the figures quoted by Whetstone[295] for cyclodiene sales in the United States in 1962 represent an interesting midterm view of the usage there and are given for interest in Table 17. In addition, the general global use of these compounds was probably somewhere near maximal in the mid-sixties.

When considering the agricultural and other uses of DDT, some mention was made of the relative practical efficiency of cyclodiene insecticides for comparison. The first detailed evaluation of the insecticidal activity of aldrin and dieldrin was reported by Kearns, Weinman, and Decker in 1949, and is particularly interesting in view of the comparisons made with DDT, lindane, chlordane, and heptachlor.[302] In tests against houseflies, heptachlor, aldrin, and dieldrin were found to have about the same order of toxicity as lindane, being about 8 to 18 times as toxic as DDT; the residual effectiveness of heptachlor and aldrin was similar to that of chlordane, but dieldrin deposits showed only a slight decrease in activity after 1 month. The speed of action in these tests appeared to be in the order, lindane > aldrin = dieldrin > heptachlor > chlordane > DDT. Thus, flies exposed to lindane were all paralyzed in 2 hr, whereas the other compounds required 3.5 to 5 hr for complete knock-down. Lindane and the cyclodienes were tested against adult American and German cockroaches by topical application in dioxan solutions, and tests for residual efficacy of these insecticides were made with adult male *Blattella germanica* using impregnated surfaces. DDT was far less toxic to cockroaches than the other compounds and was not included in the tests. Dieldrin was rather more effective against both species than lindane, aldrin, or heptachlor and was also more toxic than chlordane, although lindane showed most rapid action and gave maximum mortality in 24 hr, whereas mortality from the other treatments increased progressively for up to 5 days. A 1 mg/1,000 cm^2 deposit of dieldrin killed 90% of adult male *B. germanica* during a 48-hr exposure as long as 49 days after its formulation, whereas deposits of the other compounds at this level gave no appreciable mortality 21 days after their formulation.

The measurement of LD50's to grasshoppers, *Melanoplus differentialis*, showed that heptachlor, aldrin and dieldrin were approximately sixfold more toxic as contact insecticides and three- or fourfold more toxic than chlordane as stomach poisons. Parathion was more toxic than heptachlor, aldrin, or dieldrin as a contact toxicant, but less toxic as a stomach poison. Lindane appeared to be intermediate between these cyclodienes and parathion in either contact or stomach toxicity. In topical application tests on milkweed bugs, *Oncopeltus fasciatus*, aldrin proved to be somewhat more toxic than dieldrin, threefold more toxic than heptachlor or lindane, and about 12-fold more toxic than chlordane. Only lindane approached 3,5-dinitro-o-cresol (DNOC) in speed

TABLE 17

Estimated Sales of Hexachlorocyclopentadiene and Derived Insecticides in the United States in 1962

Compound	Approximate Quantity (million lb)	Approximate Price ($/lb)
Hexachlorocyclo pentadiene[a]	50	0.20
Chlordane	5–10	0.65
Heptachlor	<5	0.96
Aldrin	9–12	0.99
Dieldrin	5–10	1.85
Endrin[b]	5–10	2.77
Endosulfan	1–2	1–2

Data adapted from Whetstone[295]

[a]Used mainly, but not entirely for the manufacture of cyclodiene insecticides.
[b]Marketed by Shell Chemical Corporation and Velsicol Chemical Corporation.

of action toward the Chinch bug (*Blissus leucopterus*), although aldrin and dieldrin were superior in the long term. Organochlorines were shown to be poorly effective against full grown larvae of codling moth (*Carpocapsa pomonella*), although DDT at a level of 20 to 45 mg/g of larval weight, or lindane at 2 to 13 mg/g had some lethal effect; parathion, in contrast, kills at all doses greater than about 0.0075 mg/g and acts so rapidly that the toxic effect appears almost immediately following application. In these circumstances, the larvae died without spinning cocoons, whereas the action of organochlorines was so much slower that those receiving a subsequently lethal dose actually died in the pupal stage. This effect was attributed to the much faster penetration of parathion, as compared with the organochlorines, through the body wall of mature larvae.

Against the black carpet beetle, *Attagenus piceus*, only DDT and dieldrin completely prevented damage at all concentrations tested; heptachlor and aldrin were about equally effective for 2 weeks after impregnation but on subsequent reinfestation, the test material was markedly less toxic. Chlordane and lindane were relatively less effective than the other chemicals. DDT appeared to be much less effective against the clothes moth (*Tineola biselliella*) than against carpet beetles; aldrin showed the same tendency and only dieldrin was effective against this insect at all concentrations tested. Tests in which an adult male and female carpet beetle were placed in fresh insecticide treated larval food showed that lindane and parathion caused rapid paralysis of the adults at remarkably low concentrations so that the female did not deposit eggs, while DDT was about fourfold less toxic than the other compounds by this method of test.

Acetone-water suspensions of DDT, lindane, and aldrin caused a low mortality of mobile stages of the two spotted mite when sprayed on rose leaves, but the eggs and resting stages were unaffected. In these circumstances, chlordane and aldrin gave 30 to 50% kill of mobile stages and appeared to be residually toxic toward hatching larvae and newly emerged resting stages. Dieldrin gave no mortality of the mobile stages, but did have some effect on both these and the newly emerged resting stages. From these experiments it was concluded that none of the new organochlorine compounds tested showed promise as an acaricide. The general conclusion from extensive

tests on ten different insect species was that the toxicities lie in the following sequence: dieldrin > aldrin = heptachlor = lindane > chlordane > chlorinated camphene > DDT. The residual efficacy appeared to be in the order dieldrin > DDT > aldrin > heptachlor = chlordane > lindane. Thus, the newer cyclodienes were shown in laboratory experiments to be as effective as lindane and much more effective than DDT as contact poisons and more effective than lindane as residual insecticides.[302]

c. Applications in Agriculture and Against Pests Affecting Man and Animals

Aldrin, dieldrin, and endrin are nonsystemic and persistent insecticides having high contact and stomach activity towards many insects. Since aldrin is less volatile than HCH but more volatile than most other organochlorines, including dieldrin and endrin, it is less persistent than the epoxides and may be used for foliage applications when rapid kill and a short residual effect are required; dieldrin and endrin are used when a long residual effect is necessary. Its high vapor pressure gives aldrin an advantage over the epoxides, affording a slight fumigant effect which enables it to reach soil insects in habitats inaccessible to the other compounds; the long residual effect in soil makes it economical in use but means that care must be exercised not to exceed the recommended rates of application to the extent that soil insect predators are killed and pests are selected for resistance, as has frequently occurred in the past. Since the chemical evaporates rapidly from the soil surface, it is incorporated to a depth of 2 to 10 in., depending on the pest to be controlled. Dosage rates used vary according to the pest, stage of plant growth, soil type and temperature, and method used for application; as little as 0.5 to 0.75 lb per acre may be sufficient when rows of crops are treated against soil insects, but insect control is frequently best obtained by broadcast treatments of 1 to 3 lb per acre, although this means adding more insecticide to the environment and is also more costly. Foliage treatments are of the order of 0.75 to 1 lb per acre. The application of aldrin-fertilizer mixtures is a time saving and economic way of insecticide application which is widely used, although discontinued in some places, as in Great Britain.

Dieldrin has shown itself to be of great value for the control of crop destroying insects on foliage, as well as in soil, although aldrin has advantages as an internal soil insecticide, as indicated. As a soil insecticide, dieldrin is perhaps most valuable for surface treatments, where its lower volatility gives it persistence lacking in aldrin and enables it to control insects living or feeding at soil level. In the majority of uses, dieldrin applications range from 0.25 to 0.75 lb per acre, although for some purposes less than 0.1 lb per acre is sufficient. Like aldrin, it affords excellent protection as a seed dressing when used at the rate of 0.25 to 2 oz per bushel of seed either alone or in combination with a fungicide, and root dipping in 0.05% suspensions of aldrin or dieldrin wettable powders is used to protect transplant crops against wireworms, cabbage maggots, and similar pests. In addition to direct treatment of the soil surface with an aldrin or dieldrin formulation, the chemicals may be incorporated into baits; for example, 1.5 lb of either in 100 lb of bran is scattered evenly over the surface at 20 to 25 lb per acre to control leather jackets, cutworms, grasshoppers, crickets, and mole crickets. Dusts are valuable for the protection of vegetable crops since they are less persistent on plant surfaces than sprays and may consequently be used nearer to harvest. Since they require no preliminary mixing they are easy to apply, do not require expensive equipment for application and are useful in small scale applications; they are, however, somewhat less effective than sprays from the insect control standpoint. Due to the weight of the particles, granules provide a means whereby the toxicant can be carried to the leaf axils and growing points of plants and have been used successfully against corn borers and maggots infesting spinach, where the pest habitats are somewhat protected from the toxicant. For most of these applications, aldrin and dieldrin are interchangeable, except that the greater persistence of the latter makes it more useful in exposed situations.

Among the many insect pests that spend some part of their life cycle in the soil, a number of major groups can be recognized and representatives of one or more of these are to be found in most soils. The main uses of aldrin (and dieldrin, since the epoxide is formed from aldrin under natural conditions) are against the groups shown in Table 18, attacking wheat, barley, oats, and rye; maize and sorghum; potatoes and other solanaceous plants; various root crops and other vegetables; cotton, sugar cane, and tobacco. Aldrin is also recommended for use against pests of figs and dates, vines and hops, soybeans, sweet potatoes, yams, fiber crops, ornamental plants, trees and shrubs, forest trees and grassland.[305] In addition to the general pests listed in Table 18, others, specific to a particular crop or crop type, are well controlled by aldrin, and some examples are given in Table 19. Both aldrin and dieldrin are good ant and termite killers and the requirement for residual activity mainly determines which is to be used; dieldrin is generally preferred in exposed situations, whereas aldrin is used where its fumigant action is valuable, in treating nests, for

TABLE 18

Major Groups of Insects Controlled by Aldrin (and Dieldrin)

Pest group	Major genera in each pest group
Annual grubs	*Cyclocephala*
Ants	*Formicids*
Cockchafer grubs	*Amphimallon, Melolontha, Anomala, Anoxia, Serica*
Cutworms	*Agrotis, Euxoa*
Leatherjackets	*Tipulids*
Maize beetles	*Astylus, Gonocephalum*
Mole crickets	*Gryllotalpids*
Rootworms	*Diabrotica, Buphonella*
Root flies	*Hylemyia, Eroischia, Psila*
Termites	*Isopterans*
White grubs	*Lachnosterna (Holotrichia, Leucopholis) Phyllophaga, Eulepida*
Wireworms	*Agriotes, Melanotus, Eleodes*

TABLE 19

Pests That Attack Specific Crops and Are Normally Controlled by Aldrin

Crop	Pest	
Banana	Banana root borer	*Cosmopolites sordidus*
	Coccids and mealy bugs	*Coccidae*
Maize	Corn stalk borer	*Elasmopalpus lignosellus*
Orchard	Fruit fly	*Anastrephe fraterculus*
fruit	Mediterranean fruit fly	*Ceratitis capitata*
	Weevils	*Curculionidae*
		Otiorrynchus spp.
	Scale insects	*Coccus viridis*
Potatoes	Black hardback beetle	*Dyscinetus* spp.
	Potato weevil	*Premnotrypes solani*
	Potato cutworm	*Xylomyges eridania*
Sugar beet	Millipedes	*Blaniulus guttulatus*
		Bothynoderes spp.
		Cleonus spp.
		Conorrynchus spp.
		Temnorrhynchus spp.
Sugar cane	White grubs	*Allisonotum impressicolle*
		Leucopholis spp.
	Wireworms	*Limonius* spp.
	Funnel ants	*Aphaenogaster* spp.
	Cane beetles	*Dermolepida* spp.
	Root borer	*Dorysthenis* spp.
	Cane beetle	*Heteronychus* spp.
Tobacco	Crickets	*Scapteriscus* spp.

example, or for incorporation in baits which will be carried into the nests.

In principle endrin may also be used for any of these applications and especially in those for which dieldrin is favored, since its persistence (Vol. II, Chap. 3B.1b) is somewhat similar; in practice it is mainly used as a foliage insecticide. Its ability to control the Mexican bean beetle (*Epilachna varivestis*) and various aphids, which are poorly affected by other cyclodienes, was quickly noticed and is discussed in the patent by Bluestone which describes the synthesis of endrin.[190] It is now generally recognized to be complementary to dieldrin in being highly effective against caterpillars and sucking pests that are not easily controlled by the latter.[306] For this reason, endrin has proved to be exceptionally useful for the protection of tropical crops such as maize, sugar cane, rice, tobacco, coffee, and cocoa against various sucking pests such as aphids, capsids, and thrips, and against caterpillars and boring insects.[307] Insects attacking grasses, clover, and forage crops such as alfalfa, as well as orchard insects including some flies, leaf caterpillars, saw-flies, and psyllas are also controlled. For the control of sucking insects 0.1 to 0.3 lb per acre (a.i.) of endrin is usual, with higher rates for caterpillars (0.25 to 0.5 lb per acre) and for beetles and boring pests (0.5 to 0.7 lb per acre). Since endrin is persistent, one or two treatments per season are sufficient for control, although the cocoa moth caterpillar (*Acrocercops*) may require up to eight applications for full control.

The most outstanding application of endrin is against cotton pests, and it can protect the crop from sucking, leaf-eating, and boll-damaging varieties of insects throughout the growing season. At one time, the spiny bollworm (*Earias insulana*) and related species were pests of such significance in areas from the Mediterranean eastward to the Philippines and southward in Africa and Australia that cotton growing was uneconomic because of insect damage. Since the introduction of endrin, the crop can be grown in many of these areas. Various other species described as "bollworms," as well as the boll weevil (*Anthonomus grandis*), are well controlled by endrin, although the pink bollworm (*Pectinophora gossypiella*) is only partially controlled by this agent. Other cotton insect pests such as various leaf-hoppers, aphids, thrips, and plant bugs are incidentally controlled by these applications of endrin. Thrips on this crop are often controlled by applications of 0.05 lb per acre, but applications within the range 0.2 to 0.5 lb per acre are more normal for other pests; control is best achieved by frequent applications throughout the growing season, from the time when the insects first appear.

Endrin is effective against the various stages of numerous insects infesting forest trees, especially defoliating caterpillars and other leaf-eating species, sucking pests infesting buds and shoots, and stem-boring beetles. For the control of leaf-eating caterpillars, ultra low volume sprays containing 0.25% endrin have been used, the concentration being increased to 0.5% for leaf rolling species (Tortricids). In the case of bagworms, the lepidopterous larvae (and adult females) which spend their time in protected cocoons, or "bags," successful use has been made of a 0.5% spray applied at the rate of 0.12 lb per acre of endrin. Lepidopterous larvae boring into maize and sugar cane are difficult to kill because they are directly exposed to the toxicant for a very short time between hatching and reaching the shelter of leaf sheaths and stems. Nonpersistent insecticides must be applied just as the eggs are hatching, but the timing is less critical for persistent ones. Since endrin is toxic to numerous caterpillars and persists for up to 15 days on the foliage, it gives good control of these boring species and incidental control of exposed sucking insects and leaf-eating caterpillars. For the boring insects, the higher rate of spray (0.5 lb per acre) is applied so as to achieve thorough penetration; granular formulations are also rather effective for borer control because of their distribution characteristics. Similar uses are found against the boring and sucking insects attacking rice, the regular rate of treatment (0.1 to 0.3 lb per acre) being used for sucking and leaf-eating insects and 0.5 lb per acre for boring pests such as the rice stem borer (Chilo suppressalis), another lepidopterous larva.

All the main insect pests attacking tobacco (again, mainly sucking insects and various lepidopterous larvae) can be controlled with endrin. The thrips species are very susceptible and a 0.05 lb per acre application is sufficient for control; for the others 0.2 to 0.4 lb per acre is a more normal rate. Another control measure that has been used successfully involves dipping tobacco seedlings in dilute endrin emulsion

(0.02%) when transplanting, and using a similar emulsion to repel later insect attacks as the growing season advances. In addition to the range of uses already mentioned, endrin applied at 0.1 to 0.4 lb per acre will control the caterpillars, beetles and their larvae, grasshoppers, springtails, plant bugs, and aphids that infest turf and pasture; pests attacking root crops and other vegetables; caterpillars (Amsacta) on ground nuts and castor oil beans, and leaf hoppers (Idocerus) on mangoes and nut trees. Although Tetranychid mites are not killed by endrin, the toxicant is outstanding for the control of certain Tarsonemid mites, such as Hemitarsonemus latus, infesting rubber and tea, and the cyclamen mite, Steneotarsonemus pallidus, which is found on market garden crops and flowers; for such uses 0.02 to 0.05% endrin sprays are effective.

Although aldrin, dieldrin, and endrin have higher acute toxicities to some mammals than chlordane or DDT, endrin being the most toxic of the practical cyclodienes (Vol. II, Chap. 3C.2b), the residues on edible crops at harvest do not present any hazard to the consumer provided that the instructions for use are followed, particularly in regard to the interval between application and harvest; even endrin is less toxic on an acute basis than many of the organophosphorus insecticides currently available, although its residues (which usually consist at least partly of less toxic degradation products) may persist longer. Endrin is particularly notable for its high rate of disappearance from mammalian tissues, which contrast with the other cyclodienes and will be discussed more fully in relation to metabolic studies (Vol. II, Chap. 3B.1c and B.2e). Phytotoxicity appears to present few problems with these cyclodienes, if properly applied, although for some sensitive plants, deleterious effects have been traced to the solvent used in formulation and there have been indications that endrin itself can cause damage to maize.

The uses of cyclodienes discussed above represent the range of possible practical applications that emerged in the course of development of these chemicals and do not necessarily reflect present day usage, which is discussed in the following section on "current status." Figures for the amounts of endrin used in various applications up to 1958 give some idea of the usage patterns that developed for this compound.[308] These were (in million lb) cotton (6.8), fruit and vegetables

(1.0), tobacco (0.55), rice (0.40), pasture (including field rodent control) (0.4), oil seed crops (0.30), tree crops (0.20), maize (0.15), sugar cane (0.15). Endrin sales by Shell and Velsicol in the United States in 1962 totaled about half of the total of aldrin plus dieldrin, and are probably similar to the sales of chlordane in that year (Table 17);[295] its use in the U.S. is now mainly limited to nonagricultural commodities such as cotton, although it is still registered for some agricultural purposes. There is a small use of endrin-DDT formulations for control of cotton pests in certain African countries. Endrin is still an important agent for the control of rice pests, especially rice stem borer, in a number of tropical countries.

The practical applications of organochlorine insecticides may be considered both from the standpoint of their potential uses, as determined during their development, and in regard to present usage, which is limited by the general desire to reduce the levels of these compounds in the environment and by insect resistance, which destroys their efficiency and makes replacement by other compounds necessary. The potential of aldrin and dieldrin, especially the latter, as agents for the control of disease vectors and other insects of public health importance is very great, because of their residual behavior on surfaces. For certain purposes, aldrin has advantages over dieldrin because it resembles lindane to some extent in possessing an ability to exert a vapor effect at the surface of materials into which it has been absorbed. For example, the internal walls of some kinds of dwellings in the tropics absorb organochlorine insecticides rather rapidly, which tends to eliminate equally rapidly the insecticidal effect of relatively non-volatile compounds such as DDT or dieldrin applied in sprays. Nevertheless, for most purposes, dieldrin has been favored over aldrin for public health uses. It has been used for the control of mosquitoes, various species of noxious flies (houseflies, tsetse flies, *Glossina* spp; black flies, *Simulium* spp.; sandflies, *Phlebotomus, Culicoides,* etc.), fleas (oriental rat flea, *Xenopsylla* spp., etc.), bedbugs (*Cimex* spp.), Reduviid bugs (*Triatoma* spp.), ticks and mites (*Ornithodoros* spp., trombiculid mites), cockroaches (*Blattella, Blatta,* and *Periplaneta*), and scorpions.

Dieldrin wettable powders or emulsifiable concentrates are the most useful formulations for residual spraying in mosquito control and when this toxicant was introduced, an apparent eight-fold higher toxicity to mosquitoes appeared to more than offset the higher cost (according to 1967 figures, 50% dieldrin wettable powder (w.p.) cost $1.98/kg, whereas DDT 75% w.p. cost $0.40 and HCH 30% w.p. $0.50/kg).[50] In practice, it has to be applied at more than one-eighth of the normal DDT-treatments; 0.5 g/m^2 is usual, as compared with 1 to 2 g/m^2 for DDT, and either treatment normally lasts 6 to 12 months. For mosquito control, dieldrin has an advantage over DDT because it lacks DDT's tendency to excite and repel. Spray operators working inside buildings are inevitably exposed to the toxicants applied and dieldrin has been implicated in toxic effects on both men and domestic animals in those particular circumstances. This problem, together with the higher cost than DDT and especially the widespread development of resistance in anopheline mosquitoes, has led to dieldrin's abandonment for residual spray control in most countries, except those in which a species develops DDT-resistance. Anopheline larvae are normally very susceptible to dieldrin applied at 0.1 lb per acre (half the dosage of DDT required for equivalent control) to breeding areas as either a low volume spray from aircraft or as a spray from the ground; as granules which will penetrate dense vegetation to the breeding sites, or incorporated as a 0.5% solution in highly spreading larvicidal oils. Culicine mosquitoes are less easy to kill with insecticides than anophelines, but dieldrin controls them at 0.6 g/m^2, although not for as long as the other species; 0.1 to 0.25 lb per acre is a recommended rate for larvicidal applications. Dieldrin pellets, consisting of a 10 g mixture of equal proportions of 16% dieldrin w.p. with sand and cement, have been used to effect the slow release of dieldrin into water courses covered by vegetation or to prevent the breeding of *Aedes aegypti* in stagnant water.[50]

Although dieldrin-resistance (leading to generalized resistance to cyclodienes) has made the continued use of this chemical impractical for both housefly and mosquito control in many areas (Vol. II, Chap. 2C.), it is still a valuable agent for the control of some of the other unpleasant insects mentioned above which do not develop cyclodiene resistance.[44] Thus, tsetse flies, carriers of the trypanosomes affecting man and cattle, are a considerable obstacle to economic progress in large areas of tropical Africa. The technique used for control varies with the species of tsetse involved and with the season, but consists broadly in spraying the resting areas in vegetation with 1.5 to 4% dieldrin emulsions, which give effective control for about 1 month. Various species collectively known as sandflies are also controlled by dieldrin, heptachlor or lindane applied at 60 to 700 g per acre. DDT and lindane dusts (10% and 3%, respectively) have been the agents of choice for flea control, but 2% dieldrin dusts have been used for some purposes, especially in outdoor situations. DDT, dieldrin, and lindane have all been used to control bedbugs, although DDT is favored when resistance to it is absent.

Unfortunately, resistance to these chemicals has now become rather common, so that organophosphorus or carbamate insecticides have to be used. In South America, some species of reduviids (assassin bugs, kissing bugs) convey with their bite a trypanosome causing a fatal form of trypanosomiasis in man; they hide in cracks and crevices in walls, usually emerging only to feed, so that house interior spraying with dieldrin at 1 g/m^2 for malaria control kills them also.

Dieldrin, usually in the form of 0.5% oil solutions or water emulsions, is used for interior surface treatments to control various tick disease vectors, but not on animals. It is also highly effective against ticks in their outdoor habitats, and is used, except where such applications are likely to be inimical to wildlife. It is in situations such as these that the balance between preserving wildlife and preventing disease has to be weighed very carefully. The usual outdoor rates of organochlorine application present some hazard to non-target organisms, especially if the materials enter watercourses, so that less persistent organophosphorus or carbamate toxicants are used if the threat of disease is not acute. Two species of trombiculid mites (*Leptotrombidium* spp.) transmit scrub typhus, and aldrin, dieldrin, or lindane are the favored agents for application to the scrub, woodland, or open areas in which infestations occur. The chemicals are applied as sprays, dusts, or from fogging units, and dieldrin at 1,400 g per acre is considered to be most effective; aldrin and lindane are applied at 1,250 and 2,800 g per acre, respectively, and toxaphene or chlordane (1,100 g per acre) may also be used. Dieldrin applied in this fashion has given good control for more than two years and is superior in efficiency to all the other toxicants mentioned.[44]

Organochlorine insecticides are the agents of choice for cockroach control, although German cockroach resistance to chlordane, the most commonly used toxicant, has recently necessitated the use of the organophosphorus compounds malathion and diazinon and the carbamate propoxur. Dieldrin has been used effectively against cockroaches as a 0.5% spray or 1% dust, or as an ingredient of insecticidal lacquers for coating surfaces in canteens and in other appropriate situations where long lasting control is required, the effect often persisting for several years even when the lacquered surfaces are frequently washed down. It must be remembered, however, that chlordane resistance also extends to dieldrin, which will not therefore be effective against insects having this "cyclodiene" resistance. Applications of organochlorine compounds, including the cyclodienes, are further discussed from the standpoint of insect resistance (Vol. II, Chap. 2C).

d. Principles of Analysis

The total chlorine method as applied to technical aldrin, dieldrin, or endrin determines all of the chlorinated compounds present in these materials and is consequently not suitable for the determination of the content of the pure major components. It can be used, however, as a plant-control procedure for formulations of all three compounds provided these are prepared from carriers free of extraneous chlorine. In this case, the method is to determine the ratio of guaranteed aldrin, dieldrin, or endrin content to total chlorine content in each batch of technical material and to use this ratio to convert total chlorine content of derived formulation products into actual insecticide content.[116]

The well-known phenyl azide photometric method may be used for the determination of both HHDN and HEOD in formulations or in residues.[116] Wettable powders, dusts, granules, inorganic fertilizer mixtures, or liquid formulations are first extracted or diluted with a hexane-acetone (95:5) mixture and the solution is passed through a column in the case of the solid materials to effect removal of unwanted formulation ingredients. In the case of aldrin, the cleaned up solution is treated with excess phenyl azide, the solvent evaporated and the residue heated at 85°C for 35 min to form the dihydrophenyltriazole derivative (Figure 20). Excess phenyl azide is then evaporated in vacuo and the residue dissolved in isopropanol, reacted with hydrochloric acid and coupled with diazotized 2,4-dinitroaniline; the addition of 75% sulfuric acid then gives a colored solution which is used to determine aldrin content by measurement of its absorbance at 515 nm. Cleaned up solutions containing dieldrin are evaporated and the residue treated with acetic acid and hydrobromic acid to give the corresponding bromoacetate. This compound may be converted into an aldrin-analogue (probably lacking one methano-bridge chlorine atom) by treatment with zinc in acid solution, and the product is then determined with phenylazide as before. The method has been used to estimate μg quantities of HEOD in cleaned up solutions.

In another method which involves opening of the epoxide ring of dieldrin, the cleaned up solutions of toxicant obtained from formulations or other sources are evaporated and the residue, dissolved in dioxan, is treated with excess of anhydrous hydrogen bromide at ambient temperature for 2 hr. The excess hydrogen bromide remaining after formation of the bromohydrin is then titrated with standard alcoholic sodium hydroxide using thymol blue as indicator.[116] A second colorimetric method for dieldrin, proposed by Skerrett and Baker, involves rearranging it to the corresponding ketone (Figure 20, structure 12) with a Lewis acid such as boron trifluoride, the amount of the ketone then being determined through the formation of its colored 2,4-dinitrophenylhydrazone and absorbance measurements at 440 nm. A similar method has been proposed for endrin by the same authors, the 2,4-dinitrophenylhydrazone of the half-cage ketonic rearrangement product (Figure 21, structure 10) being used for determination.[309]

Infrared analysis is the most favored method for the determination of all three of these compounds in the technical materials and in formulations. HHDN has a suitable absorption maximum at 12.01 μ, and the absorbance of cleaned up extracts from such materials is measured using a baseline drawn between 11.85 and 12.24 μ. A similar technique is used for HEOD in dieldrin and employs the absorption peak at 11.82 μ with a baseline drawn between 11.59 and 12.18 μ. For endrin the peak at 11.76 μ is used, with baseline points at 11.50 and 11.97 μ. The infrared and phenyl azide methods are fully described in the Shell Chemical Corporation *Handbook of Aldrin, Dieldrin and Endrin Formulations;* the former

is also detailed in the CIPAC *handbook* along with standard methods for insoluble material, acidity and various physical properties of the formulations.[115,116]

The introduction of gas-liquid chromatographic analysis with electron-capture detection in 1960 by Goodwin, Goulden, and Reynolds revolutionized the analytical procedures for these compounds, since it provided a highly sensitive means for simultaneous separation and quantitation.[135] Following the first description of the use of this technique in the analysis of mixed cyclodienes, lindane, and DDT, it was rapidly taken up by a great many laboratories and it must be stated that the pitfalls attending its use for the analysis of mixed pesticides in residues were not fully appreciated in many quarters. The technique was probably something of an art at that time and the fabrication of "home made" electron-capture detectors became a popular pastime in laboratories associated with pesticide research. This almost overnight transition from an ability to measure μg (10^{-6} g) quantities of DDT analogues, lindane, and cyclodienes to the easy measurement of extraordinarily low levels of from nanograms (10^{-9} g) down to picograms (10^{-12} g) proved to be a traumatic experience with implications perhaps not fully appreciated in 1961. Traces of organochlorine compounds (and other compounds, since the method will detect a number of groups other than halogens having electron affinity) that had formerly been very much "out of sight and out of mind" suddenly assumed great prominence in the public mind and it has taken all of the intervening time to the present day to bring these observations into correct perspective. The significance of the quantity "10^{-10} grammes of a chemical" is not easy even for a scientist to visualize and so the difficulty experienced outside pesticide circles is readily understandable. Thus, a tool which has had quite remarkable value for research work has conferred mixed blessings in other ways.

Nowadays it is fairly well recognized that confirmatory procedures are required for the identification of pesticide residues by gas-liquid chromatography (GC). For this purpose it is customary to compare retention times of unknowns with reference compounds on several columns having different characteristics of polarity, etc. Advantage is also taken of any easily conducted chemical reactions that will produce a different product or products affording a new and recognizable chromatographic pattern. Additional confirmation is provided by thin-layer chromatography if sufficient material is available and, of course, by modern physical methods (Vol. I, Chap. 2C. 3b). Robinson has discussed some of the problems attending the identification of trace compounds by chromatographic techniques.[151] In the case of aldrin, dieldrin, and endrin, a number of the chemical transformations discussed in the chemical sections are used as aids in GC-identification. Aldrin may be oxidized to dieldrin, which causes an appreciable alteration in retention time; dieldrin can be converted into a halohydrin by epoxide ring opening with the appropriate halogen acid and the resulting hydroxyl group acetylated or converted to a trimethylsilyl ether with corresponding changes in chromatographic characteristics.[238]

Rearrangement to the corresponding ketone (Figure 20) with mineral or Lewis acids has been used for dieldrin confirmation, and also for endrin, in which case the half-cage pentacyclic ketonic rearrangement product is formed. Endrin undergoes thermal decomposition to the aldehyde and pentacyclic ketone of Figure 21 under some GC conditions, a fact which may sometimes help in identification.[257] If dieldrin and endrin are present together, acidic reagents give a mixture of the ketones (plus some endrin aldehyde) which is not completely resolved on a DC-200 or mixed QF1/DC11 column. In this case, resolution may be effected on a mixed 15% QF1/10% DC-200 column, or alternatively, treatment of the mixed chemicals with acetic anhydride-sulfuric acid gives endrin pentacyclic ketone and a dihydroaldrin diacetate instead of the ketone from dieldrin, and these derivatives have significantly different retention times.[238]

Yet another method involves treatment of the mixture with chromous chloride (Vol. I, Chap. 3B.4d,e), which quickly rearranges endrin to the pentacyclic ketone, then removes a bridge chlorine in the usual manner. Dieldrin, on the other hand, is deoxygenated to aldrin which then also suffers bridge monodechlorination. Based on the determination of residues in a 10 g commodity sample, these chemical confirmatory tests for aldrin, dieldrin, and endrin, which have recently been discussed by Cochrane and Chau,[64] have a sensitivity range of 0.01 to 0.05 ppm in terms of the chemical present. Since a number of such reactions have now been described for cyclodienes, there is

considerable scope for such techniques. The main drawback is that when more than one derivative is produced from each toxicant, the derivatization pattern rapidly becomes too complicated for interpretation if several pesticides are present together, so that some form of initial separation becomes necessary. Since residue analysis frequently involves the separation and identification of the components of mixtures of organochlorine insecticides, this aspect of analysis is conveniently considered for the compounds as a group in Chap. 2C., where mention is also made of the use of NMR and mass spectrometry in structure determination.

e. Current Status of Aldrin, Dieldrin and Endrin

The foregoing sections on various aspects of cyclodiene technology represent the development of these compounds to the present time, and, as in the case of DDT, their use has been clouded during the last few years by the heated controversy surrounding persistent pesticides in the environment. Clearly, the culmination of "History and Development" relates to the present and to what will happen in the future, a question which relates rather closely to world food supply. The dangers to mankind inherent in the population explosion have been widely stressed in the last few years — so much so that there is perhaps a danger that familiarity with the figures will soon breed contempt. Briefly, statistics of 1966 set the global population at 3.3 billion. By 1980, it is expected to reach 4.3 billion and to be near to 6 billion as we approach the year 2000. Thus, the global population is expected to double by that time, while half the present population already experiences either chronic hunger or serious dietary deficiencies; it is estimated that each day 10,000 people, the majority of them children, die through malnutrition in the underdeveloped countries.[306] In these countries, farm production is largely on a subsistence basis, so that more than half of the food produced is consumed where it is grown. Many of them have made great progress in increasing their food production with the assistance of the international agencies, but these advances are still out-stripped by the ever-increasing food requirements.

According to estimates made by the Food and Agriculture Organization of the United Nations in 1963, a 15% increase in global food production would be required by 1970, and a 100% increase by 1985 simply in order to maintain the present nutritional situation.[310] In fact, world food production per head actually fell about 1965–1966, and in the 1960's the situation changed to the extent of threatening even the nutritional status quo. Many commendable proposals have been made for improving this situation, such as increasing the acreage of land cultivated, rapid modernization of agricultural practices, the development of synthetic foods, and so on, but these all take much time. Meanwhile, every currently available means of increasing food production must be utilized and the greatest barrier to progress remains the enormous destruction done by insects. In 1963, the value of world crop losses due to insect pests was estimated at $12,500 million annually at farmer level. Assuming that soil pests in particular were responsible for about $4,000 million of this damage, this means that they deprived about 60 million people of a cereal based diet for one year! Crop losses averaging 10 to 15% are commonly quoted, such averages including acreages with no losses and those seriously threatened. For the seriously threatened acreages, the loss is probably nearer 25%, and these are the important areas from a practical standpoint.[311]

The cyclodienes have undoubtedly made, and continue to make, their greatest contribution as soil insecticides, and their role should be assessed in the light of the world need for grains, which require extensive protection against soil pests. The principal sources of world food supply are rice and wheat, with all kinds of grains supplying, in fact, 53% of the total food supply. While the developed nations enjoy a variety of foodstuffs, about half the global population depends upon rice as the staple food. For more than 1,400 million people in the Far East, where 90% of the world's rice is grown, this grain is the main dietary source. In Latin America, grains constitute 64% of the total protein source, while in Africa the percentage is 74%, in the Near East 72%, and in the Far East, 80%. Grains supply 63% of the protein source in Europe and the figure falls to 40% in North America, demonstrating the increasing dietary variety in these last areas.[306] This, then, is the background to the global importance of grain crops, and the corresponding present day importance of pesticides to protect them hardly needs to be stressed.

The value to any country of a vigorous pesticide chemicals industry is well illustrated by events in Japan since World War II, where rice production doubled (from 5.9 million tons) between 1945 and 1956, reached 14.5 million tons in 1967, and went into surplus in 1970, a progression due largely to sophisticated chemical pest control.[5] In Japan, the value of plant protection is estimated to be about seven times the expenditure on agrochemicals, agricultural machinery, labor, etc., and in 1965, agrochemicals added about eight times their own value to the gross national product.

To return to aldrin, dieldrin, and endrin, which, along with DDT, occupy the center of the environmental controversy, it has been estimated that the cyclodiene insecticides provide vitally needed protection for more than 65 million acres of food crops each year, the equivalent of food for 250 million people. The replacement of these compounds by other chemicals having equal efficiency for the same cost, but lower environmental persistence, has proved to be very difficult for many uses and, so far, impossible for some. Where replacements are available in principle, the cost of their use is greater and their efficiency lower; a 10 to 15% lower efficiency alone can mean loss of protection for some 6.5 to 10 million acres annually, the equivalent of a possible loss of one year's food supply for 26 to 39 million people. Economically, the global cost to farmers of using these compounds is estimated at $112 million, for a return of $1,120 million in terms of extra crop value at producer level, or $2,100 million at the retail level. Expressed as extra food supply, this represents an annual benefit of one year's food for some 30 million people.[306]

Events leading to the controversy surrounding the persistent cyclodienes stem from the almost simultaneous development of the ability to detect them in minute quantity (that is, by gas-liquid chromatography combined with electron-capture detection) and the occurrence in England in the early 1960's of the bird poisoning incidents later associated with the use of seed dressings containing aldrin, dieldrin, and heptachlor. This led to voluntary restrictions on the use of spring seed dressings containing these compounds from 1961 and was followed by voluntary withdrawal of the fertilizer mixtures containing aldrin and dieldrin that had been used in the late 1950's and early 1960's.

The 1964 report of the Advisory Committee on

Pesticides and other toxic chemicals recommended restrictions on the use of aldrin, dieldrin, and heptachlor within the United Kingdom, in order that "the present accumulative contamination of the environment by the more persistent organochlorine pesticides should be curtailed." In connection with this report, it was emphasized, as in the more recent one by the same committee in 1969,[60] that the recommendations made referred entirely to the British situation, and other countries having an entirely different background to their organochlorine usage could be expected to view matters differently.

It has been estimated that, as a result of the Advisory Committee's 1964 recommendations, there was a 25% fall in the total annual tonnages of aldrin, dieldrin, DDT, and DDD used in horticulture and agriculture between 1964 and 1969, and a further 20% reduction was anticipated as a result of the additional recommendations made in 1969.[60] The situation regarding organochlorine usage in Britain in the 1960's is summarized in Table 4, from which it is seen that the use of aldrin, dieldrin, and heptachlor fell drastically during the 1960's; heptachlor has not been used very much in Britain since the seed dressing incidents, and the sales of aldrin and dieldrin decreased by 60% during 1964 to 1967 and by 22% between 1967 and 1970.[54] The interesting environmental consequences of these changes are referred to elsewhere (Vol. II, Chap. 3B.1c). The information in the Advisory Committee's 1969 report in many ways serves to emphasize the continuing value of aldrin and dieldrin as soil insecticides in Britain despite the incidence of resistance, and it is clear that although other compounds, particularly organophosphorus insecticides, are being evaluated or used as replacements, control with the same economy will hardly be possible when the recommended withdrawals have been completely effected. According to a recent report,[305] aldrin is still used in the United Kingdom for the protection of wheat, spring barley, potatoes, brassicae, hops, narcissus, and ornamentals. The use of endrin for mite control on soft fruits (strawberries, blackcurrants, and blackberries) and on narcissus and ornamentals grown under glass continues at present because, with the exception of endosulfan, which is in a class by itself, no satisfactory replacements are available; the use of endrin on apple trees is discontinued. In regard to other uses,

aldrin and dieldrin were discontinued as sheep dips after 1966; their use in food storage practice ceased voluntarily in 1964, but dieldrin is still used as an ingredient of some permanent coatings in other situations, as in moth-proofing and for various formulations used in professional operations against ants and cockroaches. The small use of chlordane is again in domestic situations for ant and cockroach control as well as for lawn and turf pests.

A list of aldrin uses dated early 1972 shows that this chemical is not used in Scandinavia or West Germany, but is approved for various purposes, mainly agricultural, in Holland, Belgium, France, Spain, Austria, and Italy. Outside Europe, another 47 countries, including the United States, are listed as current users of aldrin, again mainly for agricultural applications.[305]

The United States is one of the world's largest grain producers, contributing about half the global maize (corn) production of 215 million tons in 1965—66. Aldrin is mainly used for the protection of this and other field crops as a soil treatment or seed protectant, although it also has some foliage applications. A useful summary of the current usages and the pests controlled was given for aldrin, dieldrin and endrin in a status report of 1967;[306] the situation had then already changed since 1964 due to voluntary withdrawals of certain usages, and it has to be remembered that it is still changing under pressure for the continuous review of these compounds. The 1972 appraisal lists aldrin uses on some 35 crop groups; no mention of dieldrin applications is made (it is, of course, formed by oxidative conversion of aldrin in the environment), and this is presumably a measure of the restrictions placed upon it since 1969. Endrin (as in November 1971) is registered for use as foliage or soil surface treatment with various field crops such as cereals and cotton (and as a seed treatment for cotton), as granules against sugar cane pests, and for post-harvest ground treatments in apple orchards. The review process has just been completed again at the time of writing, so that some further changes are likely.

In the United States, the benefit-risk concern with regard to pesticides came into prominence following the report of the President's Science Advisory Committee on the Use of Pesticides in May 1963.[312] This report pointed out that since pesticides are usually general toxicants, some of them are capable of causing illness or even death to people and wildlife if given in excessive doses. It was also emphasized that this possibility had to be balanced against the worldwide benefits arising from crop protection, increased productivity, and the control or eradication of disease-carrying insects. The crucial statement was that "Elimination of the use of persistent pesticides should be the goal." Since that time there has been a continuing confrontation between pesticide manufacturers concerned with persistent compounds and federal regulatory agencies. There is no doubt that the controversy initiated by persistent pesticides has resulted in an uncomfortable but beneficial awareness of other problems (many far more serious) related to the human environment and in this sense another debt is owed to these compounds. What seems strange is that the crusading zeal to effect total bans on the use of these compounds appears to have completely ignored, at least in the early years after 1963, the vital question of efficient and economic replacements to maintain food production and disease control at their existing levels.

A brief history of events in the United States serves to bring the current situation into perspective. In April 1963, a National Academy of Sciences — National Research Council Committee was appointed on the recommendation of the Food and Drug Administration (FDA) to review the tolerances then in force for aldrin and dieldrin residues on raw agricultural commodities. Reporting on March 25, 1965, this Committee found that certain uses of these compounds were essential in the absence of suitable substitutes, but that certain tolerances should be lowered and a number of toxicological questions answered during a period of three years ending March 1968. The Committee had recommended a gradual and orderly reduction of the tolerances in question, so that an FDA proposal (published 4 days later) to reduce aldrin and dieldrin tolerances on a large number of crops to either 0.1 ppm or zero caused considerable consternation, the immediate imposition of zero tolerances effectively constituting a ban on the practical use of these compounds for insect control on the commodities concerned. The concept of zero tolerances was actually under consideration by a National Academy of Sciences Committee at the time; this Committee reported towards the end of June 1965 and considered that " the concepts of 'no residue' and 'zero tolerance' as employed in the registration and regulation of

pesticides are scientifically and administratively untenable," and that pesticides should in future be registered on the basis of either "negligible residue" or "permissible residue," with a transition period of 5 years between the existing regulations and a new system of finite tolerances, to avoid problems arising from a sudden transition.

Before the advent of electron-capture, gas-liquid chromatography (GC) as a residue analysis method (Vol. I, Chap. 2C.3a), the limit of analytical sensitivity for endrin residues was 0.1 ppm, so that no residues below this level could be detected. The residue tolerance was set at this level in 1956, and, since higher residues had not hitherto been found at harvest, a zero tolerance was effectively established for endrin on 34 crops. All this changed following the introduction of GC with electron-capture detection; residues of endrin could now be detected below the set level of 0.1 ppm and labels registered for the use of endrin on broccoli, brussels sprouts, cabbage, and cauliflower were withdrawn in November 1963. A finite tolerance of 0.1 ppm was accordingly requested by Shell to allow for the now recognized existence of measurable residues up to this level. The FDA retorted that the data supporting the request were insufficient to permit re-establishment of the former tolerance and the issue was decided by an Advisory Committee on endrin which reported on September 14, 1966, requesting further toxicological and metabolic information about the compound in order to establish a new residue tolerance level for it. Since 1963, therefore, the exchanges between the FDA and the Shell Chemical Company in relation to aldrin, dieldrin, and endrin are best described as a series of thrusts by the former and parries by the latter, assisted at times of deadlock by appropriate Advisory Committees. During these proceedings, a great deal of information has accumulated about the behavior of these compounds in biological systems, both in response to the regulatory processes and for other, more academic reasons. A summary of the history of these events to 1967, together with outlines of the related toxicological and metabolic investigations undertaken by Shell or under the Company's direction, is given in the aldrin, dieldrin, and endrin status report of that year and summarized elsewhere by Glasser.[306],[313]

The sales peak for dieldrin occurred in 1956, and its use in the U.S. since then has steadily decreased on account of resistance in agricultural insects; there was a 70% reduction between 1956 and 1968, and forward estimates in 1969 indicated a likely further 10% decrease by 1972, the remaining uses being nonagricultural by then.[1] By 1968, aldrin sales had fallen by 30% from the peak in 1966, and forward estimates in 1969 indicated an overall fall of 60% from this peak so that the situation seems roughly parallel to that in Britain in this respect. The fate of aldrin and dieldrin in the U.S. currently depends on the result of public hearings, scheduled for late in 1973, on the benefits versus risks of their use.

The major use of endrin is against cotton pests and a decrease in this use is expected between 1969 and 1973 due to resistance problems. In other countries, the continued low cost contribution of aldrin, dieldrin, and endrin (along with DDT and HCH) to the protection of agricultural crops cannot be overestimated and is particularly significant in the developing countries of South and Central America, Africa, and Asia. The continuing value of aldrin and dieldrin, especially the latter, for the control of structural pests such as termites is also evident. In India, the area treated with pesticides increased from 10,120 ha in 1946—47 to some 6.15 million in 1961—62, and 17.4 million (representing 11.2% of the total crop area) in 1965—1966, and a report of 1967 envisaged an increase in aldrin, dieldrin, heptachlor, and chlordane usage from 90 to 1,050 tons between 1965 and 1969.[1] Insecticides are being evaluated as substitutes for endrin in many developing countries, but economic factors make other compounds unpromising; it is estimated, for example, that the use of substitutes against rice and cotton insect pests in India would nearly double the cost of control. More than 60 countries in Africa and Asia are liable to attack by locust swarms, which can consume 3,000 tons of food a day. There was a decrease in the number of swarms during 1962—68, but an outbreak in 1969, which was checked by spraying with HCH and dieldrin, showed that constant vigilance is essential; other insecticides are less effective or too expensive.

The obvious conclusion from these trends is that the advanced nations will contribute to the development of the poorer ones by providing them with the cheaper, persistent insecticides, while themselves forswearing these compounds and thus paying more for food protection employing the less persistent types. This will help to reduce the global burden of persistent pesticides, but it remains to be seen whether this situation will be entirely acceptable to the taxpayers of more advanced countries.

4. Applications of Chlordecone (Kepone), Mirex, and Pentac

An interesting range of biocidal activity residues in the structures chlordecone, mirex and "Pentac," derived from two molecules of hexachlorocyclopentadiene by different modes of dimerization (Vol. I, Chap. 3B.1). Chlordecone, introduced in 1958 by the Allied Chemical Corporation under the code "GC-1189" and trade mark Kepone®, is formulated as a 50% wettable powder or 20% emulsifiable concentrate, and as granules or dusts (5 or 10%) for various applications against leaf-eating insects and as a larvicide against flies.[40] It is also incorporated into baits for the control of ants and cockroaches, and this use (against cockroaches) is mentioned in the W.H.O. 1970 report on chemical control of public health pests.[44] Mirex, introduced by Allied as GC-1283 in 1959, has little contact insecticidal activity, its major use being as a stomach poison against ants. It is generally employed as an ingredient of baits; thus, 0.15% baits are employed against the harvester ant and 0.075% baits against the imported fire ant.[40] Pentac, introduced by the Hooker Chemical Corporation in 1960, is a "linear" dimerization product rather than a cage molecule. It has no insecticidal characteristics, but possesses pronounced acaricidal activity and is recommended for mite control on roses and other ornamentals; the standard formulation is a 50% wettable powder.[40]

Of the three compounds, mirex is particularly stable; chlordecone is also stable, but dissolves in strong aqueous alkaline solutions, as indicated previously, and the free ketone readily forms hydrates. Pentac is not decomposed under acidic or basic conditions, but suffers pronounced loss of its acaricidal activity when heated at 130°C for several hours or when exposed to UV-light or sunlight. Water solubility is low, as usual, and all are moderately soluble in aromatic hydrocarbons. Chlordecone is generally determined in formulations by IR-analysis using the carbonyl absorption frequency, and in residues by GC-analysis; IR- and GC-analysis are also used for Pentac, and GC- analysis for mirex (for both product and residue analysis).

Chapter 4

GAMMA-1,2,3,4,5,6-HEXACHLOROCYCLOHEXANE

A. ORIGINS, PREPARATION, AND PHYSICAL PROPERTIES

In many respects, γ-1,2,3,4,5,6-hexachloro-cyclohexane (γ-HCH) also called γ-benzene hexa-chloride (γ-BHC), Jacutin, Gammexane®, or lindane, empirical formula $C_6H_6Cl_6$, is the most unique of the three main sturctural types of organochlorine insecticides, since unlike DDT or aldrin, it is the only highly insecticidal representa-tive of its group; the steric requirement for toxicity appears to be so specific that other isomeric forms of 1,2,3,4,5,6-HCH exhibit either no activity, feeble activity, or physiological effects of a different type.

The stereochemistry of cyclohexane derivatives, and of the hexachlorocyclohexane isomers, is discussed fully in a subsequent section, but a few words about nomenclature are necessary here for clarification. Each double bond of benzene can add two chlorine atoms to give the mixture of isomers of hexachlorocyclohexane, which includes the γ-isomer as one constituent. The same product might be obtained by substituting one hydrogen atom on each carbon of cyclohexane by chlorine but *not* by substituting each of the six hydrogen atoms of benzene by chlorine. This last operation produces hexachlorobenzene (C_6Cl_6; sometimes called benzene hexachloride), an aromatic molecule, so that it is confusing to use the term benzene hexachloride for hexachlorocyclohexanes, which are fully reduced cyclic systems ($C_6H_6Cl_6$). Also, since chlorinated cyclohexanes may carry more than one chlorine atom on each carbon, the position of each chlorine should be stated, as in the name given at the beginning of this paragraph. In this book, the abbreviation HCH (or technical HCH for the commercial material) is used for a mixture of the isomers of 1,2,3,4,5,6-hexachloro-cyclohexane, or in cases where it is not clear whether this mixture or the pure γ-isomer was used; the latter is referred to as γ-HCH or lindane.

In a strictly historical sense, hexachlorocyclo-hexane (HCH) is the oldest of the organochlorine insecticides, since its purely chemical origins can be traced to the earliest ventures in organic chemistry. In the *Philosophical Transactions* of 1825, Faraday described how he had compressed "oil gas" to give a condensate which contained bi-carburet of hydrogen (benzene), a product which reacted with chlorine in sunlight. The products consisted of a "solid body" and a dense, viscous fluid which were soluble in alcohol, and these experiments undoubtedly comprised the first preparation of technical HCH. The discovery was taken up by Mitscherlich some years later, and in 1833 he described some chemical properties of the product, including its decomposition by bases to give trichlorobenzene. Other reports (by Peligot in 1834 and Laurent in 1836) established its con-stitution. However, interest then seems to have lapsed, for it was not until some 50 years later that Meunier discovered the α- and β-isomers, which were later investigated by Matthews (1891).[314,315] In 1912, Van der Linden con-firmed the existence of these two isomers and reported the presence of two more, thus describing the α-(mp 158°), β-(mp over 200°), γ-(mp 108 to 111°) and δ-(mp 129 to 132°) isomers.[316]

According to an account by Lambermont,[315] the material was considered as a possible fumigant during the war of 1914–18, but was never made industrially. The origins of this idea are not clear but may have been based simply on the fact that HCH is a volatile chlorination product of benzene, and that compounds such as p-dichlorobenzene were already used as fumigants; hence HCH might conceivably be expected to have such properties. However, there was no real reason to suppose that HCH had insecticidal properties until 1935 when United States patent 2,010,841 by H. Bender (to Great Western Electrochemical Co.) described a process for the manufacture of HCH and other chlorinated cyclic hydrocarbons.[317] In the course of this description, the easily overlooked comment was made that HCH and other chlorides appeared to be good insecticides because "they slowly liberate hydrochloric acid vapor," but no attempt seems to have been made to develop the material as an insecticide. Other patents describing the production of HCH were issued around that time to Stephenson and Curtis (1936), Grant (1939), and Hardie (1940), for various purposes.[318]

The outstanding insecticidal activity of HCH

appears to have been discovered almost simultaneously and independently in England and France as a result of the urgent search for new chemicals to replace the loss of overseas sources of agricultural insecticides during World War II. In France, Dupire is said to have discovered the toxicity of technical HCH to clothes moths in 1940, as a result of which finding the material was evaluated against agricultural insect pests. An account of the French developments was given by Dupire and Raucourt in November 1943.[319] This account and subsequent ones did not reach England or America for some time afterwards on account of the difficulties of wartime communications. According to an account by Bourne in 1945, HCH was used in France in large quantities during the war under the trade name "Aphtiria," a patent covering the preparation having been taken out in 1941 by Gindraux; the material was first manufactured on the Continent, where the γ-isomer later came to be known as "Isogam," by the Solvay et Cie Company.[315] In England, the toxicity of technical HCH towards the turnip flea beetle had been observed early in 1942 in the laboratories of Imperial Chemical Industries Ltd., and early in 1943 it was shown for the first time that the γ-isomer was actually responsible for the insecticidal activity of technical HCH. Wartime restrictions prevented until 1945 the announcement that the material had been used in England and that the γ-isomer was the active toxicant; the trademark Gammexane for γ-HCH was registered by the Company on February 23, 1945, and a detailed account of the associated researches was given by Slade in the course of the Hurter Memorial lecture delivered on March 8, 1945.[321]

The discovery of the insecticidal activity of γ-HCH (lindane, after Van der Linden) has its origin in a search for easily synthesized chemicals to replace the desirable (from the point of view of efficacy and low toxic hazard) but expensive, natural insecticides pyrethrum and derris (rotenone) which began in the 1930's. Dr. C. C. Tanner, who was associated with the research work in the ICI laboratories at Widnes and later at Jealott's Hill, has made his account of these events available to the author, the following narrative deriving largely from this source. The original interest in hexachlorocyclohexane was actually based on its usefulness as an intermediate in the synthesis of trichlorobenzene isomers as nonflammable dielectrics; benzene is chlorinated in the presence of UV-light to give HCH, which is then dehydrochlorinated with lime or sodium hydroxide to give trichlorobenzenes. This process was initially conducted on a pilot plant scale at ICI, Widnes, and while the photochlorination proceeded smoothly, the dehydrochlorination presented difficulties on account of the presence of unreactive material which proved to be β-HCH. Since the remainder of the material dehydrochlorinated readily it was regarded as the preponderant α-isomer, the actual composition being of no special concern since only the dehydrochlorination products were of interest at the time. Samples of crystalline HCH, presumably consisting mainly of the inactive isomers, were sent to Jealott's Hill for routine biological evaluation and were not found to be of particular interest, although according to Holmes,[322] internal records at Jealott's Hill showed that "tests in 1937 proved it to be quite active." According to Slade,[321] several thousand chemicals from the Company's various laboratories, along with compounds made specifically for evaluation as pesticides, were tested in this manner at Jealott's Hill during a 5-year period from 1934. At first the procedure was to determine median lethal doses of the chemicals for a standard range of insects, but effort was later concentrated on ways to control specific pests, such as turnip flea beetle.

In the early days of World War II, discussions on the development of substitutes for nicotine in greenhouses led to a reconsideration of some chemicals already available as possible ingredients of insecticidal smokes; a short list of 13 chemicals included the strong-smelling monochloro-HCH (heptachlorocyclohexane) and orthodichloro-HCH (octachlorocyclohexane), which had just been made for another purpose. In the event, none of these compounds was actually tested until 1942 when they were remembered in the course of the search for alternatives to derris for the control of turnip flea beetles. Extra quantities of the heptachloro- and octachlorocyclohexanes were then sent from Widnes, and, because fairly large amounts of technical HCH were available from the earlier work on nonflammable dielectrics and were kept in store, a sample of this was included. Early tests with the latter by Dr. F. Thomas were promising and by the summer of 1942 it was evident that of the derris substitutes tested, only HCH was likely to be effective. Pure samples of α- and β-HCH were now prepared and found to be inactive, so the search for the active ingredient began.

Pending the completion of a pilot plant to produce sufficient HCH for the flea beetle season of 1943, it was necessary to maintain supplies of HCH for the development work at Jealott's Hill. Something of the atmosphere surrounding these wartime operations is conveyed by Dr. Tanner's account of the arrangements. He writes that ICI chemists . . . "continued to make small quantities of the crystalline product by chlorinating benzene on the roof of a convenient building in the fitful Widnes sunlight and filtering of the solid in a large Buchner filter. When the sun went in, chlorine would dissolve in the benzene, only to react violently when the sun came out. In this way, Jealott's Hill was kept supplied."

Since the crude material contains at most only about 15% of the active γ-isomer, the biological effectiveness of any batch will be very dependent on its processing; recrystallization in particular is likely to lead to crystalline material containing an increased proportion of the less soluble, inactive α- and β-isomers. Accordingly, it is not surprising that although the field trials of 1942 were promising, the results with different batches were not always consistent. Nevertheless, a 20% powder concentrate of crude HCH and gypsum was made for dilution to "field strength" dusts; some hundreds of tons were sold and the material was at least as effective as derris powders and did not injure the plants. Early in 1943, the γ-isomer was isolated and found to be the most toxic substance to weevils so far screened by the Company. This exciting discovery showed at once that much of the active material had previously been rejected with mother liquors during the preparation of the crystalline material and explained why the technical material was so much more active than the crystalline product that only about 7 tons of it were actually needed to meet the original requirement for 30 tons set for the new pilot plant.

Following the discovery that the γ-isomer is the active component of HCH, activities were directed towards the development of full-scale manufacturing facilities and particularly towards a search for ways of increasing the lindane content of technical HCH. Methods were developed for extracting the lindane from crude HCH with solvents such as toluene or xylene (alone or mixed with light petroleum), or with trichloroethylene, cyclohexane, methyl acetate, or glacial acetic acid

as detailed in patents of 1944.[314] In this way, most of the lindane content would be recovered in a form suitable for incorporation into insecticidal formulations, but the use of such solvents is expensive. A final product containing virtually 100% of lindane can be obtained by further recrystallization of the material isolated in this manner, but this degree of purity is not required for most purposes.

News of the new Swiss insecticide (which later proved to be DDT) reached the ICI scientists around Christmas of 1943. No information regarding chemical structure was available, and the reported properties of the Swiss product seemed so similar to those of HCH that speculation arose as to whether the two were variants of the same chemical. At a later date the Widnes laboratory was to become involved in DDT synthesis. The most remarkable aspect of the Gammexane story relates to the difficulties created by the low content of the active isomer in the technical product, a situation which is quite different from that attending the synthesis of DDT or the cyclodienes. Had the insecticidal properties of HCH not been confined to the γ-isomer, it seems evident that the discovery of its insecticidal properties would have preceded the discovery of those of DDT by several years.

Very soon after Slade's presentation of the Hurter lecture in 1945, extensive investigations of HCH were begun in the United States, the earliest work being done with various technical grades of the material obtained from England.[318] All of these had the penetrating musty odor which became a famous characteristic of HCH. The γ-isomer content of different preparations used varied from 10 to 35%, and was initially determined by bioassay, no easily conducted chemical or physical methods of analysis having been described at that time. Since an outstanding feature, at least of technical HCH, is its ease of manufacture from chlorine and benzene, production was soon undertaken in America, and as many as 12 chemical companies were involved by 1947. Indeed, the simplicity of the route to the technical grade of HCH, and its value as an insecticide, even with the low lindane content, resulted in a number of countries manufacturing this product before they were able to produce DDT. Unfortunately, the use of such material in practice means that much chlorinated material (that is, the biologically inactive remainder, constituting 85% or more of the total) entered the environment to no purpose.

For example, in Japan, technical HCH has been extensively used for economic reasons and the high levels of the chemically rather inert β-isomer that have accumulated in human adipose tissue, and in the environment generally in that country led to a recent total ban on the use of HCH products.[323] If the inevitably more expensive lindane could have been used from the beginning, the current situation would no doubt be rather different. This is a good example of an individual country making the necessary adjustments in pesticide usage to meet its own particular circumstances; the γ-isomer is extensively used elsewhere in the world, and is not usually regarded as a problem compound.

The general pattern with regard to HCH production appears to be one of continuous decline. Lindane formulations are produced and used in Britain but neither technical HCH nor lindane are now made there. The 12 American manufacturers mentioned by Haller and Bowen in 1947[318] evidently increased to 16 by 1959, only to decline sharply to three by 1962, their total annual production capacity at that time being given as 21,000 metric tons.[314] This represents a considerable decline from the capacity available in 1952, and the reverse seems to have begun in 1954 when the use of HCH against plant bugs on forage crops was banned by the U.S. Government. The 1971 revision of Kenaga's list of organic insecticides cites the Hooker and Olin Mathieson Chemical Corporations and the Stauffer Chemical Company as manufacturers in the United States, and Cela Landwirtschaftliche Chemikalien Gesellschaft in West Germany.[324] In Japan, total HCH production increased from 15,400 metric tons in 1960 to nearly 46,000 in 1968, of which about 2,500 were exported. About 18% of the annual production there is processed for lindane extraction, production of this isomer having more than doubled between 1964 and 1968. The 1969 figures show a decline in production (to 35,400 metric tons) and this reduction predates the ban on HCH for home use. Four companies, Mitsubishi, Mitsui Toatsu, Mitsubishi Edogawa, and Nippon Soda, are listed in 1970 as manufacturers of HCH, and its production for export continues.[325,326]

In principle, hexachlorocyclohexane can be made in three different ways; cyclohexane itself may be chlorinated, cyclohexene may undergo both additive and substitutive chlorination to give the same product, or benzene may add three molecules of chlorine. The first two methods are of theoretical interest only, and the third is the well-used method for HCH preparation. Reaction can occur between benzene and chlorine to give either substitution products ranging from monochlorobenzene to hexachlorobenzene (in which all the hydrogen atoms are replaced by chlorine), or addition products, the hexachlorocyclohexanes, depending on the conditions employed. Substitution occurs in the presence of suitable catalysts (iodine, iron, ferric chloride, and aluminum chloride), while addition is effected in the presence of UV light or sunlight. Addition can also be promoted in other ways, as by the addition of organic peroxides or certain unsaturated compounds or by γ-irradiation from a ^{60}Co source. With the theoretical amount of chlorine (3 molecules per molecule of benzene), the reaction proceeds rapidly to completion but an oily mixture of the various isomers of tetrachlorocyclohexane can be obtained by partial additive chlorination.[327] The chain reaction leading to HCH is first order with respect to chlorine and benzene and is initiated by the addition of atomic chlorine to one double bond of the benzene ring:[314]

$$Cl_2 \xrightarrow{h\nu} 2Cl^\bullet$$

$$C_6H_6 + Cl^\bullet \longrightarrow C_6H_6Cl^\bullet \xrightarrow{Cl_2} C_6H_6Cl_2 + Cl^\bullet$$

$$C_6H_6Cl_2 + Cl_2 \longrightarrow C_6H_6Cl_4 \begin{array}{c} \xrightarrow{Cl^\bullet} C_6H_6Cl_5^\bullet \xrightarrow{Cl_2} C_6H_6Cl_6 + Cl^\bullet \\ \xrightarrow{Cl_2} C_6H_6Cl_6 \end{array}$$

This reaction leads to the mixture of isomers of hexachlorocyclohexane and some other products that is known as technical HCH. Commercially, the chlorination is usually carried out using excess benzene as solvent, but processes have been described that utilize methylene chloride, ethylene dichloride, chloroform, carbon tetrachloride, tetrachloroethane, acetone, or acetyl chloride for this purpose. Reaction temperatures commonly vary between 5 and 60°C, and care is taken to exclude any materials that might catalyze substitutive chlorination to give chlorobenzenes. The original patent of Bender[317] describes the chlorination of benzene at −15°C in the dark, for which method the free radical chlorination mechanism given above is unlikely to hold. Since benzene solidifies at temperatures below 6°C, other solvents have to be utilized as diluents. High concentrations of chlorine in the reaction mixture are said to favor the formation of the γ-isomer in the technical product, and the chlorination may be batchwise or continuous. In batch processes, chlorination may be effected at 15 to 20°C in vessels lined with lead, and is continued until 12 to 15% of HCH is present in the reaction mixture; higher concentrations result in the deposition of the less soluble isomers. Excess chlorine is then removed by blowing air through the mixture and the benzene is evaporated and recovered for further use.

In a process devised by the Stauffer Chemical Company, benzene and chlorine are continuously fed into a lead lined chlorinator and the reaction is conducted at 40 to 60°C and atmospheric pressure with frequent checks of specific gravity to determine the HCH content; the benzene is mostly evaporated at 85 to 88°C at the end of the chlorination and returned to the chlorinators after a cleanup. The residual liquid deposits mainly the less soluble α- and β-isomers when cooled to 35 to 40°C, and these are filtered off and the liquid residue allowed to solidify in shallow pans after a further steam distillation to remove remaining benzene. Such samples may contain up to 25% of lindane. Besides the chlorination in homogeneous phase, benzene has been chlorinated in aqueous or aqueous alkaline dispersion; this process is said to be more difficult to control than the other one but the technical product has been claimed to contain up to 18% of the γ-isomer. This method was used by Gunther, who prepared small batches of HCH by the exhaustive chlorination of benzene emulsified in an equal volume of 2% aqueous sodium hydroxide. The mixture was irradiated for 75 hr with a powerful UV-source giving wavelengths between 290 nm and the visible, and the 295 g of semicrystalline material produced contained 31.2, 0.9, 41.7, and 5.7%, respectively, of the α, β, γ, and δ-isomers.[329]

The addition at the chlorination stage of various compounds such as carboxylic acids or their anhydrides, nitrotoluenes or the chlorides of sulfur, selenium, or tellurium, has been claimed to improve the yield of γ-isomer in technical HCH, but there is little real evidence that the improvements effected in the initial proportion of this isomer present are more than marginal, and so the manufacture has always been overshadowed by the presence of 80% or more of unwanted material in the product. Various proposals have been made to utilize this remaining material, as, for example, in a process for 1,2,4-trichlorobenzene that can be used subsequently in the manufacture of the acaricide tetradifon. In principle, the conversion to chlorinated benzenes and their derivatives provides a route to a number of useful synthetic intermediates and crop protection chemicals, and a list of these is given in the book by Melnikov,[330] but it is difficult to say how many such products actually arise from this source. Examples are hexachlorobenzene (selective fungicide), chlorinated phenoxyacetic acids (herbicides), various chlorinated nitro- and dinitro-benzenes useful as fungicides, pentachlorophenol (herbicide), and so on.

Technical HCH is an off-white to brown amorphous powder with a very characteristic penetrating musty odor, described by one source as being reminiscent of the odor of phosgene. It begins to melt at about 65°C and consists of a mixture of the α-(55 to 70%), β-(5 to 14%), γ-(10 to 18%), δ-(6 to 10%) and ε-(3 to 4%) isomers, together with up to 4% of heptachlorocyclohexane and up to 1% of octachlorocyclohexane, the last two compounds being at least partly responsible for the characteristic odor. The η-isomer has also been isolated from technical HCH. From what has been said previously, the composition of the technical material is inevitably somewhat variable, depending on the nature of the manufacturing process used. Some physical constants of the major constituents, including the low values obtained for water solubility, are given in Table 20 and the solubilities of the major HCH-isomers in various organic solvents in Table 21. The

TABLE 20

Physical Properties of the Major Constituents of Technical Hexachlorocyclohexane

Isomer	Alpha	Beta	Gamma	Delta	Epsilon	γ-Heptachloro-cyclohexane	o-Octachloro-cyclohexane
mp(°C)[a]	159–160°(159.2)	309–310°(311.7)	112–113°(112.9)	138–139°(140.8)	219–220°(218.2)	85–86°	147–149°
Vapor pressure[b] (mm Hg at 20°C)	2.5×10^{-5}	2.8×10^{-7}	9.4×10^{-6}	1.7×10^{-5}			
Heat of combustion[c] Kcal/mol	659.01	657.25	662.32	659.24			
Refractive index (n_D^{25})	1.630 ± 0.002[c]	1.633 ± 0.004[c]	1.644 ± 0.002[c]	$n_D^{20} 1.576–1.674$[d]			
Crystalline form[c,d]	Monoclinic prisms	cubic (octahedral)	Tablets, plates prisms	Fine plates	Monoclinic needles or hexagons		
Dipole moment[c] (D) in benzene	2.16	0	2.84	2.24	0		
Solubility in water at 25°[e] (ppm)	1.63	0.70	7.90	21.3			

[a] Data adapted from Metcalf;[41] figure in parentheses from Hardie.[314]
[b] Data adapted from Balson;[333] much higher values are given by Slade.[321]
[c] Data adapted from Hardie.[314]
[d] Data adapted from Ulmann.[340a]
[e] Data adapted from Kanazawa et al.[334]

TABLE 21

Solubilities of the Major HCH Isomers in Various Organic Solvents at 20°C(g/100 g Solution)

Solvent	α	β	γ	δ
Acetic acid (glacial)	4.2	1.0	12.8	25.6
Acetone	13.9	10.3	43.5	71.1
Dioxan	33.6	7.8	31.4	58.9
Methanol	2.3	1.6	7.4	27.3
Ethanol	1.8	1.1	6.4	24.2
Chloroform	6.3	0.3	24.0	13.7
Carbon tetrachloride	1.8	0.3	6.7	3.6
Trichloroethylene	3.7	0.3	14.7	7.6
Ether	6.2	1.8	20.8	35.4
Chlorobenzene	7.4	0.4	23.4	21.4
Benzene	9.9	1.9	28.9	41.1
Xylene	8.5	3.3	24.7	42.1
Petrol ether (40–60°C)	0.7	0.1	2.1	1.6
Heavy naphtha (230–270°C)	5.8	1.5	18.1	30.4
Odorless distillate (198–257°C)	0.8	0.02	2.0	1.1
Diesel oil	1.5	0.3	4.1	9.2

(Data adapted from Slade.[321])

solubilities have been very fully investigated in connection with attempts to discover economic routes for the isolation of lindane from the technical material, and Slade lists solubilities of the α, β, γ, and δ-isomers in 45 different solvents.[321] They are generally in the order $\delta > \gamma > \epsilon > a > \beta$, which facilitates the separation of lindane from the technical material. This was done originally by treating technical HCH with the minimum amount of methanol to extract mainly the γ- and δ-isomers.[321] The insoluble a- and β-isomers are removed, and the methanol filtrate evaporated to give nearly pure γ, followed by successive crops containing γ plus β, and mixtures of γ,β, and δ. The γ-isomer crop is recrystallized from chloroform, and the δ-isomer is less easily obtained from the final crop by fractional precipitation from methanolic solution with light petroleum, followed by recrystallization from chloroform. The a- and β-isomers are readily separated from their mixture on account of the poor solubility of the latter in most solvents. On the commercial scale, technical HCH may be ground up and extracted with an aromatic solvent such as toluene or xylene, cyclohexane, glacial acetic acid, methyl acetate, etc., to remove lindane from the mixture, as described in the original patents by ICI.[331,332] Another method involves countercurrent extraction by partitioning the material between immiscible solvent pairs such as hexane and nitromethane, and in the laboratory, adsorption chromatography has frequently been used for the separations. Anthranilic acid forms a complex with lindane which can be used to remove this isomer from the mixture; decomposition of the complex with warm aqueous methanol then regenerates lindane, and this process is the subject of a German patent of 1959.[335] With regard to the practical grades of HCH, the W.H.O. specifications distinguish between the technical product containing at least 12 and not more than 16% γ-HCH, refined HCH with a γ-content between 16.1 and 98.9%, and lindane itself, containing not less than 99% of the γ-isomer.[86]

The hexachlorocyclohexanes are rather stable chemically, and unlike DDT, do not lose hydrogen chloride when heated with traces of iron and other metals and their salts. They are also stable to light and oxidation; recrystallization from hot nitric acid leaves the material unchanged. Boiling water or steam liberates only traces of hydrochloric acid, but at 200°C in sealed tubes, 1,2,4-trichlorobenzene is formed. To inhibit the decomposition of HCH stored for long periods at temperatures somewhat above ambient (for example, at 50°C), the incorporation of 1% of sodium thiosulfate is said to be effective, but storage of technical HCH for reasonable periods under normal conditions does not require stabilizers. The lability in the presence of bases is, however, well known; lime water at 60°C or even at ordinary temperature will dehydrochlorinate HCH isomers (β-HCH being the most stable) to mixed trichlorobenzenes. On the other hand, dehydrochlorination does not occur in the presence of calcium carbonate either dry or in the presence of water at 60°C, and 0.2 to 5% of this compound has been incorporated with HCH to inhibit its attack on paper sacks used to contain it.

The odor of technical HCH has always been a problem and has been attributed to the presence in it of one or more heptachlorocyclohexanes, an octachlorocyclohexane and an enneachlorocyclohexane. Purified lindane usually has only a slight odor and according to another source the smell appears when this compound is stored for a long time in the presence of light and moisture and traces of bases, and is due to the slow formation of pentachlorocyclohexene and tetrachlorocyclohexadiene by dehydrochlorination.[330] Various ways of reducing the odor have been described. They include low temperature chlorination (-8°C), heating the technical product with aluminum chloride followed by extraction of the organic material with light petroleum, steam distillation of the benzene solution, treating the ground solid with oleum, or shaking it with calcium carbonate or manganese dioxide in air for several hours.[314] A related problem, and one which has always attended the use of HCH for agricultural purposes, is the "taint" or "off-flavor" imparted to some edible crops that have been treated with it for insect control. The isolation of lindane and the development of formulations using it greatly improved this situation, but the problem of taint is still mentioned in connection with present day uses of HCH. "Taint" has been attributed to the presence in HCH of the previously mentioned impurities, but since lindane is now mostly used it seems likely that any off-flavor arises from breakdown products of this isomer on the crop. Traces of chlorinated phenols, for example, might be expected to have such an effect. Further mention of this problem is made in connection with agricultural applications.

1. Applications in Agriculture and Against Pests Affecting Man and Animals

By the end of World War II, much work had been done on the development of HCH for the control of a wide range of insect pests in agriculture and other areas and some of the results are summarized in the Hurter lecture of 1945, and in reports of evaluations subsequently made in America.[336] As far as formulations for practical use are concerned, the principles generally applicable to other organochlorine compounds and described previously for DDT and the cyclodienes apply also to the formulation of HCH; all the grades can be readily incorporated into solutions, emulsifiable concentrates, wettable powders and dusts, as in the other cases. However, it has to be remembered when formulating the material that HCH is readily converted into inactive isomers of trichlorobenzene in the presence of bases, in some cases even at room temperature. The 20% dust concentrate made with gypsum was mentioned earlier.[321] Practical solutions were made by diluting upwards of 5% lindane in decalin, xylene, carbon tetrachloride, or perchloroethylene with kerosene or other oils. The addition of emulsifying agents such as Turkey red oil to the solution concentrates gave emulsifiable concentrates suitable for subsequent dilution with water at the point of application. The volatility of lindane and its exceptional stability at elevated temperatures made it suitable for evaporation from thermal vaporizers or dispersal in the form of a smoke.

Compared with most other organochlorine insecticides, lindane is outstanding for its speed of action as well as high acute toxicity as a stomach contact or fumigant poison. Indeed, having regard to the fact that it is quite readily degraded in many insects, it is probably the most toxic of the practically used compounds in the organochlorine group in terms of actual concentration of toxicant required at the site of action. For this reason, it is frequently used as a standard for comparative evaluation of other insecticides and a number of uses have been mentioned already in connection with DDT, with which it is frequently coformulated, and when discussing (Vol. I, Chap. 3C.3b) the early evaluation of aldrin and dieldrin made by Kearns and his colleagues.[302] The discovery that even technical HCH was more effective than derris dust against turnip flea beetle came at a time when replacements for the natural insecticide were urgently needed and hundreds of tons of HCH dust were used annually in Britain against this pest after the initial discovery in 1943. Subsequently, the range of activity was found to extend to, among others, Orthoptera (locusts, crickets, cockroaches), Lepidoptera (the numerous leaf-eating larvae attacking orchard and agricultural crops; various armyworms), Coleoptera (mustard beetle, blossom beetle; various weevils such as apple blossom weevil, pea and bean weevil; stored product pests such as grain weevils and hide beetles), and various Hymenoptera and Diptera.[321]

Early tests on locusts indicated that HCH was remarkably toxic when incorporated into locust baits;[24] even the technical material containing only 10 to 12% of lindane proved to be substantially more toxic than sodium arsenite, and excellent results were obtained in the field with HCH at tenfold lower concentration than the sodium arsenite in the treatment used for comparison. However, the effectiveness of HCH dusts appeared to vary with the locality (probably on account of differences in climatic conditions, since HCH is more effective at higher temperatures) and they were more effective than chlordane dusts against grasshoppers (*Melanoplus* spp.) in Illinois, but inferior to the cyclodiene against this species in Colorado. Aphids are also quite well controlled by lindane, and the advantages conferred by the fumigant action are seen in situations where these and other pests can hide in cracks and crevices on the plant and so escape the normal contact action of most toxicants. A good example is provided by the boll weevil, infesting cotton. DDT is relatively ineffective against this insect; the adults are difficult to kill and the larvae, which remain within the cotton bolls, are protected from contact with toxicants. HCH, presumably assisted by its fumigant action, can poison larvae within the bolls and is also toxic to adults. Thus, in the early 1950's, 3% lindane dust (applied at 10 lb per acre), or dusts containing 3% lindane plus 5% DDT (also formulated with added sulfur), accounted for a substantial part of the control measures against cotton pests, since lindane is effective against most species except bollworm (*Heliothis* spp.). During the sulfur shortage of the early 1950's there was, however, a move towards the use of toxaphene and the cyclodienes in place of lindane.

The fumigant action also makes the chemical a useful soil insecticide, although for some other field uses, especially in exposed situations, the relatively high volatility results in poor performance on crops. The other problem in these uses is that of tainting of crops mentioned previously. In Britain, one of the largest uses of HCH has been for controlling wireworms (*Agriotes* spp.) attacking cereals and sugar beets. Cereals do not appear to acquire taint, nor does the refined sugar from beets, although there are indications that the fresh beets may sometimes acquire it. However, potatoes are especially liable to acquire taint when they are grown on land which was previously treated with HCH, and the effect can persist for some years after the original application. Much depends on the mode of application, which, if uneven, may cause patches of taint in the crop. Surface applications, as for the control of flea beetle (*Phyllotetra* spp.) are a smaller problem in this regard than applications for which the same amount of chemical is worked into the soil. Under British conditions, it was discovered that technical HCH could be used for the control of orchard insects on fruit without tainting problems, but blackcurrants are particularly susceptible to this phenomenon so that HCH cannot be used in orchards where this soft fruit is grown under the trees. This question of taint is discussed in detail by Holmes,[322] who comments that crops such as tomatoes and mushrooms do not appear to taint, whereas onions, with a strong natural flavor, can be tainted rather easily. Taint may arise either from HCH remaining on a crop, or from HCH or its transformation products within the crop; the variation between species perhaps suggests some association with varying routes of toxicant degradation in vivo. For crops such as cereals, apples, plums, tomatoes, strawberries, and mushrooms, which appear not to absorb the material, tainting is unlikely if sufficient time is allowed for volatilization and weathering of spray deposits before harvest.

The production of substantially pure lindane at reasonable cost presented great difficulties for some time and it is inevitably more expensive than the technical material. Accordingly, the technical material had to be used for operations such as mosquito or locust control, in which low cost is paramount and tainting problems do not normally arise. This situation led to some benefits in terms of improved pesticide technology, because of the necessity for new techniques providing for economic and effective use of minimal amounts of lindane. It transpired, for example, that a minimum of 0.5 lb per acre of pure γ-HCH was required for effective wireworm control during potato growing, and this was likely to cause taint unless the treatment was spread over two seasons. A major step towards the solution of both the economic and tainting problems was therefore made by incorporating lindane into organomercurial seed dressings, since in this way as little as 8 to 12% of the previously used amount of toxicant gave nearly complete protection of cereal crops against wireworm attack. In Britain, this discovery reduced the cost of an effective application about sixfold compared with that of a broadcast treatment with the toxicant.[322]

Prior to the introduction of the organochlorine insecticides, the control of orchard fruit pests required a complex assembly of traditional control agents. Among some nine major groups of pests attacking apple in Britain, apple blossom weevil, woolly aphis, and sawfly were traditionally difficult to control. The apple blossom weevil problem was largely solved by DDT, and 50% wettable powders of technical HCH were later found to control a wider range of these insects than had been possible with any previous chemical control program. Sawfly infestations were controlled more efficiently and with less critical timing than hitherto by application of 0.2 and 0.1% suspensions of HCH at the "green cluster" and "petal fall" stages, respectively.

Neither HCH nor DDT is recommended against the codling moth because they do not control red spider mites and their inimical effects on mite predators are now known to be responsible for the severe infestations observed in orchards where the compounds have been used. For this reason, the use of organophosphorus or carbamate alternatives has increased considerably in recent years. HCH is not normally phytotoxic when applied to plants at rates which are insecticidal, except, as in the case of DDT, to cucurbits. However, cucurbits which were susceptible to technical and partly refined grades of HCH appeared not to be harmed by preparations containing more than 90% of γ-HCH, and of the individual isomers, only δ- proved to be toxic to leaves and stems.[337] Seedlings of various root crops, kale, and spinach are said to be damaged by wettable powders or dusts containing more than 0.04% of lindane, so care needs to be exercised in using the material, but most crops appear to grow without harm in soil containing much higher levels than are produced by applica-

tions at normal rates. Evidence has been advanced that HCH resembles colchicine in inducing chromosomal polyploidy in growing seedlings. A comprehensive review of the earlier literature on both phytotoxicity and taint has been given by Brown.[24]

The comparisons with DDT given in Table 14 show that in the public health area, HCH is generally a better toxicant from the standpoint of acute effects. It has therefore been a valuable addition to the agents available for the control of insects of public health importance. Unfortunately, resistance to cyclodienes is accompanied by resistance to HCH, and many former uses of HCH are no longer practicable on this account, which is why DDT has proved to be the highly important mainstay of the malaria control and eradication programs.

HCH is the obvious replacement for DDT for housefly control in cases in which DDT resistance has arisen and has been widely used until in turn overtaken by resistance. Residual sprays containing 0.5% of lindane are used; in the U.S. lindane and methoxychlor are the only compounds permitted for this purpose in dairy barns, and even lindane is barred from use in poultry houses. For outdoor use, sprays containing up to 2% of lindane may be applied to rubbish dumps and other breeding sites (it is considerably more toxic than DDT to housefly larvae) with replacements by organophosphorus compounds when organochlorine resistance is present. HCH smokes have been used for tsetse fly control, and the chemical has also been used for aerial applications. Lindane has the advantage over DDT of being toxic to pupae as well as larvae of blackflies (*Simulium* spp.) but DDT is the preferred larvicide and control agent. Lindane will control blood-sucking flies such as *Phlebotomus* and *Culicoides* (sandflies), but lacks the persistence of DDT for prolonged effectiveness. Various blowflies attacking sheep are well controlled by HCH; 1% emulsions or solutions applied to breech or poll prevent "strike" and also kill established maggots.[24,44]

Where cyclodiene resistance is absent, HCH has some advantages over DDT in residual sprays for house interiors in mosquito control. Unlike DDT, it does not excite the insects to leave the deposit before acquiring a lethal dose, and furthermore, the fumigant action enables it to remain toxic even when covered with dust or smoke, which is not the case with DDT. Thus, the sorption of HCH by surfaces is advantageous, since the effect is prolonged, whereas there is rapid loss from nonsorptive ones due to volatilization. The usual rate of application for this purpose is 500 mg/m^2 and Burnett found that on highly sorptive mud walls, a 240 mg/m2 treatment with lindane killed 90% of *Anopheles gambiae* 6 months afterwards and the toxic effect was still very evident after 9 months,[338] whereas such surface deposits normally last about half as long as those of DDT or dieldrin. The ideal control agent would combine the high toxicity of HCH with the persistence of DDT or dieldrin, and attempts have been made to create such a formulation by combining HCH with various chlorinated biphenyls to reduce the volatility on nonsorptive surfaces; the persistence on such surfaces has been increased to at least 6 months by this technique, which in effect creates a physical barrier to volatilization around the HCH particles.[339] The normal toxic effects shown on sorptive surfaces were not impaired. Lindane is also used at the rate of 50 to 100 g per acre as a fog, mist or dust for external space treatments against adults. As a mosquito larvicide, it is generally less toxic than DDT, and dieldrin, which is several times more toxic, is preferred when DDT-resistance is present. The rate of treatment of breeding sites depends on the species, the dosage required being higher for *Culex* than for *Anopheles,* for example, since the former breed in polluted water in which insecticide persistence is likely to be lower; for lindane it is normally about 500 g per acre. Granules or briquettes are preferred over liquids or dusts for larval control in rice fields.

In contrast to DDT, HCH is an excellent control agent for various mites infesting man and livestock. Thus, the itch mite, *Sarcoptes scabei,* which causes human scabies and sarcoptic mange on cattle is completely controlled by sprays containing only 0.036% of lindane, although the effect is not so durable as when sulfur preparations are used. HCH is more effective than DDT against chiggers (*Trombiculid* spp.), the vectors of scrub typhus, but is not sufficiently resistant to washing to be an effective clothing impregnant.[340] However, it is excellent for control of these insects in the scrub areas of their habitat. Although slower in action, DDT is more valuable for control of chicken mite, since it combines persistence with no risk of tainting hen eggs. Before the appearance of resistance, HCH showed great promise as a

cattle tick control agent, being effective in suspensions or emulsions at concentrations down to 0.003% of lindane. Females of the blue tick, *Boophilus decoloratus,* take a long time to die, but do not feed or oviposit during this period.[24] HCH and lindane are preferred control agents for the *Ornithodoros* tick vectors of relapsing fever, against which houses and exterior habitats are treated with 0.5% lindane dust at 15 to 10 g/m^2, or lindane suspension at 0.1 to 12.5 g/m^2. The treatment rates vary widely in rate and frequency, depending on the situation; *O. moubata,* which frequents houses, is well controlled, but *O. tholozani,* which frequents animal burrows and caves has proved much more recalcitrant. Lindane is used in the form of 0.5% water emulsions or oil solutions for surface treatments in dwellings, or as 0.03% liquid preparations for animal treatments to control various ixodid ticks which are carriers of many human bacterial, rickettsial, and viral diseases such as encephalitis, haemorrhagic fevers, American spotted fever, Colorado tick fever, and other unpleasant afflictions in Europe, America, Africa, and Asia. All the organochlorine compounds are effective against these species and several organophosphorus and carbamate compounds are alternatives in places where wildlife appears to be threatened. Thorough and heavy coverage of forest areas is sometimes required for control, as in the case of *Ixodes persulcatus* in the U.S.S.R., but lindane is normally applied at about 280 g per acre and the other compounds at two to four times this rate.[44]

Many species of biting and sucking lice are readily killed by HCH, which is frequently more toxic than DDT but lacks its staying power on garments and tends to develop odor, so that the human use is mainly as a 0.2% aqueous emulsion shampoo for the control of head lice. Thus, DDT dust (10% DDT) is the control agent of choice, but 1% lindane or malathion dust is an alternative against resistant strains of body lice. Resistance to all organochlorines is now widespread among bedbugs (*Cimex* spp.), but 0.5% lindane sprays (solutions or emulsions) can be used when only DDT-resistance is present, and lindane is considerably more toxic than DDT to normal strains of these insects. Again, resistance permitting, lindane is used as a 3% dust to control plague fleas by placing this where the rodent carriers can pick it up on their fur. Similar dusts (1%) are used against human, cat, or dog fleas, as well as oil-based sprays

(1%) for interior and exterior applications around dwellings. DDT is ineffective in residual applications against reduviid bugs such as *Triatoma* and *Rhodnius*, for which the interiors of dwellings and associated buildings are treated at 0.5 g/m² with residual sprays of HCH (about half the corresponding rate for dieldrin). A deposit at this rate remains effective for 2 months or more, depending on the nature of the surface, and fortunately, there is so far no evidence of resistance in these insects.[44] HCH dust is about as effective as chlordane for cockroach control but the odor makes the material undesirable for household use, and chlordane resistance in the German cockroach precludes the use of HCH for such strains in any case. A number of ant species are susceptible to HCH and the fumigant action of the toxicant is particularly useful when application is directly into nests. Some species are said to be repelled by HCH so that its application to trails is less successful on this account and also because of volatility.

The comparative tests conducted by Kearns[302] during the original evaluation of aldrin and dieldrin showed that only DDT and dieldrin completely protected the test fabric from damage by the black carpet beetle, *Attagenus piceus.* Lindane and chlordane were less effective, although in other tests in which toxicants were incorporated into artifical food, lindane showed toxicity towards adults at very low concentrations. In these tests, lindane was also less effective than dieldrin in preventing damage by clothes moths. Like DDT, HCH is highly effective for controlling a number of stored product insects when incorporated into stored grain; the early work at Jealott's Hill showed that 0.4 ppm of its weight of lindane incorporated into grain gave a 50% kill of grain weevils in 5 days.[321] Although the larval forms of some species are naturally rather tolerant to HCH, and there is now evidence of widespread resistance to it among stored product insects (Vol. II, Chap 2C.3c), the material still appears to be quite widely used for control of these pests.

Current patterns of HCH usage* are not easy to discern, apart from the obvious fact that its production in the United States declined greatly some years ago and that of Japan has now taken a downward turn. In Britain, the use of this material in food storage, in ships in British ports, and for industrial and public health purposes amounts to no more than a few thousand pounds. In 1962–64, the annual use of HCH was about 100 tons, constituting about one-sixth of the total organochlorine usage (Table 4). The considerable reduction in cyclodiene usage since then, together with some reduction in DDT usage, which appears to have been procured with difficulty, suggests that HCH usage may be about the same or may even have increased to compensate for the other reductions, since soil insecticides are much needed and the replacement problem is by no means solved. However, the problem of taint is an automatic restraint on some uses. It must also be noted that HCH is not restricted in Britain, except in relation to its use in thermal vaporizers.[60]

2. Stereochemistry, Constitution, and Reactions

The benzene ring, with its conjugated double-bond system, is necessarily planar; the cyclohexane ring, on the other hand, has six carbon atoms with nearly tetrahedral valencies, with the result that the ring is not planar. This situation is seen most readily with the aid of skeletal atomic models of the Dreiding type and is illustrated in Figure 23. The ring system can adopt two configurations (Figure 23, structures 1 and 2), the first being called the "step" or "chair" form, and the second the "boat" form, for obvious reasons. The energy difference between the two is not large enough to permit their isolation, but there is much evidence that the chair conformation (1) is more stable and is the one normally adopted since in it the distance between non-bonded atoms is greatest and their interactions minimal. Two important consequences arise from this lack of planarity and seem to have been ignored prior to the now classical work of Hassel.[341] In the first place, three of the carbon atoms of the chair form lie in one horizontal plane (as the molecule is depicted in Figure 23) and the other three in a parallel plane about 0.5Å lower; secondly, six of the carbon-hydrogen bonds are directed parallel to the vertical symmetry axis of the molecule, and are therefore called axial (a), while the other six lie approximately in the general horizontal plane of the ring (actually at an angle of ± 19.5° to it) and

---

*Since this account was written, a useful compilation of information about lindane has been published in the form of a multi-author book (E. Ulmann, Editor) entitled *Lindane, Monograph of an Insecticide.*[340a] Contents include a list of countries using lindane and an extensive list of its applications.

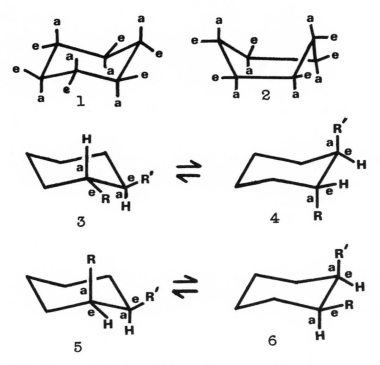

FIGURE 23. The chair and boat forms of cyclohexane and the conformations of simple derivatives.

are called equatorial. The two types are clearly distinct, and, in the absence of strong electrostatic effects, the most stable conformation of a molecule composed of six-membered alicyclic rings will consist of chair forms with their larger substituents in the equatorial positions.

If the chair form of cyclohexane is considered for a moment to be rigid, it is seen that a single substituent can be placed in either an axial or an equatorial position; since isomerism is not found in such molecules it follows that the two possible forms must be freely interconvertible without the breaking of carbon-carbon bonds. The interconversion is probably associated with the normal thermal vibrations of the carbon atoms of the cyclohexane ring. This situation complicates the simple, classical picture of *cis-trans* isomerism in cyclohexane derivatives, which were depicted as deriving from planar rings. A *trans*-1,2-disubstituted cyclohexane will exist in the chair form (Figure 23, structure 3) with R and R′ equatorial, rather than the alternative one (4) in which these two groups are axial; however, both forms are *trans*, as is more obvious in (4), and ring conversion converts one into the other, so that only one

compound can be isolated in practice. On the other hand, an adjacent axial and equatorial bond are *cis* and a *cis*-1,2-disubstituted cyclohexane in which two substituents of different sizes are present will have the configuration shown (Figure 23, structure 5), in which the larger group R′ occupies the equatorial position, rather than the alternative (6) in which R′ is axial. Again, the compound is actually a nonseparable equilibrium mixture in which the molecule (5) predominates. Both molecules are depicted by the same classical structure in which a planar cyclohexane ring has the substituents R and R′ attached to the same side of it (1,2-*cis*). When the molecule is depicted in this way, without the complication of the conformational possibilities, it is readily seen that it has no elements of symmetry; its mirror image cannot be superimposed upon it, so that in principle, a separation into two optical isomerides (each having two possible conformations) is possible. The planar depiction of these molecules is very useful for determining the number of possible stereoisomers but unfortunately gives no idea of the true spatial arrangements of the substituents. However, the true spatial configuration of a

197

molecule can be derived from the planar structure (assumed to be in the chair form) if the actual conformation of one substituent is known, since a *cis*-substituent adjacent to an axial one must be equatorial, while an adjacent *trans*-substituent must be axial, and so on. In this way one can work round the ring drawn in planar fashion from the substituent of known conformation, assigning the correct conformations to all the substituents.

From what has been said so far, it will be evident that the stereochemistry of cyclohexane derivatives becomes increasingly complicated with increasing substitution, especially if the substituents are different. Allowing for all possible modes of substitution of chlorine in the cyclohexane molecule (including for example those with two chlorines on one carbon atom) there are 18 possible position isomers of hexachlorocyclohexane.[342] The photochemical substitution chlorination of cyclohexane, for example, gives representatives of only two of these types, namely, one of the isomers of 1,1,2,4,4,5-hexachlorocyclohexane and α-HCH; the present account is mainly concerned with representatives of the latter class, and on the basis of a planar cyclohexane ring, eight geometrical isomers are possible, of which one (the α-isomer) can be separated into a pair of enantiomorphs. From a conformational point of view, the situation is more complex because of the isomerism that is possible in the conformational structures of each of these geometrical isomers derived on the basis of the planar formulae. The conformational possibilities are shown in Table 22, and the first five-ring conversion forms (in column 2) are no doubt unstable because they are sterically strained. Ring conversion of the γ- and ι-isomers give identical molecules and for the η-isomer gives its mirror image. The α-conformation is asymmetrical and its ring conversion product is different and also asymmetrical, so that each of these conformations corresponds to a different enantiomorphic pair. This gives a total of 10 ordinary conformational isomers plus three existing as optically active pairs; a grand total of 16 steric and optical isomers.

If ring conversion did not occur readily, the thirteen stereoisomers should each be capable of separate existence, but there is strong evidence that it does occur, so the number of separable conformations reduces to those in column one of Table 22, of which only the α-form exists as a dl pair, making a total of nine possible isomeric

TABLE 22

Conformation of the Isomers of 1,2,3,4,5,6-Hexachlorocyclohexane

Isomer	Conformation based on chair form of cyclohexane		Optical activity
	1	2	
β	eeeeee	⇌ aaaaaa	O
δ	aeeeee	⇌ eaaaaa	O
α[a]	aaeeee	⇌ eeaaaa	±
θ	aeaeee	⇌ eaeaaa	O
ε	aeeaee	⇌ eaaeaa	O
γ[b]	aaaeee	⇌ eeeaaa (identical)	O
η[b]	aeaaee	⇌ eeaaea (dl isomers)	±
ι	aeaeae	⇌ eaeaea (identical)	O

[a] α is asymmetrical and is capable of existing in enantiomorphic forms; the ring conversion product in column 2 represents a different pair of dl isomers.
[b] η is asymmetrical and this conformation represents a dl pair; however, ring conversion gives the mirror image conformation and hence results in racemisation, so that separation is impossible.

forms. In theory, each of the eight isomers may also exist in one or more boat forms, but there is no evidence for the presence of such conformations at room temperature. Five of the isomers, including α, are relatively strain-free and might be expected to predominate in technical HCH. In fact, the α-isomer is by far the major component of the technical material with β, γ, and δ-isomers constituting most of the remainder. The η- and ε-isomers have also been isolated from technical HCH[343,344] but it seems likely that the structure of the eighth isomer is too strained for it to exist. A report of 1969 states that the mother liquors following the separation of lindane and the other isomers from technical HCH contains unidentified components, one of which is a hexachlorocyclohexane that shows appreciable toxicity to flour beetles and "differs chromatographically from all known components."[345] Figure 24 shows the major isomers of HCH in both conformational form, as they exist naturally, and as derivatives of a planar cyclohexane ring. Cristol showed that the α-isomer is in fact a racemate by using the alkaloid brucine to dehydrochlorinate one enantiomorphic form selectively to trichlorobenzene, a process which left the laevorotatory form intact:[346]

$$2C_6H_6Cl_6 + 3 \text{ brucine} \rightarrow (-) C_6H_6Cl_6$$
$$+ 3 \text{ brucine.HCl} + C_6H_3Cl_3$$

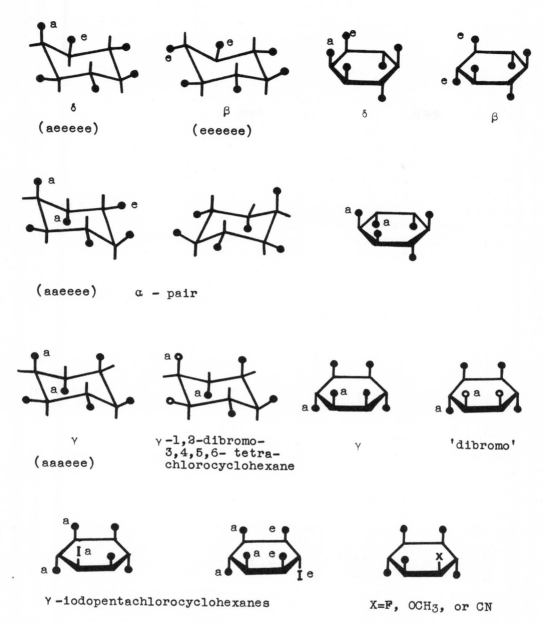

FIGURE 24. Conformations of the major isomers of hexachlorocyclohexane, and of a weakly insecticidal dibromo-analogue of γ-HCH. Also shown are the same structures derived from a "planar" cyclohexane ring. In the planar structures the plane of symmetry (absent in α-HCH) is in the plane of the paper. Two corresponding atoms are designated in the planar and conformational structures for orientation. The iodo-compound on the left may be in the ring conversion form in which the bulky iodine atom is equatorial.

The centrosymmetrical structure of β-HCH was demonstrated by X-ray analysis in 1928. Since that time, X-ray diffraction and electron diffraction, supported by dipole moment measurements, chemical evidence from chlorination, dehydrochlorination reactions and other considerations, have been used to establish the presently accepted structures of the α, β, γ, δ-, and ε-isomers.[347-350] As an example of the synthetic method for constitution determination, the all-equatorial β-HCH gives α-heptachlorocyclohexane ($C_6H_5Cl_7$, mp 153 to 154°C) and two isomeric octachlorocyclohexanes ($C_6H_4Cl_8$) on chlorination.[342] One of the octachloro-compounds (called o-octa) is also obtained from o-dichlorobenzene and the other from p-dichlorobenzene (called β-p-octa) by additive photochlorination. Since it arises from βHCH (eeeee), α-$C_6H_5Cl_7$ must have the (ae)eeeee configuration (brackets indicating that the carbon atom concerned carries two chlorines) and the derivation of the other compounds indicates them to be (ae)(ae)eeee and (ae) ee(ae)ee, respectively. Employing the fact that both α and δ-HCH give o-octa on chlorination, while only δ- gives α-$C_6H_5Cl_7$, it follows that δ-HCH must be aeeeee, while α- must have the other possible structure, aaeeee, which can give rise to o-octa. On chlorination, ε-HCH gives β-p-octa and so must have the aeeeaee structure, which accords with its zero dipole moment. Consideration of the interrelations between the remaining isomers and the chlorination products of the γ-isomer, along with dehydrochlorination experiments and other information, lead to the aaaeee structure for this compound and to the current structural assignments for the other known isomers. Small amounts of θ-HCH mp 124 to 125°C, arise from the chlorination of γ- or ε-tetrachlorocyclohexene.[351]

The chlorination of HCH isomers to give mixtures of hepta- and octachloro-derivatives may be effected in carbon tetrachloride, while liquid chlorine gives higher substitution products. When heated above 140°C with anhydrous ferric chloride in sealed tubes, the individual HCH isomers isomerize; α- is most stable and converts slowly into a mixture of β-, γ-, δ-, and ε. Under these conditions β-HCH readily affords the α-isomer, while γ- gives mainly δ-, the latter itself being isomerized to α- and traces of γ. The maximum amount of γ-isomer formed in these isomerizations is only a few percent.[352] These reactions are reminiscent of the isomerization of β-dihydrohep-

tachlor and its derivatives under similar conditions (Vol. I, Chap. 3B.2a). Numerous reaction conditions convert HCH isomers into various chlorobenzenes.[314] HCH vapour is converted into a mixture of pentachloro- and hexachloro-benzenes when mixed with air or oxygen and chlorine and heated to 470°C in the presence of a catalyst, and may be catalytically dehydrochlorinated at elevated temperatures (280 to 350°C) in the presence of chlorine, iron, or aluminum and their salts, to give mixtures of trichlorobenzenes (which are partly chlorinated when chlorine is the initiator). Excess chlorine gives excellent yields of hexachlorobenzene. Treatment of HCH with zinc dust in acids dechlorinates the material to benzene. Although generally inert to strong acids, the isomers are decomposed by chlorosulfonic acid or sulfur trioxide to give benzene, among other products. p-Dichlorobenzene, 1,2,4,5-tetrachlorobenzene, and small amounts of 1,2,3- and 1,3,5-trichlorobenzenes are also formed when HCH is heated with sulfur at 240 to 290°C. Sodium sulfide reacts with two molecules of HCH to eliminate two molecules of sodium chloride, forming a sulfur birdge between two pentachlorocyclohexyl nuclei;[353] sodium disulfide reacts similarly:

$$2C_6H_6Cl_6 + Na_2S \rightarrow C_6H_6Cl_5 \cdot S \cdot C_6H_6Cl_5$$

$$2C_6H_6Cl_6 + Na_2S_2 \rightarrow C_6H_6Cl_5S \cdot SC_6H_6Cl_5$$

These compounds, as well as various derivatives containing oxidized sulfur were prepared by French investigators and some of them have been applied as pest control agents; the sulfide, for example, has been used to control May beetles and grasshoppers.

Of the reactions of HCH, the best known one is that which occurs under alkaline conditions; technical HCH gives mainly 1,2,4-trichlorobenzene (60 to 80%), with roughly equal amounts of 1,2,3- and 1,3,5-trichlorobenzenes, totaling 4 to 10%, when treated with strong alkalis. The pure α-, γ-, δ-, and ε-isomers dehydrochlorinate readily in alcoholic sodium or potassium hydroxide giving 1,2,4-trichlorobenzene (65 to 86%); 1,2,3-trichlorobenzene (5 to 15%), and 1,3,5-trichlorobenzene (6 to 15%). At room temperature, β-HCH reacts very slowly, but produces 1,2,4-trichlorobenzene nearly exclusively at reflux temperature.[354,355] The elimination of hydrogen chloride is an E_2-type mechanism involving the liberation of hydrogen and chlorine in *trans*-relation to one another. Inspection of the structures shows that all isomers except β have *trans*-related adjacent hydrogen and chlorine to facilitate dehydrochlorination; in β-HCH, all chlorines are equatorial and therefore *trans-* to each other, so that adjacent hydrogen and chlorine are in all cases in *cis*-relationship. The dehydrochlorination rates are in the order α->δ>γ>ε>β, and Cristol found the rate constants at 20°C for elimination of the first chlorine from α, γ, and β to be 0.169, 0.045, and 3×10^{-6} liters/sec/mol, respectively.[354] The differential rates of dehydrochlorination have been employed as a method of analysis for lindane; at 0°C only the α- and δ-isomers are dehydrochlorinated by 1 N alcoholic potassium hydroxide in 15 min, but in 50 min, α, δ, and γ are all dehydrochlorinated so that the difference between the two determinations gives an estimation of the γ-isomer present.[356]

The conversion of HCH into benzene by zinc in acid media has already been mentioned; the chromous chloride reagent employed in the derivatization of cyclodiene compounds for gas-chromatographic analysis (Vol. I, Chap. 3C.3d) reacts rapidly with γ-HCH and much less rapidly with the γ and β-isomers, but the products have not so far been identified.[232] γ-HCH does not normally react with sodium borohydride, but under the conditions used for the bridge monodechlorination (Vol. I, Chap. 3B.2b) of cyclodienes, namely, with this reagent in a solvent, plus a catalyst such as a cobalt, nickel, or palladium salt, adjacent equatorial chlorine atoms are removed at 0 to 20°C; a hitherto unknown 1,2,3,4-tetrachlorocyclohexane is formed, mp 141 to 143°C, and has been assigned the 1e, 2a, 3a, 4a configuration with respect to the chlorine substituents.[220] Other reactions of HCH relate mainly to its biological degradation, which resembles the chemical conversion insofar as chlorinated benzenes are produced in some organisms. However, there are additional oxidations and conjugation reactions which lead to phenolic and thiophenolic products, and these are discussed in connection with metabolism (Vol. II, Chap. 3B.2b and B.3b).

A common feature of the different groups of organochlorine insecticides is that the molecules, at least when drawn in planar representation, appear to be rather simple, whereas in fact the chemistry is rather complex. This has become increasingly evident from research in recent years

on metabolism and the nature of organochlorine residues in the environment. The potential complexity of the stereochemistry of polyhalogenated cyclohexanes, especially when the molecules contain more than one type of halogen, will be evident from the previous account, and toxicity seems, according to information available at the present time, not only to be limited to molecules having the "aaaeee" configuration of lindane, but to be outstanding above all other molecules in this particular hexachloro-derivative. This statement may require qualification in relation to some quite recent investigations, but it may be noted that of a large number of compounds examined by Riemschneider in the 1950's, including tetrachlorocyclohexenes, heptachlorocyclohexanes, octachlorocyclohexanes, hexa- and heptabromocyclohexanes, and numerous mixed halogen compounds, only γ-$C_6H_6Cl_4Br_2$, in which one axial and an adjacent equatorial chlorine of lindane are replaced by bromine (configuration 1e2a(Br)3a4a-5e6e(Cl); Figure 24), showed slight toxicity to insects.[357] This finding has been confirmed recently, but it is found that an asymmetrical monobromo-compound having the γ-configuration is more active.[358] Although various claims have been made in the past regarding the toxicity of analogues, as in the case of certain hepta- and octahalogen derivatives for example, the situation seems to be somewhat different from the DDT and cyclodiene series (especially the latter), where a number of readily available compounds show considerable toxicity. For this reason, no detailed account of the early synthetic work on HCH analogues in this area is given here. Some mention of the methods used was made in connection with structural investigations of the HCH isomers; they include the addition of chlorine, bromine, and iodine chloride to benzene and its derivatives, and similar addition to various cyclohexane derivatives, as well as the halogenation of cyclohexane and related compounds. For details, reference may be made to the numerous original papers of Riemschneider, Nakajima, and others.[41,314,357-361]

However, recent work of Nakajima and his colleagues at Kyoto University is of interest. They have described the synthesis of the isomeric forms of 3,4,5,6-tetrachlorocyclohex-1-ene (TCH) from the α-isomer, which is the major component of the isomeric mixture of these compounds obtained by the partial additive chlorination of benzene.[327,]

^{360,361} Of these isomers, γ-TCH (3a4e5e6e) is particularly important, since it affords lindane on *trans*-addition of chlorine and offers a route to other analogues having the same configuration as lindane. The α-isomer adds iodine chloride in the *trans*-manner to give a mixture of pentachloro-monoiodocyclohexanes, one of which reverts to the starting material with sodium iodide in dimethylsulfoxide, while the other gives a mixture of α-, β-, and γ-TCH from which the latter can be isolated in 43% yield under optimal conditions. With iodine chloride, this isomer gives by addition a mixture of two of the four possible *trans*-addition products (Figure 24) which are diastereoisomeric forms of 6-iodo-1,2,3,4,5-pentachlorocyclohexane. Tests against mosquitoes show that these compounds are no more effective than the weak excitant α-HCH. If a fluorine atom or a cyano- or methoxy-group is included in the molecule (presumably by adding the elements of ClF, ClOCH₃ or ClCN to γ-TCH) so as to give the γ-configuration (Figure 24, last structure; X = F, OCH₃ or CN), the first two compounds are said to be highly insecticidal and the cyano-analogue somewhat less toxic towards mosquitoes (*Culex pipiens pallens*). In the structure shown, when the X is varied, activity toward this insect decreases in the order Cl (lindane) > OCH₃ > F > H > Br > OC₂H₅ > I > CN, and highly active compounds are also produced when one or both of the chlorine atoms 1 and 4 (when X = Cl) are replaced by bromine.[657]

In principle, the HCH isomers may be radiolabeled with tritium, chlorine-36 or carbon-14, the last form of labeling being generally preferred. In a typical preparation described by Bridges,[14]C-labeled benzene, diluted with about twice its volume or pure, dry methylene chloride, is irradiated at – 35°C with an intense source of UV light while chlorine is passed through the apparatus previously swept with nitrogen.[362] To ensure complete chlorination, passage of chlorine is continued for 2 to 3 hr after solid material begins to separate and irradiation in nitrogen is continued for a little time after the chlorine flow is cut off. The crude product (82% based on the benzene used) contains α-HCH (59%), γ-HCH (25%), δ-HCH (16%) and none of the β-isomer; separation on a silicic acid column gives 21% of the pure γ-isomer. The method may, of course, be adapted to prepare the ^{36}Cl-labeled compound. ^{14}C-lindane is currently available commercially at specific activities of 20 to 40 mCi/mmol.[362]

3. Analysis

The analytical methods used for the determination of lindane, technical HCH, or formulations containing them alone, as well as methods for their determination in formulations containing other pesticides, are described in detail in the *CIPAC Handbook.*[115] For lindane and technical HCH, total chlorine may be determined by the potassium-xylene or sodium-isopropanol (Stepanow) dechlorination method discussed for other organochlorines. Total hydrolyzable chlorine is determined by heating under reflux for 30 min with 1N ethanolic potassium hydroxide, followed by Volhard or electrometric silver nitrate titration methods for the liberated chloride. The melting point of pure γ-HCH is lowered 0.7° by 1% of an impurity, so that the setting point of a sample of lindane (determined by the plateau on the cooling curve of the molten material) gives the γ-isomer content according to the equation:

Mol % γ – isomer = antilog [2–(0.00643T)],

where T = 112.86 – setting point of sample

For formulations, the active materials are extracted and the above methods or infrared analysis are used when lindane is involved; the total or hydrolyzable chloride method is applied to the mixed isomers. With mixed formulations, other pesticides (as also emulsifiers, etc.) are frequently separated from HCH or lindane by adsorption chromatography on alumina. Depending on the ratio of the compounds present, DDT can be separated from HCH or lindane by partition chromatography on silica gel impregnated with nitromethane, using hexane saturated with nitromethane as mobile solvent. The separated pesticides may then be determined by methods appropriate to each one. This method is also used to separate the HCH isomers in the technical material or in HCH extracted from formulations. If a sufficiently large sample is available, the separated γ-HCH is weighed directly; otherwise this isomer may be determined by polarography, which can be employed so as to eliminate interference from other derivatives such as hepta- and octachlorocyclohexanes.

Determination of the recovered materials by infrared spectroscopy also presents no difficulty at this stage, if pure isomers are available as standards, since the spectra show a number of prominent absorption bands in CCl_4 or CS_2 that are suitable for measurement.[113] IR-analysis is widely used for the technical materials and formulations because of its speed and convenience; the wavelengths commonly used are 12.58 μ, 13.46 μ, 13.22 μ, 11.81 μ, 13.96 μ, for the α-, β-, δ-, γ-, and ϵ-isomers, respectively. When mixtures of isomers are present, a method of successive approximations is used in which reference solutions are prepared containing the isomers that are not of analytical interest, and are adjusted in concentration until they compensate these isomers in the sample. The absorbance of the required isomer is then read at the chosen wavelength. In this method the precision is ±0.5% for isomers other than γ-HCH, but ±2.0% for the latter.

Several colorimetric methods have been used for HCH estimation, the most widely used one being that of Schechter and Hornstein,[363] which is valuable because it can be used on crude samples containing HCH. The HCH is dechlorinated to benzene with zinc in boiling acetic acid and the benzene produced is nitrated to m-dinitrobenzene, which, when dissolved in methyl ethyl ketone and treated with strong alkali, gives a violet color. The absorbance is read at 565 nm after 20 min. All isomers are determined by this method, although the conversion of β- and δ-HCH to benzene requires a longer treatment with zinc dust than for α- and γ-HCH. The method appears to be specific for the HCH isomers, although any related compound which produces benzene in the process will give the same color. In the determination of HCH in contaminated air by this method, an accuracy of ±2% has been achieved and the method has obvious applications in residue analysis.

As with other chlorinated insecticides, the combination of electron capture detection with gas-liquid chromatography enables picogram (10^{-12}g) levels of the HCH isomers to be separated and determined with relative ease, and this method is very widely used today for the analysis of HCH residues. The electron affinity of the isomers is in the order $\alpha > \delta > \gamma > \epsilon > \beta$ and appropriate isomers may be used as internal standards for one another in quantitative determinations.[360]

Chapter 5

POLYCHLOROTERPENE INSECTICIDES (TOXAPHENE)

A. ORIGINS, PREPARATION, PHYSICAL PROPERTIES

Each of the groups of organochlorine insecticides so far discussed has a character of its own, and the compounds produced by chlorinating various naturally occurring terpenes are no exception. The mixtures of compounds that comprise this group are easy to prepare, but beyond that are the most enigmatic of all the organochlorines used in insect control. Since the products are largely defined by their chlorine content and the nature of the materials chlorinated, the designations employed are not easy to understand and, because skeletal rearrangements of the starting hydrocarbons occur readily, there is likely to be overlap between the constituents of the various mixtures anyway. Basically, these materials arise by the chlorination of α-pinene or camphene, and the insecticidal properties of the products increase with the chlorine content, from which empirical formulae can be derived. GC-analysis shows, however, that the higher chlorination products are complex mixtures and, as in the case of technical HCH, the possibility exists that the insecticidal activity may reside in some relatively minor constituent, since little information is available regarding the nature of individual components.

Various products may be obtained by the chlorination of a mixture of terpenes consisting mainly of α-pinene and camphene in the presence of light and free radical initiators. Thus, the product known as Strobane® contains not less than 66% of chlorine, and is said to be identical with the chloropinene produced in the USSR by incorporating 66 to 68% of chlorine into "pinene." Another product, obtained by chlorinating pinene from turpentine oleoresin, contains 55% of chlorine and has been used as an insecticide in Russia although it is more phytotoxic than the Strobane® composition and unsuitable for plant protection. Melnikov[330] describes another product, chemically similar to toxaphene, which is obtained by photochemical or free radical initiated chlorination of bornyl chloride to a chlorine content of 64 to 67%. This material, called polychloropinene or chlorothene, is used in

agriculture in the USSR, although it is more phytotoxic than toxaphene and also more toxic to the honeybee. The name derives from its origin; manufacture begins with the addition of hydrogen chloride to a pinene fraction of turpentine oleoresin to give bornyl chloride, with molecular rearrangement (Figure 25). The best known and most widely used member of this series of products is prepared by photochemical chlorination of camphene, a structural isomer of α-pinene, to a 67 to 69% chlorine content in the presence of free radical initiators. It is variously called octachlorocamphene, toxaphene (ISO approved name), camphechlor (BSI approved name), polychlorocamphene (USSR), or simply chlorinated camphene.[40,330] Since the free radical chlorination process may proceed through a variety of normal or rearranged structures (some possibilities are shown in Figure 25), the complexity of the product is not surprising; a typical GC-analysis shows it to contain at least 12 components.[365]

Figure 25 summarizes the relationships that appear to exist between these different products according to various statements that appear in the literature, and it is evident that the properties of these materials are very dependent on the degree of chlorination and the nature of the starting material, even when the degree of final chlorination appears similar. This is borne out by the data of Busvine, who examined the insecticidal potency of a number of these products towards houseflies (*Musca domestica*) and blowflies (*Lucilia cuprina*).[218] Of the compounds examined in this investigation, only 2,6-dichlorocamphane was a single compound. Two other samples examined had empirical formulae $C_{10}H_{15}Cl_3$ and chlorine contents of 43.8% and 44.3%, respectively, and their potencies to houseflies were somewhat different. The less potent material was obtained by the chlorination of bornyl chloride in carbon tetrachloride exposed to sunlight (a procedure claimed by Desalbres and Rache[366] to give 2,6,7-trichlorocamphane, mp 60°C, having good insecticidal properties) and contained about eight constituents; the more potent sample was obtained by the chlorination of 2,6-dichlorocamphane. Busvine's experiments showed that a sample of

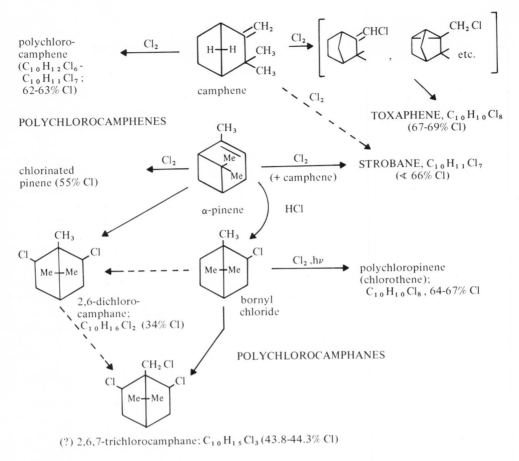

POLYCHLOROCAMPHENES

POLYCHLOROCAMPHANES

FIGURE 25.   Insecticidal mixtures obtained by the chlorination of camphene and α-pinene.[41,218,330]

polychlorocamphene containing 62 to 63% of chlorine (Figure 25) was nearly as toxic as Strobane to houseflies and blowflies, 2,6-dichloro-camphane being very much less toxic and toxaphene rather more toxic than Strobane. A particularly interesting feature of this investigation was the results obtained with a specimen of adamantane ($C_{10}H_{16}$; a cage molecule effectively consisting of four fused cyclohexane chairs) chlorinated to contain 66% of chlorine. This material, containing an average of seven chlorine atoms per molecule, consisted of about 12 compounds according to GC-analysis and was several times more potent than toxaphene to houseflies and blowflies.[218] Again, the nature of the individual constituents is open to speculation.

A number of small molecules of terpenoid origin show evidence of slight insecticidal activity, and Russian workers, Khanenia and Zhuravlev,[367] who in 1944 were seeking agents to control body

lice, found that the mild toxicity of terpenes contained in turpentine could be greatly enhanced by chlorination. The insecticidal properties of toxaphene were first described in 1947, and the product was introduced by Hercules Powder Company in 1948, under the code name Hercules 3956®.[40] Its discovery was undoubtedly the product of the intense interest in chlorinated cyclic hydrocarbons generated by the earlier introduction of lindane (announced by Slade in 1945) and chlordane (discussed by Kearns[265] without indication of structure in 1945). Strobane was introduced by the B. F. Goodrich Chemical Company in 1951 and is listed in the second (1971) edition of Martin's *Pesticide Manual* as being of little current commercial interest.[40] A survey by the U. S. Department of Agriculture in 1964 revealed that the use of toxaphene slightly exceeded that of DDT, and the two together accounted for 46% of the total pesticides used that year;[1] toxaphene is still widely used at the present time.

Technical toxaphene is an amber, waxy solid, softening range 70 to 95°C, d^{27} 1.66, obtained by chlorinating camphene obtained from the southern pine to a chlorine content of 67 to 69%, corresponding to an average empirical formula of $C_{10}H_{10}Cl_8$ and corresponding molecular weight of 414.[40,364] The photochemical chlorination is frequently conducted in carbon tetrachloride to reduce the viscosity of the reaction mixture and proceeds vigorously and exothermically at first, but the final stages of chlorination are difficult and require heat; temperatures may not be too high, however, since hydrogen chloride is slowly lost above 120°C. The product has a mild terpene odor, and the recorded vapor pressure of 0.17 to 0.40 mm Hg at 25°C (3.0 to 4.0 mm at 90°C) is very much higher than that of the other organochlorines previously discussed. Its solubility in water, about 3 ppm at ambient temperature, is also higher than that of other organochlorines except lindane. Like chlordane, its excellent solubility in most organic solvents makes formulation easy. The solubilities (g/100 ml at 27°C) in acetone, benzene, carbon tetrachloride, ethylene dichloride, toluene, and xylene amount to complete miscibility and are high for turpentine (350 to 400), kerosene (more than 280) and fuel oil (250 to 275). Solubilities in alcohols such as isopropanol (15 to 18) and 95% ethanol (10 to 13) are low. Other information given by the manufacturer includes heat of fusion (0.39 cal/g), specific heat (0.258 cal/g/°C at 41°C) and viscosity (1.4 poises at 100°C). It is interesting to compare these properties with some of those given for Strobane, which is described as a viscous, straw-colored liquid with a mild aromatic odor, d^{20} 1.60; n_D^{20} 1.580 and vapor pressure 3×10^{-7} mm Hg at 20°C. It is insoluble in water and sparingly soluble in alcohols, but readily soluble in most other solvents. There seems to be considerable variation in the vapor pressures recorded in the literature for individual organochlorine compounds, presumably a reflection of variability in the techniques used for measurement; the difference between these values for Strobane and toxaphene seems rather extreme.

Toxaphene dehydrochlorinates on heating above 120°C or when exposed to ultraviolet light or strong sunlight. Dehydrochlorination is accelerated by alkalis and is catalyzed by iron, so that stainless steel equipment is recommended when the technical material is melted during formulation processes.[364]

B. APPLICATIONS IN AGRICULTURE AND AGAINST INSECTS AFFECTING MAN AND ANIMALS

Toxaphene is supplied as the technical material, as a 40% dust concentrate and as a viscous liquid containing 10% of xylene;[364] this last preparation can decompose by an exothermic reaction between xylene and toxaphene that is catalyzed by iron or aluminum at temperatures as low as 70°C. An aluminum alloy is recommended for storage tanks, although stainless steel and brass can be used, and dry conditions are essential. The dust concentrate should also be stored in a cool, dry place.

Since toxaphene dehydrochlorinates in the presence of alkalis, it is incompatible with pesticides giving an alkaline reaction, but otherwise compatible with many other agricultural chemicals. Emulsifiable concentrates containing 4, 6, or 8 lb of technical toxaphene/gal are formulated with xylene, 0.5% epichlorohydrin as stabilizer, and an emulsifier, for admixture with water in field use. For the control of ectoparasites on animals, high-grade emulsion concentrates are made for dilution to 0.5% solutions usable as dips or sprays. Toxaphene solutions (5% in oil) are also applied to cattle rubbing devices for the same purpose. Other formulations include a 40% wettable powder for use on vegetables and other crops; 5, 10, and 20% dusts (often admixed with sulfur for mite control in cotton applications) and 5, 10, and 20% granules frequently applied from aircraft for the control of armyworms, various inaccessible boring larvae (corn borer), and budworms.

Certain chlorinated terpenes such as 2,6-dichlorocamphane are synergistic with DDT, and the discovery of synergistic action between toxaphene and DDT when in 2:1 ratio led to a number of formulations of this type for use where resistance to one or other of the constituents has developed.[364] Thus, an emulsifiable concentrate containing 4 lb toxaphene and 2 lb DDT/gal proved to be highly effective against cotton boll weevil in unusually difficult situations, methyl parathion sometimes being added to give rapid knock-down of boll weevils, and parathion for aphid control. Dust formulations incorporating the 2:1 ratio are also used on cotton and have found certain applications on vegetables. The coformulation of toxaphene with parathion combines the activity spectrum of both toxicants and is effective by stomach, contact and fumigant action against a diversity of lepidopterous larvae, leaf miners, aphids, scales, and other pests, some of which are unaffected by toxaphene alone. Limited recourse has also been had to combinations with various miticides, fungicides (maneb, zineb), organophosphorus insecticides (dioxathion, diazinon, malathion, etc.), endrin, endosulfan, and *Bacillus thuringiensis* for specific purposes. In some cases, coformulation of toxaphene with other materials has caused varying degrees of phytotoxicity, or the residual toxic effect of the organochlorine has been significantly reduced. Thus, admixture with bordeaux mixture, calcium arsenate, ferrous sulfate, lime, lime-sulphur, or summer oils is not recommended; mixtures with oil particularly have caused damage to fruit tree foliage. However, investigations of the effect of toxaphene on tree foliage carried out in 1948 showed that 0.4% aqueous suspensions were harmless to some 70 species of trees and shrubs and the material appears to be nonphytotoxic in most combinations, except to cucurbits, which it may damage extensively.[24] Although there have been some indications that low concentrations of it may depress seedling growth, it did not injure crops when present in soil at 25 lb/per acre.

A list of permitted residue tolerances for toxaphene of 1962 gives values for 19 fruit crops (usually 7 ppm), six main cereal crops (5 ppm), fat of meat, fish and poultry (7 ppm), cottonseed (5 ppm), four kinds of tree nuts (7 ppm), and 28 varieties of vegetables (7 ppm) and serves to illustrate the wide range of uses of this material.[364] More recent lists indicate similar uses, and the permitted tolerances have remained unchanged. The material has proved valuable for a number of purposes for which DDT is not suited and in some situations has advantages over chlordane, although its general toxicity is lower than that of either of these toxicants. Toxaphene has the advantage over HCH of not causing tainted crops. Some of the uses depend upon its ability to reach insects in inaccessible situations that appear to defeat DDT, and it is often effective when target insects are heavily sclerotized. Thus, its combined stomach and contact poisoning action

makes it an effective control agent for grasshoppers (*Melanoplus* spp.) in the field, although in laboratory tests it appears to be about sevenfold less effective than chlordane by both stomach and contact action.[368] Various pentatomid and mirid plant bugs are better controlled by toxaphene than by DDT, although the cyclodienes are usually superior, and several species of thrips, especially when found on cotton, are also well controlled.[24]

It is evident that in field practice, toxaphene ranks with DDT and other organochlorines as a toxicant for many of the lepidopterous larvae infesting agricultural crops, although the amounts required are usually, but, surprisingly, not always higher. Toxaphene has proved to be effective against tortricid caterpillars on fruit (as a 0.1% spray), but is less efficient against red banded leaf roller, for example, than DDD applied as a 0.05% emulsion, and will not control codling moth. The granules and dusts have given adequate control of various boring larvae, such as the sugar cane borer, and have the advantage over DDT of not causing aphid infestations (presumably arising because DDT kills their natural predators). Toxaphene as a 0.2% spray gives good control of psychid moth larvae (known as bagworms on account of their protective cocoons); parathion is better, but, of course, much more hazardous in use.[24] As a soil insecticide it appears to be generally inferior to the cyclodienes or HCH, probably because it tends to disappear from soil more rapidly than the other organochlorine insecticides; nevertheless, a report of 1949 indicated an application of 25 lb per acre to be still effective against larvae of the Japanese beetle after 18 months, as were HCH and chlordane.[369]

A table compiled by Brown in 1950[24] summarizes the status of organochlorines as agents for weevil control at that time, and shows that toxaphene is more effective than DDT against the six species listed, being sometimes equally effective and sometimes less effective than HCH or chlordane; it is more effective than any of the others against the Cowpea curculio, *Chalcodermus seneus*. At this time, aldrin and dieldrin were still being evaluated and a footnote indicates that dieldrin had proved superior to chlordane and toxaphene against alfalfa weevil, and these in turn are more effective than DDT. Although generally less effective than HCH as an acute toxicant, toxaphene has the virtue of an ability to control most of the main cotton pests moderately well and has proved very valuable for this purpose, especially in combination with DDT, which by itself is relatively ineffective against boll weevil; the 2:1 toxaphene-DDT combination already mentioned has proved very effective (at 1 to 6 lb per acre) for the control of boll weevil, leaf eating caterpillars, various flea hoppers, plant bugs, and thrips. Along with other polychloroterpenes, toxaphene possesses some degree of acaricidal activity, but it is frequently formulated with sulfur for cotton applications, due to its tendency to provoke *Tetranychus* mite infestations, presumably through predator reduction. It is quite effective against various species of orchard mites, whereas chlordane and other cyclodienes, except dieldrin, show little miticidal activity.

The importance of the cotton application is illustrated by a U.S. Department of Agriculture survey of 1964,[1] which indicated that toxaphene was the pesticide used in largest volume in the United States and that 69% of this usage was on cotton. This use is paralleled in Southern and Central America, Africa, and other cotton growing areas of the world.[370] For such applications, the toxicant may be applied from the air using conventional emulsifiable concentrates diluted with water, or much more economically using special low-volume concentrates which can be applied in about one tenth the volume from a greater altitude. Information available from trials conducted in Guatemala in 1967-68 with toxaphene-DDT and toxaphene-methyl parathion shows that cotton yields in 1967 were significantly higher with low-volume than with conventional formulations; the crop value was nearly double the 1964-67 average for the area and the net profit in dollars per acre about fourfold greater.

As a mixture of compounds containing chlorinated bridged rings, toxaphene is frequently classified with the cyclodienes, although its origins are really quite different. Nevertheless, there is clearly a general resemblance between such molecules and the hexachloronorbornene nucleus of the cyclodienes, and this seems to be reflected in the fact that cyclodiene-resistance extends to toxaphene. Toxaphene resistance in the boll weevil and some other cotton pests became widespread from about 1955 onward, which is one reason why the synergistic combination with DDT has proved so valuable.

Since toxaphene is readily soluble in most commercial solvents of the type used in household pest control formulations, it requires no auxiliary low flash point solvent for incorporation into such preparations, and this property, combined with relative safety to warm-blooded animals at the concentrations used and good residual efficiency, has made it a useful material for such purposes. Furthermore, there is no discoloration of surfaces when the solvents themselves are colorless and the material itself has a pleasant odor. Accordingly, the technical mixture has been incorporated in space sprays consisting of 1% toxaphene, 2% isobornyl thiocyanoacetate plus related terpenoids (Thanite®) and 97% refined petroleum distillate; or 1% toxaphene, 0.1% pyrethrins, 0.5% piperonyl butoxide made up with the petroleum distillate. In these combinations the Thanite® or pyrethrum (with pyrethrum synergist) is added to give a rapid knock-down effect. For use as a household aerosol spray, a typical combination is 2% toxaphene, 0.2% pyrethrins, 0.8% piperonyl butoxide, 12% refined petroleum distillate, and 85% of a propellant consisting of a mixture of the Freons normally used for this purpose. These combinations are used for the control of houseflies, mosquitoes and other flying insects, spiders, silverfish, cockroaches, ants, fleas, bedbugs, moths, and carpet beetles. Early tests against fabric pests showed that 0.8% of fabric weight of toxaphene gave effective protection against black carpet beetle, furniture beetle, and clothes moth, although the toxicant was readily removed when the impregnated materials were treated with dry-cleaning agents.[364]

Frequent mention of the use of toxaphene for the control of insects affecting man and animals is to be found in the section on resistance, for in many places it took its turn with the cyclodienes as a replacement for DDT when resistance to this toxicant appeared. Unfortunately, the efficiency of toxaphene was usually short-lived when cyclodiene resistance was already present and in some cases the failure was immediate, as with its use against the cattle tick (*Boophilus microplus*) in Queensland in 1954. Thus, although this material has been used at one time or another for most of the usual public health applications, its use has probably declined with the rise of cyclodiene resistance, and it may be significant that in the account of chemical control methods for vectors and pests of public health importance given by W.H.O. in 1970, toxaphene is only specifically mentioned in connection with the control of chigger infestations, and then as a less satisfactory alternative to dieldrin, aldrin or lindane.[44]

C. ANALYSIS

Methods for the analysis of technical toxaphene and the toxaphene content of its various formulations are described in detail in the *CIPAC Handbook* and in the manufacturer's technical bulletins.[115,364] When toxaphene is the only chlorinated material present, the usual methods for total chlorine estimation are applied (the sodium-isopropanol or sodium biphenyl methods), the liberated chloride precipitated with silver nitrate being estimated gravimetrically or the excess silver nitrate by the usual titration methods (see DDT, for example). The sodium biphenyl method is used both for technical toxaphene and toxaphene dust. Total chlorine may also be determined by the Parr bomb oxidation method in appropriate cases.

The infrared spectrum of toxaphene shows prominent absorption bands in the 6 to 8 μ region and the specified absorbance at 7.2 μ of 0.0177 (maximum) is used to characterize the material (by comparison with an authentic specimen). For its determination in formulations such as dusts or emulsifiable concentrates, the carriers or dispersing agents are removed by passing the material, extracted into benzene, through columns of acid alumina. The residue from solvent evaporation is then subjected to IR-analysis in carbon tetrachloride by comparison of the difference in absorbance at 7.2 and 7.3 μ (that is, the difference between the maximum and minimum absorbances which occur at these wavelengths) with corresponding value for reference material. Benzene extraction in these procedures dissolves only a little of any sulfur present in the formulation. A related method uses the same cleanup procedure and employs the absorbance difference between 7.7 and 5.8 μ; if DDT is also present, it is determined simultaneously by comparing the absorbance difference between 9.1 and 5.8 μ with a calibration curve prepared using various concentrations of standard DDT.

Sodium reduction methods for total chlorine (Vol. I, Chap. 1C.) may be applied in residue analysis when the history of the samples is known so that any extraneous chlorine containing materials can be allowed for. The sodium-liquid ammonia technique for total chlorine described by Beckman has been adapted for the determination of toxaphene residues in crop extracts and cattle dips;[371] chloride determination is by automatic titration using coulometrically generated silver ion. The Schoniger oxygen flask method is also applicable and enables multiple analyses to be conducted simultaneously. Adsorption chromatography on florisil is frequently used to free samples from plant pigments and waxes before further treatment, and extraction of hexane solutions with mixed concentrated and fuming sulfuric acids has been used to free toxaphene from fats and oils.[364]

A spectrophotometric procedure for toxaphene residues involves heating the cleaned up residue with diphenylamine and zinc chloride at 205°C to give a greenish blue complex with an absorption maximum at 640 nm.[372] This method enables 20 to 700 μg to be detected and has been applied to extracts of crops and foods and to various formulations, including those containing DDT or sulfur. Sulfur interferes with the determination but contamination with it can be largely avoided if methanol is used to extract the crop or sample. Cyclodiene compounds can interfere but this possibility does not normally raise serious problems. Small quantities of waxes and pigments interfere severely, so that rigorous cleanup is required.

A typical gas chromatogram of toxaphene exhibits about 12 peaks, of which three are particularly prominent.[365] In principle, toxaphene residues can be analyzed, as in the case of chlordane, by comparing these "signature peaks" with those in a known amount of reference sample, but this will give an exaggerated value for the residue, since other components disappear at a greater rate. In fact, toxaphene residues seem to disappear rather rapidly; a field study of the weathering of toxaphene residues on kale by Klein and Link showed that after 14 days, only 1% of the original residue of more than 100 ppm was still present.[365]

REFERENCES

1. Department of Health, Education, and Welfare, Report of the Secretary's Commission on Pesticides and Their Relationship to Environmental Health, Parts I and II, U.S. Govt. Print. Off., Washington, D.C., 1969, 44.
2. O'Brien, R. D., *Insecticides, Action and Metabolism,* Academic Press, New York, 1967, 165.
3. Perkow, W., *Die Insektizide,* Huther Verlag, Heidelberg, 1956, 360.
4. von Rumker, R., Guest, H. R., and Upholt, W. M., The search for safer, more selective and less persistent pesticides, *Bioscience,* 20, 1004, 1970.
5. *Agrochemical Industry in Japan,* published by Japan Agricultural Chemicals Overseas Development Commission (JACODEC), Tokyo, 1971.
6. Pesticides and Health, Royal Dutch Shell Group Briefing Service, February, 1967.
7. Lauger, P., Martin, H., and Muller, P., Uber Konstitution und toxische wirkung von naturlichen und neuen synthetischen insektentotenden stoffen, *Helv. Chim. Acta,* 27, 892, 1944.
8. Zeidler, O., Verbindungen von chloral mit brom-und chlorbenzol, *Chem. Ber.,* 7, 1180, 1874.
9. Lauger, P., Uber neue, sulfogruppenhaltige mottenschutzmittel, *Helv. Chim. Acta,* 27, 71, 1944.
10. Chattaway, F. D. and Muir, R. J. K., The formation of carbinols in the condensation of aldehydes with hydrocarbons, *J. Chem. Soc. (Lond.),* 701, 1934.
11. Muller, P., Uber zusammenhange zwischen konstitution und Insektizider wirkung I, *Helv. Chim. Acta,* 29, 1560, 1946.
12. West, T. F. and Campbell, G. A., *DDT and Newer Persistent Insecticides,* revised 2nd ed., Chapman and Hall, London, 1950.
13. Busvine, J. R., New synthetic contact insecticides, *Nature (Lond.),* 158, 22, 1946.
14. Barlsch, E., Eberle, D., Ramsteiner, K., Tomann, A., and Spindler, M., The carbinole acaricides: Chlorobenzilate and chloropropylate, *Residue Rev.,* 39, 1, 1971.
15. Martin, H., Ed., *Pesticide Manual,* 2nd ed. British Crop Protection Council, 1971, 495.
16. Busvine, J. R., Mechanism of resistance to insecticides in house flies, *Nature (Lond.),* 168, 193, 1951.
17. Miles, J. W., Goette, M. B., and Pearce, G. W., A new high temperature test for predicting the storage stability of DDT water-dispersible powders, *Bull. W.H.O.,* 27, 309, 1962.
18. Hadaway, A. B. and Barlow, F., Some physical factors affecting the efficiency of insecticides, *Trans. R. Soc. Trop. Med. Hyg.,* 46, 236, 1952.
19. van Tiel, N., On 'Supona', a new type of DDT suspension, *Bull. Entomol. Res.,* 43, 187, 1952.
20. Philips, N. V., Gloeilampen fabricken (Stabiliser for aqueous pesticide emulsions), Netherland Patent 6809975, June 1970.
21. Wiesmann, R., cited in *DDT, the Insecticide Dichlorodiphenyltrichloroethane and its significance,* Muller, P., Ed., Birkhauser Verlag, Basel, 1955, 14.
22. van Tiel, N., Influence of Resins on Crystal Size and Toxicity of Insecticidal Residues, Ph.D. Thesis, University of Leyden, 1955.
23. Hayhurst, H., The action on certain insects of fabrics impregnated with DDT, *J. Soc. Chem. Ind.,* 64, 296, 1945.
24. Brown, A. W. A., *Insect Control by Chemicals,* John Wiley, New York, 1951.
25. Carruth, L. A. and Howe, M. L., Chlorinated hydrocarbons and cucurbit foliage, *J. Econ. Entomol.,* 41, 352, 1948.
26. Cullinan, F. P., Effect of DDT on plants, *Agric. Chem.,* 2, 18, 1947.
27. Wilson, J. K. and Choudhri, R. S., DDT and soil microflora, *J. Econ. Entomol.,* 39, 537, 1946.
28. Edwards, C. A., Soil pollutants and soil animals, *Sci. Am.,* 220, 88, 1969.
29. Linsley, E. G. and MacSwain, J. W., DDT dusts in alfalfa, and bees, *J. Econ. Entomol.,* 40, 358, 1947.
30. Stahl, C. F., Relative susceptibility of *Protoparce sexta* and *P. quinquemaculata, J. Econ. Entomol.,* 39, 610, 1946.
31. Dowden, P. B., DDT to control *Lymantria dispar,* cited in Brown, A. W. A., *Insect Control by Chemicals,* John Wiley, New York, 1951, 603.
32. Hofmaster, R. N. and Greenwood, D. E., Organic insecticides for fall armyworm, *J. Econ. Entomol.,* 42, 502, 1949.
33. Blanchard, R. A. and Chamberlin, T. R., Insecticides for corn earworm and armyworm, *J. Econ. Entomol.,* 41, 928, 1948.
34. Bishopp, F. C., New insecticides to date, *J. Econ. Entomol.,* 39, 449, 1946.
35. Weinman, C. J., Chlorinated hydrocarbons for codling moth control, *J. Econ. Entomol.,* 40, 567, 1947.
36. Sun, Y-P., Rawlins, W. A., and Norton, L. B., Toxicity of chlordane, BHC and DDT, *J. Econ. Entomol.,* 41, 91, 1948.
37. Plumb, G. H., DDT sprays against Dutch elm disease, *J. Econ. Entomol.,* 43, 110, 1950.
38. Kapoor, I. P., Metcalf, R. L., Nystrom, R. F., and Sangha, G. K., Comparative metabolism of methoxychlor, methiochlor and DDT in mouse, insects, and in a model ecosystem, *J. Agric. Food Chem.,* 18, 1145, 1970.
39. E.I. du Pont de Nemours & Co., Industrial and Biochemicals Department, Technical Data Sheets, 1969, etc.
40. Martin, H., Ed., *Pesticide Manual,* 2nd ed., British Crop Protection Council, 1971, 428.
41. Metcalf, R. L., *Organic Insecticides,* John Wiley (Interscience) New York, 1955.
42. Brown, A. W. A., Insect resistance in arthropods, *W.H.O. Monogr. Ser.,* No. 38, 1958.

43. **Brown, A. W. A. and Pal, R.,** Insect Resistance in Arthropods, 2nd ed., W.H.O., Geneva, 1971.
44. Insecticide resistance and vector control, *W.H.O. Monogr. Ser.,* No. 443, 1970.
45. **Niswander, R. E. and Davidson, R. H.,** Chlordane and lindane versus DDT for roaches, *J. Econ. Entomol.,* 41, 652, 1948.
46. **Busvine, J. R.,** DDT and analogues for Pediculus and Cimex, *Nature (Lond.),* 156, 169, 1945.
47. **Busvine, J. R.,** Review of lousicides, *Br. Med. Bull.,* 3, 215, 1945.
48. **Busvine, J. R. and Pal, R.,** The impact of insecticide-resistance on control of vectors and vector-borne diseases, *Bull. W.H.O.,* 40, 731, 1969.
49. **Busvine, J. R.,** DDT and BHC for the bedbug and louse, *Ann. Appl. Biol.,* 33, 271, 1946.
50. **Pampana, E.,** *A Textbook of Malaria Eradication,* 2nd ed., Oxford University Press, 1969.
51. **Peffly, R. L. and Gahan, J. B.,** Residual toxicity of DDT analogues, *J. Econ. Entomol.,* 42, 113, 1949.
52. **Metcalf, R. L., Kapoor, I. P., and Hirwe, A. S.,** Biodegradable analogues of DDT, *Bull. W.H.O.,* 44, 363, 1971.
53. **Bruce, W. N. and Decker, G. C.,** Chlordane, DDD and DDT for fly control, *J. Econ. Entomol.,* 40, 530, 1947.
54. **Brooks, G. T.,** Pesticides in Britain, in *Environmental Toxicology of Pesticides,* Matsumura, F., Boush, G. M., and Misato, T., Eds., Academic Press, New York and London, 1972, 61.
55. The place of DDT in operations against malaria and other vector-borne diseases, *Off. Rec. W.H.O.,* No. 190, 1971, 176.
56. Food and Agriculture Organization, *F.A.O. Production Year Book,* 1969, 23, 496.
56a. **Carson, R.,** *Silent Spring,* Houghton-Mifflin Co., Boston, 1962, 368.
57. Department of Health, Education, and Welfare, *Report of the Secretary's Commission on Pesticides and Their Relationship to Environmental Health,* Pts. I and II, U.S. Govt. Print. Off., Washington, D.C., 1969.
58. **Matsumura, F.,** Current pesticide situation in the United States, in *Environmental Toxicology of Pesticides,* Matsumura, F., Boush, G. M., and Misato, T., Eds., Academic Press, New York and London, 1972, 33.
59. **Edwards, C. A.,** CRC Uniscience Series, *Persistent Pesticides in the Environment,* Butterworths, London, 1970.
60. Department of Education and Science, Further review of certain persistent organochlorine pesticides used in Great Britain, *Report by the Advisory Committee on Pesticides and Other Toxic Chemicals,* HMSO London, 1969.
61. **Muller, P.,** Physike und chemie des DDT-insektizides, in *DDT, the Insecticide Dichlorodiphenyltrichloroethane and its Significance,* Muller, P., Ed., Birkhauser Verlag, Basel, 1955.
62. **Haller, H. L.,** The chemical composition of technical DDT, *J. Am. Chem. Soc.,* 67, 1591, 1945.
63. **Forrest, J., Stephenson, O., and Waters, W. A.,** Chemical investigations of the insecticide 'DDT' and its analogues, *J. Chem. Soc. (Lond.),* 333, 1946.
64. **Cochrane, W. P. and Chau, A. S. Y.,** Chemical derivatization techniques for confirmation of organochlorine residue identity, in Pesticides Identification at the Residue Level, *Adv. Chem. Ser.,* No. 104, 11, 1971.
65. **Rogers, E. F., Brown, H. D., Rasmussen, I. M., and Heal, R. E.,** The structure and toxicity of DDT insecticides, *J. Am. Chem. Soc.,* 75, 2991, 1953.
66. **Skerrett, E. J. and Woodcock, D.,** Insecticidal activity and chemical composition. II. Synthesis of some analogues of DDT, *J. Chem. Soc. (Lond.),* 2718, 1950.
67. **Metcalf, R. L. and Fukuto, T. R.,** The comparative toxicity of DDT and analogues to susceptible and resistant houseflies and mosquitos, *Bull. W.H.O.,* 38, 633, 1968.
68. **Holan, G.,** 1,1-Bis(p-ethoxyphenyl) dimethylalkane Insecticides, German patent 1,936,494, 1970; *C.A.,* 72, 90045, 1970.
69. **Kaluszyner, A., Reuter, S., and Bergmann, E. D.,** Synthesis and biological properties of diaryl (trifluoromethyl) carbinols, *J. Am. Chem. Soc.,* 77, 4165, 1955.
70. **Wiles, R. A.** (to Allied Chemical Corporation), U.S. patent 3,285,811, 1966; *C.A.,* 60, 18572, 1967.
71. **Holan, G.,** Diphenyldichlorocyclopropane Insecticides, Australian patent 283,356, 1968; *C.A.,* 71, 12815, 1969.
72. **Muller, P.** (to Geigy AG), Insecticidal Compounds, U.S. patent 2,397,802; *C.A.,* 40, 3849, 1946.
73. **Haas, H. B., Neher, M. B., and Blickenstaff, R. T.,** *Ind. Eng. Chem.,* 43, 2875, 1951.
74. **Holan, G.,** 1,1-Bis(p-Ethoxyphenyl)dimethylalkane Insecticides, German patent 1,936,495, 1970; *C.A.,* 72, 100259, 1970.
75. **Grummitt, O., Buck, A. C., and Becker, E. I.,** 1,1-Di(p-chlorophenyl)ethane, *J. Am. Chem. Soc.,* 67, 2265, 1945.
76. **Bergmann, E. D. and Kaluszyner, A.,** Di(p-chlorophenyl)trichloromethylcarbinol and related compounds, *J. Org. Chem.,* 23, 1306, 1958.
77. **Hinsberg, O.,** Uber die bildung von saureestern und saure amiden bei gegenwart von wasser und alkali, *Chem. Ber.,* 23, 2962, 1890.
78. **Huisman, H. O., Uhlenbroek, J. H., and Meltzer, J.,** Preparation and acaricidal properties of substituted diphenylsulfones, diphenylsufides and diphenylsulfoxides, *Recl. Trav. Pays-Bas Belg.,* 77, 103, 1958.
79. **Monroe, E. and Hand, C. R.,** Some bis-(aryloxy)methanes, *J. Am. Chem. Soc.,* 72, 5345, 1950.
80. **Hilton, B. D. and O'Brien, R. D.,** A simple technique for tritiation of aromatic insecticides, *J. Agric. Food Chem.,* 12, 236, 1964.
80a. **Kapoor, I. P., Metcalf, R. L., Hirwe, A. S., Coats, J. R., and Khalsa, M. S.,** Structure activity correlations of biodegradability of DDT analogs, *J. Agric. Food Chem.,* 21, 310, 1973.

81. **Fields, M., Gibbs, J., and Walz, D. E.,** The synthesis of 1,1,1-trichloro-2,2-bis(4'-chlorophenyl-4'-C^{14}) ethane, *Science,* 112, 591, 1950.
82. **Dachauer, A. C., Cocheo, B., Solomon, M. G., and Hennessy, D. J.,** The synthesis of tertiary carbon deuterated DDT and DDT analogs, *J. Agric. Food Chem.,* 11, 47, 1963.
83. **Sumerford, W. T.,** A synthesis of DDT, *J. Am. Pharm. Assoc.,* 34, 259, 1945.
84. **Rueggeberg, W. H. C. and Torrans, D. J.,** Production of DDT, *Ind. Eng. Chem.,* 38, 211, 1946.
85. **Mosher, S. H., Cannon, M. R., Conroy, E. A., Van Strien, R. E., and Spalding, D. P.,** Preparation of technical DDT, *Ind. Eng. Chem.,* 38, 916, 1946.
86. *Specifications for Pesticides used in Public Health,* 3rd ed., W.H.O., Geneva, 1967
87. **Gunther, F. A.,** cited in West, T. F. and Campbell, G. A., *DDT and Newer Persistent Insecticides,* 2nd ed., Chapman and Hall, London, 1950, 43.
88. **Gooden, E. L.,** Optical crystallographic properties of DDT, *J. Am. Chem. Soc.,* 67, 1616, 1945.
89. **Wild, H. and Brandenberger, E.,** Zur kristallstruktur des p,p'-dichlorodiphenyl trichlorathan, *Helv. Chim. Acta,* 28, 1692, 1945.
90. **Balson, E. W.,** Vapour pressures of DDT, BHC, and DNOC, *Trans. Faraday Soc.,* 43, 54, 1947.
90a. **Spencer, W. F. and Claith, M. M.,** Volatility of DDT and related compounds, *J. Agric. Food Chem.,* 20, 645, 1972.
90b. **Claith, M. M. and Spencer, W. F.,** Dissipation of pesticides from soil by volatalization of degradation products. I. Lindane and DDT, *Environ. Sci. Technol.,* 6, 910, 1972.
91. **Kenaga, E. E.,** Factors related to bioconcentration of pesticides, in *Environmental Toxicology of Pesticides,* Matsumura, F., Boush, G. M., and Misato, T., Eds., Academic Press, New York and London, 1972, 193.
92. **Riemschneider, R.,** Chemical structure and activity of DDT analogues with special consideration of their spatial structures, *Adv. Pest. Control Res.,* 2, 307, 1958.
93. **Bradlow, H. and Van der Werf, C.,** Composition of Gix, *J. Am. Chem. Soc.,* 69, 662, 1947.
94. **Schneller, G. H. and Smith, G. B. L.,** 1-Trichloro -2,2-bis(p-methoxyphenyl)ethane (methoxychlor), *Ind. Eng. Chem.,* 41, 1027, 1949.
95. **Harris, L. W.,** Condensation of aldehydes, U.S. patent 2,572,131; *C.A.,* 46, 5088, 1952.
96. **Adams, C. E.,** British patent 709,564; *C.A.,* 49, 10376, 1955.
97. **Grummitt, O.,** Di (p-chlorophenyl) methylcarbinol, a new miticide, *Science,* 111, 361, 1950.
98. **Barker, J. S. and Maughan, F. R.,** Acaricidal properties of Rohm and Haas FW-293, *J. Econ. Entomol.,* 49, 458, 1956.
99. **Gatzi, K. and Stammbach, W.,** Uber die natur der nebenprodukte in technischen p,p'- dichlorodiphenyl-trichlorathan, *Helv. Chim. Acta,* 29, 563, 1946.
100. **Schechter, M. S. and Haller, H. L.,** Colorimetric tests for DDT, and related compounds, *J. Am. Chem. Soc.,* 66, 2129, 1944.
101. **Balaban, I. E. and Sutcliffe, F. K.,** Thermal stability of DDT, *Nature (Lond.),* 155, 755, 1945.
102. **Scholefield, P. G., Bowden, S. T., and Jones, W. J.,** The thermal decomposition of 1,1,1-trichloro-2,2-bis-(p-chlorophenyl)ethane (DDT) and some of its phase equilibria, *J. Soc. Chem. Ind.,* 65, 354, 1946.
103. **Fleck, E. E. and Haller, H. L.,** Catalytic removal of hydrogen chloride from some substituted *a*- trichloroethanes, *J. Am. Chem. Soc.,* 66, 2095, 1944.
104. **Gunther, F. A., Blinn, R. C., and Kohn, G. K.,** Labile organo-halogen compounds and their gas chromatographic detection and determination in biological media, *Nature (Lond.),* 193, 537, 1962.
105. **Fleck, E. E., Preston, R. K., and Haller, H. L.,** Sym-tetraphenylethane from DDT and related compounds, *J. Am. Chem. Soc.,* 67, 1419, 1945.
106. **Cristol, S. J.,** Kinetic study of the dehydrochlorination of substituted 2,2-diphenylchloroethanes related to DDT, *J. Am. Chem. Soc.,* 67, 1494, 1945.
107. **England, B. D. and McLennan, D. J.,** Kinetics and mechanism of the reaction of DDT with sodium benzenethiolate and other nucleophiles, *J. Chem. Soc. (Lond.),* B7, 696, 1966.
108. **Cristol, S. J. and Haller, H. L.,** Dehydrochlorination of 1-trichloro-2-o-chlorophenyl-2-p-chlorophenylethane (o,p'-DDT isomer), *J. Am. Chem. Soc.,* 67, 2222, 1945.
109. **McKinney, J. D., Boozer, E. L., Hopkins, H. P., and Suggs, J. E.,** Synthesis and reactions of a proposed DDT metabolite, 2,2-bis(p-chlorophenyl) acetaldehyde, *Experientia,* 25, 897, 1969.
110. **Plimmer, J. R., Klingebiel, U. I., and Hummer, B. E.,** Photooxidation of DDT and DDE, *Science,* 167, 67, 1970.
111. Institut für ökologische Chemie, *Jahresbericht 1970,* Klein, W., Weisgerber, I., and Bieniek, D., Eds., Gesellschaft für Strahlen und Umweltforschung mbH, München, 1971, 48.
112. **McKinney, J. D. and Fishbein, L.,** DDE formation, dehydrochlorination or dehypochlorination, *Chemosphere,* [2], 67, 1972.
113. **Zweig, G., Ed.,** *Analytical Methods for Pesticides, Plant Growth Regulators and Food Additives,* Vols. I–V, Academic Press, New York and London, 1963, et seq.
114. **Gunther, F. A., Ed.,** *Residue Reviews,* Springer-Verlag, New York.
115. **Raw, G. R., Ed.,** *CIPAC Handbook, Vol. I, Analysis of Technical and Formulated Pesticides,* Collaborative International Pesticides Analytical Council Ltd., Harpenden, U. K., 1970.
116. Shell Chemical Corporation, *Handbook of Aldrin, Dieldrin and Endrin Formulations,* 2nd ed., Agricultural Chemicals Division, New York, 1959, et seq.

117. **Castillo, J. C. and Stiff, H. A.,** Application of the xanthydrol-KOH-pyridine method to the determination of 2,2-bis(p-chlorophenyl)-1,1,1-trichloroethane (DDT) in water, *Mil. Surg.,* 97, 500, 1945; *C.A.,* 2561, 1946.

118. **Claborn, H. V.,** Determination of DDT in the presence of DDD (1,1-dichloro-2,2-bis(p-chlorophenyl)ethane (also called TDE), *J. Assoc. Offic. Agric. Chem.,* 29, 330, 1946.

119. **Bradbury, F. R., Higgons, D. J., and Stoneman, J. P.,** Colorimetric method for the estimation of 2,2-bis(p-chlorophenyl)-1,1,1-trichloroethane (DDT), *J. Soc. Chem. Ind.,* 66, 65, 1947.

120. **Miskus, R.,** Analytical methods for DDT, in *Analytical Methods for Pesticides, Plant Growth Regulators and Food Additives,* Vol. II, Zweig, G., Ed., Academic Press, New York and London, 1964, 97.

121. **Claborn, H. V. and Beckman, H. F.,** Determination of 1,1,1-trichloro-2,2-bis(p-methoxyphenyl) ethane in milk and fatty materials, *Anal. Chem.,* 24, 220, 1952.

122. **Gillett, J. W., Van der Geest, L. P. S., and Miskus, R. P.,** Column partition chromatography of DDT and related compounds, *Anal. Biochem.,* 8, 200, 1964.

123. **Jones, L. R. and Riddick, J. A.,** Colorimetric determination of 2-nitro-1,1-bis(p-chlorophenyl)alkanes, *Anal. Chem.,* 23, 349, 1951.

124. **Blinn, R. C. and Gunther, F. A.,** The utilisation of infrared and ultra violet spectrophotometric procedures for assay of pesticide residues, *Residue Rev.,* 2, 99, 1963.

125. **Gilbert, G. G. and Sader, M. H.,** Fluorimetric determination of pesticides, *Anal. Chem.,* 41, 366, 1969.

126. **Allen, P. T.,** Polarographic methods for pesticides and additives, in *Analytical Methods for Pesticides, Plant Growth Regulators and Food Additives,* Vol. V, Zweig, G., Ed., Academic Press, New York, 1966, 67.

127. **Gajan, R. J.,** Analysis of pesticide residues by polarography, *J. Assoc. Offic. Agric. Chem.,* 48, 1027, 1965.

128. **Gajan, R. J. and Link, J.,** An investigation of the oscillography of DDT and certain analogues, *J. Assoc. Offic. Agric. Chem.,* 47, 1118, 1964.

129. **Cisak, A.,** Study of the electroreduction mechanism of halogen derivatives of cyclohexane, *Rocz. Chem.,* 37, 1025, 1963.

130. **Cisak, A.,** Polarographic properties of aldrin, isodrin, a-chlordane, β-chlordane, dieldrin and endrin, *Rocz. Chem.,* 40, 1717, 1966.

131. **Gajan, R. J.,** Applications of polarography for the detection and determination of pesticides and their residues, *Residue Rev.,* 5, 80, 1964.

132. **Gajan, R. J.,** Recent developments in the detection and determination of pesticides and their residues by oscillographic polarography, *Residue Rev.,* 6, 75, 1964.

133. **Casida, J. E.,** Radiotracer studies on metabolism, degradation, and mode of action of insecticide chemicals, *Residue Rev.,* 25, 149, 1969.

134. **Gorsuch, T. T.,** Radioactive Isotope Dilution Analysis, Review No. 2, The Radiochemical Centre, Amersham, U.K.

135. **Goodwin, E. S., Goulden, R., Richardson, A., and Reynolds, J. G.,** The analysis of crop extracts for traces of chlorinated pesticides by gas-liquid partition chromatography, *Chem. Ind. (Lond.),* 1220, 1960.

136. **Goodwin, E. S., Goulden, R., and Reynolds, J. G.,** Rapid identification and determination of residues of chlorinated pesticides in crops by gas-liquid chromatography, *Analyst,* 86, 697, 1961.

137. **Widmark, G.,** Modern trends of analytical chemistry as regards environmental chemicals, in *Environmental Quality and Safety; Chemistry, Toxicology and Technology,* Vol. I, Coulston, F. and Korte, F., Eds., Georg Thieme Verlag, Stuttgart, Academic Press, New York, 1972, 78.

138. **Mitchell, L. C.,** Separation and identification of chlorinated organic pesticides by paper chromatography, *J. Assoc. Offic. Agric. Chem.,* 41, 781, 1958.

139. **McKinley, W. P.,** Paper chromatography, in *Analytical Methods for Pesticides, Plant Growth Regulators and Food Additives,* Vol. I, Zweig, G., Ed., Academic Press, New York, 1963, 227.

140. **Winteringham, F. P. W., Harrison, A., and Bridges, R. G.,** Radioactive tracer-paper chromatography techniques, *Analyst,* 77, 19, 1952.

141. **Abbott, D. C. and Thomson, J.,** The application of thin-layer chromatographic techniques to the analysis of pesticide residues, *Residue Rev.,* 11, 1, 1965.

142. **Mangold, H. K.,** Isotope technique, in *Thin Layer Chromatography, a Laboratory Handbook,* 2nd ed., Stahl, E., Ed., Allen and Unwin, London, 1969.

143. **Burchfield, H. P. and Storrs, E. E.,** *Biochemical Applications of Gas Chromatography,* Academic Press, New York and London, 1962.

144. **Coulson, D. M. Cavanagh, L. A., de Vries, J. E., and Walther, B.,** Microcoulometric gas chromatography of pesticides, *J. Agric. Food Chem.,* 8, 399, 1960.

145. **Cassil, C. C.,** Pesticide residue analysis by microcoulometric gas chromatography, *Residue Rev.,* 1, 37, 1962.

146. **Lovelock, J. E. and Lipsky, S. R.,** Electron affinity spectroscopy — a new method for the identification of functional groups in chemical compounds separated by gas chromatography, *J. Am. Chem. Soc.,* 82, 431, 1960.

147. **de Faubert Maunder, M. J., Egan, H., and Roburn, J.,** Some practical aspects of the determination of chlorinated pesticides by electron-capture gas chromatography, *Analyst,* 89, 157, 1964.

148. **Clark, S. J.,** Quantitative determination of pesticide residues by electron absorption chromatography: characteristics of the detector, *Residue Rev.,* 5, 33, 1964.

149. **Burke, J. A. and Holswade, W.,** A gas chromatographic column for pesticide residue analysis: retention times and response data, *J. Assoc. Offic. Agric. Chem.,* 49, 374, 1966.
150. **Ott, D. E. and Gunther, F. A.,** DDD as a decomposition product of DDT, *Residue Rev.,* 10, 70, 1965.
151. **Robinson, J.,** Organochlorine compounds in man and his environment, *Chem. Br.,* 7, 472, 1971.
152. **Bowman, M. C. and Beroza, M.,** Extraction p values of pesticides and related compounds in six binary solvent systems, *J. Assoc. Offic. Agric. Chem.,* 48, 943, 1965.
153. **Bache, C. A. and Lisk, D. J.,** Selective emission spectrometric determination of nanogram quantities of organic bromine, chlorine, iodine, phosphorus and sulfur compounds in a helium plasma, *Anal. Chem.,* 39, 787, 1967.
154. **Keith, L. H., Alford, A. L., and Garrison, A. W.,** The high resolution NMR spectra of pesticides II; DDT-type compounds, *J. Assoc. Offic. Agric. Chem.,* 52, 1074, 1969.
155. **Baldwin, M. K., Robinson, J., and Carrington, R. A. G.,** Metabolism of HEOD (dieldrin) in the rat: examination of the major faecal metabolite, *Chem. Ind. (Lond.),* 595, 1970.
156. **Cochrane, W. P., Forbes, M., and Chau, A. S. Y.,** Cyclodiene chemistry. IV. Assignment of configuration of two nonachlors via synthesis and derivatisation, *J. Assoc. Offic. Agric. Chem.,* 53, 769, 1970.
157. **Mackenzie, K. and Lay, W. P.,** Fragmentation of tetrahydromethanonaphthalenes and their diaza- and benzo-analogues, *Tetrahedron Lett.,* No. 37, 3241, 1970.
158. **McKinney, J. D., Keith, L. H., Alford, A., and Fletcher, C. E.,** The proton magnetic resonance spectra of some chlorinated polyclodiene pesticide metabolites, rapid assessment of stereochemistry, *Can. J. Chem.,* 49, 1993, 1971.
159. **Keith, L. H. and Alford, A. L.,** Long-range couplings in the chlorinated polycyclodiene pesticides, *Tetrahedron Lett.,* 2489, 1970.
160. **Nagl, H. G., Klein, W., and Korte, F.,** Uber das reaktions verhalten von dieldrin in losung und in der gasphase, *Tetrahedron,* 26, 5319, 1970.
161. **Roll, D. B. and Biros, F. J.,** Nuclear quadrupole resonance spectrometry of some chlorinated pesticides, *Anal. Chem.,* 41, 407, 1969.
162. **Damico, J. N., Barron, R. P., and Ruth, J. M.,** The mass spectra of some chlorinated pesticidal compounds, *Org. Mass Spectr.,* 1, 331, 1968.
163. **Scopes, N. E. A. and Lichtenstein, E. P.,** The use of *Folsomia fimetaria* and *Drosophila melanogaster* as test insects for the detection of insecticide residues, *J. Econ. Entomol.,* 60, 1539, 1967.
164. **Sun, Y-P.,** Bioassay-insects, in *Analytical Methods for Pesticides, Plant Growth Regulators and Food Additives,* Vol. I, Zweig, G., Ed., Academic Press, New York and London, 1963, 399.
165. **Alder, K. and Stein, G.,** The course of the diene synthesis, *Angew Chem.,* 50, 510, 1937.
166. **Bergmann, F. and Eschinazi, H. E.,** Sterical course and the mechanism of the diene reactions, *J. Am. Chem. Soc.,* 65, 1405, 1943.
166a. **Gill, G. B.,** The application of the Woodward-Hoffman orbital symmetry rules to concerted organic reactions, *Q. Rev. Chem. Soc. (Lond.),* 22, 338, 1968.
167. **Buchel, K. H., Ginsberg, A. E., Fischer, R., and Korte, F.,** β-Dihydroheptachlor, ein insektizid mit sehr niedriger warmblüter-toxizitat, *Tetrahedron Lett.,* No. 33, 2267, 1964.
168. **Soloway, S. B.,** Stereochemistry of Bridged Polycyclic Compounds, Ph.D. Thesis, University of Colorado, 1955.
169. **Benson, W. R.,** Note on nomenclature of dieldrin and related compounds, *J. Assoc. Offic. Agric., Chem.,* 52, 1109, 1969.
169a. **Bedford, C. T.,** Von Baeyer – IUPAC Names and Recommended Trivial Names of Dieldrin, Endrin, and 24 Related Compounds, Shell Research Ltd., Tunstall Laboratory Research Report, November 1972.
170. **Bird, C. W., Cookson, R. C., Crundwell, E.,** Cyclisations and rearrangements in the isodrin-aldrin series, *J. Chem. Soc. (Lond.),* 4809, 1961.
171. **Parsons, A. M. and Moore, D. J.,** Some reactions of dieldrin and the proton magnetic resonance spectra of the products, *J. Chem. Soc. (Lond.),* C., 2026, 1966.
172. **Eckroth, D. R.,** A method for manual generation of correct von Baeyer names of polycyclic hydrocarbons, *J. Org. Chem.,* 32, 3362, 1967.
172a. International Union of Pure and Applied Chemistry, *IUPAC Nomenclature of Organic Chemistry,* 3rd ed., Butterworth, London, 1971.
173. **Straus, F., Kollek, L., and Heyn, W.,** Uber den ersatz positiven wasserstoffs durch halogen, *Chem. Ber.,* 63B, 1868, 1930.
174. **Kleiman, M.,** Chlorination of cyclopentadiene, U.S. patent 2,658,085, 1953; *C.A.,* 48, 12798, 1954.
175. **Lidov, R. E., Hyman, J., and Segel, E.,** Diels-Alder adducts of hexahalocyclopentadiene with quinones, U.S. patent 2,584,139, 1952; *C.A.,* 46, 9591, 1952.
176. **Frensch, H.,** Entwicklung und chemie der dien-gruppe, einer neuen klasse biozider wirkstoffe, *Med. Chem.,* 6, 556, 1957.
177. **Ungnade, H. E. and McBee, E. T.,** The chemistry of perchlorocyclopentenes and cyclopentadienes, *Chem. Rev.,* 58, 249, 1958.
178. **Gilbert, E. E. and Giolito, S. L.,** Pesticides, U.S. patent 2,616,928; *C.A.,* 47, 2424, 1953.
179. **McBee, E. T., Roberts, C. W., Idol, J. D., and Earle, R. H.,** An investigation of the chlorocarbon, $C_{10}Cl_{12}$, mp485° and the ketone, $C_{10}Cl_{10}$, mp349°, *J. Am. Chem. Soc.,* 78, 1511, 1956.

180. **Gilbert, E. E., Lombardo, P., Rumanowski, E. J., and Walker, G. L.,** Preparation and insecticidal evaluation of alcoholic analogs of kepone, *J. Agric. Food Chem.,* 14, 111, 1966.
181. **Allen, W. W.,** The effectiveness of various pesticides against resistant two-spotted spider mites on green house roses, *J. Econ. Entomol.,* 57, 187, 1964.
182. **McBee, E. T. and Smith, D. K.,** The reduction of hexachlorocyclopentadiene, 1,2,3,4,5-pentachlorocyclopentadiene, *J. Am. Chem. Soc.,* 77, 389, 1955.
183. **McBee, E. T., Meyers, R. K., and Baranauckas, C. F.,** 1,2,3,4-Tetracyclopentadiene. 1. The preparation of the diene and its reaction with aromatics and dienophiles, *J. Am. Chem. Soc.,* 77, 86, 1955.
184. **Soloway, S. B.,** Correlation between biological activity and molecular structure of the cyclodiene insecticides, *Adv. Pest Control Res.,* 6, 85, 1965.
185. **Volodkovich, S. D., Melnikov, N. N., Plate, A. F., and Prianishnikova, M. A.,** Reaction of 1,1-difluorotetrachloro-cyclopentadiene with some unsaturated compounds, *J. Gen. Chem. (U.S.S.R.),* 28, 3153, 1958.
186. **Riemschneider, R. and Kuhnl, A.,** Zur chemie von polyhalocyclopentadienen und verwandten verbindungen, *Monatsh. Chem.,* 86, 879, 1953.
187. **Banks, R. E., Harrison, A. C., Haszeldine, R. N., and Orrell, K. G.,** Diels-Alder reactions involving perfluorocyclo-pentadiene, *Chem. Commun.,* 3, 41, 1965.
188. **Riemschneider, R.,** The chemistry of the insecticides of the diene group, *World Rev. Pest Control,* 2, 29, 1963.
189. **Riemschneider, R. and Nehring, R.,** Ein Thiodan-Analoges aus trimethyltrichlorocyclopentadienen, *Z. Naturforsch. Teil. B.,* 17b, 524, 1962.
190. **Bluestone, H.,** 1,2,3,4,10,10-Hexachloro-6,7-Epoxy-1,4,4a,5,6,7,8,8a-Octahydro-1,4,5,8- Dimethanonaphthalene and Insecticidal Compositions Thereof, U.S. patent 2,676,132, 1954; *C.A.,* 48, 8474, 1954.
191. **Mackenzie, K.,** The reaction of hexachlorobicyclo [2,2,1] heptadiene with potassium ethoxide, *J. Chem. Soc. (Lond.),* 86, 457, 1962.
192. **Brooks, G. T. and Harrison, A.,** The effect of pyrethrin synergists, especially sesamex, on the insecticidal potency of hexachlorocyclopentadiene derivatives ('cyclodiene' insecticides) in the adult housefly, *Musca domestica,* L., *Biochem. Pharmacol.,* 13, 827, 1964.
193. **Riemschneider, R., Gallert, H., and Andres, P.,** Uber der herstellung von 1,4,5,6,7,7-hexachlorobicyclo [2.2.1] hepten-5-bishydroxymethylen-2,3, *Monatsh. Chem.,* 92, 1075, 1961.
194. **Riemschneider, R. and Kotzsch, H. J.,** Hexachlorocyclopentadiene and allylglykolather, *Monatsh. Chem.,* 91, 41, 1960.
195. **Riemschneider, R., Herzel, F., and Koetsch, H. J.,** 2-Methylen-1,4,5,6,7,7-hexachlorbicyclo [2.2.1] hepten-(5), *Monatsh. Chem.,* 92, 1070, 1961.
196. **Riemschneider, R.,** Penta- und hexachlorocyclopentadien als philodiene komponenten, *Botyu-Kagaku,* 28, 83, 1965.
197. **Forman, S. E., Durbetaki, A. J., Cohen, M. V., and Olofson, R. A.,** Conformational equilibria in cyclic sulfites and sulfates. The configurations and conformations of the two isomeric thiodans, *J. Org. Chem.,* 30, 169, 1965.
198. **Feichtinger, H. and Linden, H. W.,** Telodrin, its synthesis and derivatives, *Chem. Ind. (Lond.),* 1938, 1965.
199. **Gross, H.,** Darstellung und reaktionen des 2,5-dichlor-tetrahydrofurans, *Chem. Ber.,* 95, 83, 1962.
200. **Hyman, J.,** Improvements In or Relating to a Method of Forming Halogenated Organic Compounds and the Products Resulting Therefrom, British patent 618,432, 1949.
201. **Herzfeld, S. H. and Ordas, E. P.,** 1-Hydroxy-4, 7-methano-3a,4,7,7a-tetrahydro-4, 5,6,7,8,8-hexachloroindene and method of preparing same, U.S. patent 2,528,656, 1950.
202. **Goldman, A., Kleiman, M., and Fechter, H. G.,** Production of Halogenated Polycyclic Alcohols, U.S. patent 2,750,397, 1956.
203. Arvey Corporation, Improvements in or Relating to Halogenated Dicyclopentadiene Epoxides, British patent 714,869, 1954.
204. **Buchel, K. H., Ginsberg, A. E., and Fischer, R.,** Isomierung und chlorierung von dihydroheptachlor, *Chem. Ber.,* 99, 416, 1966.
205. **Buchel, K. H., Ginsberg, A. E., and Fischer, R.,** Synthese und struktur von heptachlor-methano-tetrahydroindanen, *Chem. Ber.,* 99, 405, 1966.
206. **Buchel, K. H., Ginsberg, A. E., and Fischer, R.,** Synthese und struktur von isomeren des chlordans, *Chem. Ber.,* 99, 421, 1966.
207. **March, R. B.,** The resolution and chemical and biological characterisation of some constituents of technical chlordane, *J. Econ. Entomol.,* 45, 452, 1952.
208. **Cochrane, W. P., Forbes, M., and Chau, A. S. Y.,** Assignment of configuration of two nonachlors via synthesis and derivatisation, *J. Assoc. Offic. Agric. Chem.,* 53, 769, 1970.
209. **Hyman, J., Freireich, E., and Lidov, R. E.,** Bicyclo-heptadienes and process of preparing the same, British patent 701,211, 1953.
210. **Plate, A. F. and Pryanishnikova, M. A.,** Preparation of bicyclo [2.2.1] hepta-2,5-diene by the condensation of cyclopentadiene with acetylene, *Izvest. Akad. Nauk SSSR., Otdel Khim. Nauk,* 741, 1956.
211. **Lidov, R. E. and Soloway, S. B.,** Improvements In or Relating to Methods of Preparing Insecticidal Compounds and the Insecticidal Compounds Resulting From Said Methods, British patent 692,546, 1953.

212. **Lidov, R. E. and Soloway, S. B.,** Process of Preparing Polycyclic Compounds and Insecticidal Compositions Containing the Same, British patent 692,547, 1953.
213. **Winteringham, F. P. W. and Harrison, A.,** Mechanisms of resistance of adult houseflies to the insecticide dieldrin, *Nature, (Lond.),* 184, 608, 1959.
214. **Kuderna, J. G., Sims, J. W., Wilkstrom, J. R., and Soloway, S. B.,** The preparation of some insecticidal chlorinated bridged phthalazines, *J. Am. Chem. Soc.,* 81, 382, 1959.
215. **Kleiman, M.,** Pesticidal Compounds, U.S. patent 2,655,513, 1953.
216. **Bellin, R. H.,** Endrin Stabilization Using Inorganic and Organic Nitrite Salts, U.S. patent 2,768,178, 1956.
217. **Nagl, H. G., Klein, W., and Korte, F.,** Uber das reaktionsverhalten von dieldrin in losung und in der gasphase, *Tetrahedron,* 26, 5319, 1970.
218. **Busvine, J. R.,** The insecticidal potency of γ-BHC and the chlorinated cyclodiene compounds and the significance of resistance to them, *Bull. Entomol. Res.,* 55, 271, 1964.
219. **Adams, C. H. M. and Mackenzie, K.,** Dehalogenation of isodrin and aldrin with alkoxide base, *J. Chem. Soc. (Lond.),* C, 480, 1969.
220. **Bienieck, D., Moza, P. N., Klein, W. and Korte, F.,** Reduktive dehalogenierung von chlorierten cyclischen kohlen wasserstoffen, *Tetrahedron Lett.,* 4055, 1970.
221. **Volodkovich, S. D., Vol'Fson, L. G., Kuznetsova, K. V., and Mel'nikov, N. N.,** Synthesis of α-oxides by the oxidation of polycyclic haloderivatives with hydrogen peroxide, *J. Gen. Chem. (U.S.S.R.),* 29, 2797, 1959.
222. **Brooks, G. T.,** The synthesis of ^{14}C-labelled 1:2:4:10:10-hexachloro-6:7-epoxy-1:4:4a:5:6:7:8:8a-octahydro-*exo*-1:4-exo-5:8-dimethanonaphthalene, *J. Chem. Soc. (Lond.),* 3693, 1958.
223. **Korte, F.,** Metabolism of Chlorinated Insecticides, Paper Presented to the Joint FAO/IAEA Division of Atomic Energy in Agriculture, Panel on the uses of Radioisotopes in the detection of Pesticide Residues, Vienna, April 1965.
224. **Korte, F. and Rechmeier, G.,** Mikrosynthese von Aldrin-^{14}C und Dieldrin-^{14}C, *Justus Liebigs Ann. Chem.,* 656, 131, 1962.
225. **McKinney, R. M. and Pearce, G. W.,** Synthesis of carbon-14-labeled aldrin and dieldrin, *J. Agric. Food Chem.,* 8, 457, 1960.
226. **Thomas, D. J. and Kilner, A. E.,** The synthesis of the insecticides aldrin and dieldrin labelled with carbon-14 at high specific activity, in *Radioisotopes in the Physical Sciences and Industry,* International Atomic Energy Agency, Vienna 1962.
227. **Korte, F. and Stiasni, M.,** Mikrosynthese von ^{14}C-markiertem Telodrin, *Justus Liebigs Ann. Chem.,* 656, 140, 1962.
228. **Black, A. and Morgan, A.,** Herstellung einiger Cl-38-markierten chlorierten Kohlen wasserstoffe durch neutronenbestrahlung und gaschromatographie, *Int. J. Appl. Radiat. Isot.,* 21, 5, 1970.
229. **Dailey, R. E., Walton, M. S., Beck, V., Leavens, C. L., and Klein, A. K.,** Excretion, distribution, and tissue storage of a ^{14}C-labeled photoconversion product of ^{14}C-dieldrin, *J. Agric. Food Chem.,* 18, 443, 1970.
230. **Riemschneider, R.,** Zur chemie von polyhalocyclopentadienen und verwandten verbindungen, IV. Thermische spaltung und oxydation des adduktes $C_{10}H_6Cl_6$, *Chem. Ber.,* 89, 2697, 1956.
231. **Lidov, R. E. and Bluestone, H.,** Cyclobutano-compounds, U.S. patent 2,714,617; *C.A.,* 50, 5756, 1956.
232. **Cochrane, W. P. and Forbes, M. A.,** Practical Applications of Chromous Chloride to the Confirmation of Organochlorine Pesticide Residue Identity, Paper presented at the 2nd International Congress of Pesticide Chemistry, Tel Aviv, Israel, 1971.
233. **Davidow, B. and Radomski, J. L.,** Isolation of an epoxide metabolite from fat tissues of dogs fed heptachlor, *J. Pharmacol. Exp. Ther.,* 107, 259, 1953.
234. **Brooks, G. T. and Harrison, A.,** The toxicity of α-DHC and related compounds to the housefly (*M. domestica*) and their metabolism by housefly and pig liver microsomes, *Life Sci.,* (Oxford), 6, 1439, 1967.
235. **Brooks, G. T.,** unpublished results.
236. **Chau, A. S. Y. and Cochrane, W. P.,** Cyclodiene chemistry. I. Derivative formation for the identification of heptachlor, heptachlor epoxide, cis-chlordane, trans-chlordane and aldrin pesticide residues by gas chromatography, *J. Assoc. Offic. Agric. Chem.,* 52, 1092, 1969.
237. **Cochrane, W. P.,** Cyclodiene chemistry II. Identification of the derivatives employed in the confirmation of the heptachlor, heptachlor epoxide, cis-chlordane, and trans-chlordane residues, *J. Assoc. Offic. Agric. Chem.,* 52, 1100, 1969.
238. **Chau, A. S. Y. and Cochrane, W. P.,** Cyclodiene chemistry III. Derivative formation for the identification of heptachlor, heptachlor epoxide, cis-chlordane, trans-chlordane, dieldrin, and endrin pesticide residues by gas chromatography, *J. Assoc. Offic. Agric. Chem.,* 52, 1220, 1969.
239. **Cochrane, W. P. and Forbes, M. A.,** Isomerisation of 1-exo-4,5,6,7,8,8-heptachloro-2,3-endo-epoxy-3a,4,7,7a-tetrahydro-4,7-methanoindane with base, *Can. J. Chem.,* 49, 3569, 1971.
240. **Polen, P. B.,** personal communication, 1970.

241. **Rosen, J. D.,** Conversion of pesticides under environmental conditions, in *Environmental Quality and Safety,* Vol. 1, Coulston, F. and Korte, F., Eds., Georg Thieme Verlag, Stuttgart; Academic Press, New York, 1972.

242. **McGuire, R. R., Zabik, M. J., Schuetz, R. D., and Flotard, R. D.,** Photolysis of 1,4,5,6,7,8,8-heptachloro-3a,4,7,7a-tetrahydro-4,7-methanoindene (cage formation vs. photodechlorination), *J. Agric. Food Chem.,* 18, 319, 1970.

243. **Brooks, G. T. and Harrison, A.,** Structure-activity relationships among insecticidal compounds derived from chlordene, *Nature (Lond.),* 205, 1031, 1965.

244. Institut fur Okologische Chemie, Jahresbericht 1969, Klein, W. and Drefahl, B., Eds., Gesellschaft für Strahlen und Umweltforschung mbH, München, 1970.

244a. **Ivie, G. W., Knox, J. R., Khalifa, S., Yamamoto, I., and Casida, J. E.,** Novel photoproducts of heptachlor epoxide, *trans*-chlordane, and *trans*-nonachlor, *Bull. Environ. Contam. Toxicol.,* 7, 376, 1972.

245. **Vollner, L., Klein, W., and Korte, F.,** Photoumlagerung der komponenten des technischen chlordans, *Tetrahedron Lett.,* 2967, 1969.

246. **Schwemmer, B., Cochrane, W. P., and Polen, P. B.,** Oxychlordane, animal metabolite of chlordane: isolation and synthesis, *Science,* 169, 1087, 1970.

247. **Polen, P. B., Hester, M., and Benziger, J.,** Characterisation of oxychlordane, animal metabolite of chlordane, *Bull. Environ. Contam. Toxicol.,* 5, 521, 1971.

248. **Maier-Bode, H.,** Properties, effect, residues and analytics of the insecticide endosulfan, *Residue Rev.,* 22, 1, 1968.

249. **Chau, A. S. Y. and Cochrane, W. P.,** Cis-opening of the dieldrin oxirane ring, *Chem. Ind. (Lond.),* 1568, 1970.

250. **Adams, C. H. M. and Mackenzie, K.,** Dehalogenation of isodrin and aldrin with alkoxide base, *J. Chem. Soc. (Lond.),* C, 480, 1969.

251. **Robinson, J., Richardson, A., Bush, B., and Elgar, K.,** A photoisomerisation product of dieldrin, *Bull. Environ. Contam. Toxicol.,* 1, 127, 1966.

252. **Benson, W. R.,** Photolysis of solid and dissolved dieldrin, *J. Agric. Food Chem.,* 19, 66, 1971.

253. **Cookson, R. C. and Crundwell, E.,** Transannular reactions in the isodrin series, *Chem. Ind. (Lond.),* 703, 1959.

254. **Soloway, S. B., Damiana, A. M., Sims, J. W., Bluestone, H., and Lidov, R. E.,** Skeletal rearrangements in reactions of isodrin and endrin, *J. Am. Chem. Soc.,* 82, 5377, 1960.

255. **Weil, E. D., Colson, J. G., Hoch, P. E., and Gruber, R. H.,** Toxic chlorinated methanoisobenzofuran derivatives, *J. Heterocycl. Chem.,* 6, 643, 1969.

256. **Bruck, P., Thompson, B., and Winstein, S.,** Dechlorination of isodrin and related compounds, *Chem. Ind. (Lond.),* 405, 1960.

257. **Phillips, D. D., Pollard, G. E., and Soloway, S. B.,** Thermal isomerisation of endrin and its behaviour in gas chromatography, *J. Agric. Food Chem.,* 10, 217, 1962.

258. **Rosen, J. D., Sutherland, D. J., and Lipton, G. R.,** The photochemical isomerisation of dieldrin and endrin and effects on toxicity, *Bull. Environ. Contam. Toxicol.,* 1, 133, 1966.

259. **Zabik, M. J., Schuetz, R. D., Burton, W. L., and Pape, B. E.,** Studies of a major photolytic product of endrin, *J. Agric. Food Chem.,* 19, 308, 1971.

260. **Brooks, G. T.,** Degradation of Organochlorine Insecticides, Problems and Possibilities, in Proceedings of the 2nd International IUPAC Congress of Pesticide Chemistry 1971, Vol. 6; *Fate of Pesticides in Environment,* Tahori, A. S., Ed., Gordon and Breach, London, 1972, 223.

261. **Martin, H. and Wain, R. L.,** Insecticidal action of DDT, *Nature (Lond.),* 154, 512, 1944.

262. **Prill, E. A.,** Diels-Alder syntheses with hexachlorocyclopentadiene, *J. Am. Chem. Soc.,* 69, 62, 1947.

263. **Riemschneider, R.,** Neue kontakt-insetizide der halogenkohlen wasserstoffe-klasse (M410, M344), *Mitt. Physiol. Chem. Inst. Univ. Berlin,* R8, March 1947; *C.A.,* 48, 2973, 1954.

264. **Riemschneider, R. and Kuhnl, A.,** *Mitt. Physiol. Chem. Inst. Univ. Berlin,* R.11, October 1947 (see also reference 186).

265. **Kearns, C. W., Ingle, L., and Metcalf, R. L.,** New chlorinated hydrocarbon insecticide, *J. Econ. Entomol.,* 38, 661, 1945.

266. **Hyman, J.,** personal communication, 1972.

267. Velsicol Corporation, Improvements in the Production of New Diels-Alder Adducts, British patent 614,931, 1948; *C.A.,* 43, 4693, 1949.

268. **Hyman, J.,** Improvements in or relating to method of forming halogenated organic compounds and the products resulting therefrom, British patent 618,432, 1949; *C.A.,* 43, 5796, 1949.

269. **Ingle, L.,** *A Monograph On Chlordane. Toxicological and Pharmacological Properties,* University of Illinois, 1965.

270. Velsicol Chemical Corporation, *Chlordane Formulation Guide,* Bulletin No. 502–35R, 1970.

271. Velsicol Chemical Corporation, Standard for Technical Chlordane, August 1971.

272. **Polen, P. B.,** Chlordane: Composition, Analytical Considerations and Terminal Residues, paper presented to a meeting of the IUPAC Commission on Terminal Residues, Geneva, Switzerland, 1966.

273. **Riemschneider, R. and Graviz, B. B.,** Uber den raumlichen Bau einiger Addukte aus hexachlorocyclopentadien und ungesattigen verbindungen (Dien-gruppe), *Botyu-kagaku,* 25, 123, 1960.

274. **Vogelbach, C.,** Isolation of new crystalline substances from technical grade chlordan, *Angew Chem.,* 63, 378, 1951.

275. **Harris, C. R.,** Factors influencing the biological activity of technical chlordane and some related components in soil, *J. Econ. Entomol.,* 65, 343, 1972.

276. **Saha, J. G. and Lee, Y. W.,** Isolation and identification of the components of a commercial chlordane formulation, *Bull. Environ. Contam. Toxicol.,* 4, 285, 1969.
277. Belt-plus Insecticide, Velsicol Chemical Corporation Development Bulletin no. 01-054-601, April 1971.
278. **Freeman, S. M. D.,** cited in Ingle, L., *A Monograph on Chlordane. Toxicological and Pharmacological Properties,* University of Illinois, 1965, 12.
279. **Herzfeld, S. H. and Ordas, E.,** Improvements In or Relating To Processes For the Production of Derivatives of Diels-Alder Adducts of Hexachlorocyclopentadiene and Cyclopentadiene, British patent 652,300, 1951.
280. **Whaley, R. E.,** Heptachlor Insecticide, U.S. patent 3,541,163, 1970; *C.A.,* 74, 53177, 1971.
281. Velsicol Chemical Corporation, Heptachlor Formulation Guide, Bulletin no. 504-16, 1964.
282. **Busvine, J. R.,** The newer insecticides in relation to pests of medical importance, in symposium on insecticides, *Trans. R. Soc. Trop. Med. Hyg.,* 46, 245, 1952.
283. Velsicol Chemical Corporation, Chlordane Federal Label Acceptances, Bulletin no. 502-32, March 1968.
284. **Miles, J. R. W., Tu, C. M., and Harris, C. R.,** Metabolism of heptachlor and its degradation products by soil microorganisms, *J. Econ. Entomol.,* 62, 1334, 1969.
285. **Carter, F. L., Stringer, C. A., and Heinzelman, D.,** 1-Hydroxy-2,3-epoxychlordene in Oregon soil previously treated with technical heptachlor, *Bull. Environ. Contam. Toxicol.,* 6, 249, 1971.
286. **Bonderman, D. P. and Slach, E.,** Appearance of 1-hydroxychlordene in soil, crops and fish, *J. Agric. Food Chem.,* 20, 328, 1972.
287. **Miles, J. R. W., Tu, C. M., and Harris, C. R.,** Degradation of heptachlor epoxide and heptachlor by a mixed culture of soil microorganisms, *J. Econ. Entomol.,* 64, 839, 1971.
288. **Harris, C. R.,** personal communication, 1972.
289. **Davidow, B.,** A spectrophotometric method for the quantitative estimation of technical chlordan, *J. Assoc. Offic. Agric. Chem.,* 33, 886, 1950.
290. **Ordas, E. P., Smith, V. C., and Meyer, C. F.,** Spectrophotometric determination of heptachlor and technical chlordane on food and storage crops, *J. Agric. Food Chem.,* 4, 444, 1956.
291. Velsicol Chemical Corporation, Evaluation of Technical Heptachlor Formulations, Bulletin No. 504-17, undated.
292. **Dorn, W.,** Statement of the State Secretary on a request of MdB(MP) Prinz Zu Sayn Wittgenstein et al. regarding the fish kill in the Rhine, given at 20.2. 1970 in the Deutsche Bundestag (Parliament), 1970. Cited in *Pesticides in the Modern World,* Cooperative Programme of Agro-Allied Industries, with FAO and other U.N. Organisations at the Newgate Press Ltd., London, 1972, 5.
293. **Farbwerke Hoechst, AG,** personal communications, 1972.
294. **Feichtinger, H. and Tummes, H.,** Verfahren Zur Hestellung Von 4,5,6,7,10,10-Hexachlor-4,7-Endomethylen-4,7,8,9-Tetrahydrophthalan, German patent 960,284, 1957.
295. **Whetstone, R, R.,** Chlorinated derivatives of cyclopentadiene, in Kirk-Othmer: *Encyclopedia of Chemical Technology,* Vol. 5, 2nd ed., McKetta, J. J. and Othmer, D. F., Eds., John Wiley (Interscience), 1964, 240.
296. **Farbwerke Hoechst, AG,** *Thiodan and the Environment,* technical bulletin translated by Hoechst U.K., 1971; *Thiodan,* [an undated bulletin on formulations and applications of Thiodan], Farbwerke Hoechst AG, Frankfurt a. Main.
297. **Frensch, H., Goebel, H., and Czech, M.,** Agents for killing undesired fish, U.S. patent 2,799,685; *C.A.,* 3852, 1962.
298. **Gorbach, S. and Knauf, W.,** Endosulfane and the environment in *Environmental Quality and Safety,* Vol. I, Coulston, F. and Korte, F., Eds., Georg Thieme Stuttgart, Academic Press, New York, 1972, 250.
299. **Schoettger, R. A.,** Toxicology of Thiodan in several fish and aquatic invertebrates, *Investigations in Fish Control 35,* United States Department of the Interior, Bureau of Sport Fisheries and Wildlife, U.S. Govt. Print. Off., Washington D.C., 1970
300. **Zweig, G. and Archer, T. E.,** Quantitative determination of Thiodan by gas chromatography, *J. Agric. Food Chem.,* 8, 190, 1960.
301. Shell Research Ltd., cited in Phillips, F. T., The rates of loss of dieldrin and aldrin by volatilisation from glass surfaces, *Pestic. Sci.,* 2, 255, 1971.
302. **Kearns, C. W., Weinman, C., and Decker, G. C.,** Insecticidal properties of some new chlorinated organic compounds, *J. Econ. Entomol.,* 42, 127, 1949.
303. **Jager, K. W.,** *Aldrin, Dieldrin, Endrin and Telodrin, an Epidemiological and Toxicological Study of Long-term Occupational Exposure,* Elsevier, Amsterdam, 1970.
304. **Park, K. S. and Bruce, W. N.,** The determination of water solubility of aldrin, dieldrin, heptachlor and heptachlor epoxide, *J. Econ. Entomol.,* 61, 770, 1968.
304a. **Delacy, T. P. and Kennard, C. H. L.,** Crystal structures of endrin and aldrin, *J. Chem. Soc. (Lond.),* Perkin II, 2153, 1972.
305. Aldrin and Dieldrin, An Appraisal of Data Submitted to International Regulatory Agencies, Shell International Chemical Company Ltd., Regulatory Affairs Division, January 1972.
306. Summary of Technical Information, Registered Label Uses and Label Uses Withdrawn Since 1964, in; Aldrin, Dieldrin, Endrin, a Status Report, Shell Chemical Company, Agricultural Chemicals Division, September 1967.
307. *Endrin, the Foliage Insecticide,* Shell International Chemical Company Ltd., Bulletin, 1964.
308. *The Safe Handling and Toxicology of Endrin,* Shell International Chemical Company Ltd., undated bulletin.

309. **Skerrett, E. J. and Baker, E. A.,** A new colour reaction for dieldrin and endrin, *Chem. Ind. (Lond.)*, 539, 1959.

310. Third World Food Survey, Basic Study No. 11; Food and Agriculture Organisation, 1963.

311. Pesticides and Health, Shell briefing service, Royal Dutch Shell Group, February 1967.

312. The Use of Pesticides, President's Scientific Advisory Committee Report, Weisner, J. B., Chairman, Govt. Print. Off., Washington, D.C., May 15, 1963.

313. **Glasser, R. F.,** The actions taken by Shell in response to recommendations of advisory committees and regulatory agencies, in Symposium on the Science and Technology of Residual Insecticides in Food Production with Special Reference to Aldrin and Dieldrin, Shell Oil Company, 1968, 225.

314. **Hardie, D. W. F.,** Benzene Hexachloride, in *Kirk-Othmer Encyclopedia of Chemical Technology,* Vol. 5, 2nd ed., McKetta, J. J. and Othmer, D. F., Eds., John Wiley (Interscience), 1964, 267.

315. **Lambermont, F.,** L'Hexachlorcyclohexane et son isomère gamma, paper presented at the 8th International Congress on Agriculture, Brussels, July 12, 1950.

316. **Van der Linden, T.,** Uber die benzol-hexachloride und ihren zerfall in trichlor-benzole, *Chem. Ber.,* 45, 231, 1912.

317. **Bender, H.,** Chlorinating benzene, toluene, etc., U.S. patent 2,010,841; 1935. *C.A.,* 29, 6607, 1935.

318. **Stephenson, H. P., Curtis, A. L., Grant, A. E., and Hardie, T.,** cited in Haller, H. L., and Bowen, C. V., Basic facts about benzene hexachloride, *Agric. Chem.,* 2, 15, 1947.

319. **Dupire, A. and Raucourt, M.,** Un insecticide nouveau: l'hexachlorure de benzene, *C. R. Acad. Agric. Fr.,* 29, 470, 1943.

320. **Bourne, L. B.,** Hexachlorocyclohexane as an insecticide, *Nature (Lond.),* 156, 85, 1945.

321. **Slade, R. E.,** The γ-isomer of hexachlorocyclohexane (gammexane), Hurter Memorial Lecture, *Chem. Ind. (Lond.),* 64, 314, 1945.

322. **Holmes, E.,** Recent British developments in taint-free use of BHC, *Agric. Chem.,* 6,[12], 31, 1951.

323. **Ishikura, H.,** Impact of pesticide usage on the Japanese environment, in *Environmental Toxicology of Pesticides,* Matsumura, F., Boush, G. M., and Misato, T., Eds., Academic Press, New York and London, 1972, 1.

324. **Kenaga, E. E. and Allison, W. E.,** Commercial and experimental organic insecticides (1971 revision), *Bull. Entomol. Soc. Am.,* 16, 68, 1970.

325. **Ishii, K.,** Chlorinated pesticides, in *Japan Pesticide Information,* No. 3, Japan Plant Protection Association, 1970, 11.

326. **Shindo, N.,** The present situation of agricultural chemicals in the chemical industry of Japan, in *Japan Pesticide Information,* No. 7, Japan Plant Protection Association, 1971, 10.

327. **Kurihara, N., Sanemitsu, Y., Kimura, T., Kobayashi, M., and Nakajima, M.,** Stepwise synthesis of 3,4,5,6-tetrachlorocyclohexene-1 (BTC) isomers, *Agric. Biol. Chem.* (Tokyo), 34, 784, 1970.

328. Cited in Hardie, D. W. F., Benzene hexachloride, in *Kirk-Othmer Encyclopedia of Chemical Technology,* Vol. 5, 2nd ed., Mcketta, J. J. and Othmer, D. F., Eds., John Wiley (Interscience), 1964, 274.

329. **Gunther, F. A.,** Chlorination of benzene, *Chem. Ind. (Lond.),* 399, 1946.

330. **Melnikov, N. N.,** *Chemistry of Pesticides,* Gunther, F. A. and Gunther, J. D., Eds., *Residue Rev.,* Vol. 36, Springer-Verlag, New York, 1971.

331. **Cooke, W. H. and Smart, J. C.,** Insecticide, British patent 586,439, 1944; *C.A.,* 41, 7640, 1947.

332. **Hay, J. K. and Webster, K. C.,** Insecticide, British patent 586,442, 1944; *C.A.,* 7640, 1947.

333. **Balson, E. W.,** Studies in vapour pressure measurement, *Trans. Faraday Soc.,* 43, 54, 1947.

334. **Kanazawa, J., Yashima, T., and Kiritani, K.,** Contamination of ecosystem by pesticides II, *Kagaku (Science),* 41, 383, 1971.

335. **Furst, H. and Praeger, K.,** Isolation of pure γ-isomer from crude hexachlorocyclohexane mixtures, German patent 1,093,791, 1959; *C.A.,* 55, 19124, 1961.

336. **Bishopp, F. C.,** The insecticide situation, *J. Econ. Entomol.,* 39, 449, 1946.

337. **Rohwer, S. A.,** Effect of individual BHC isomers on plants, *Agric. Chem.,* 4, 75, 1949.

338. **Burnett, G. F.,** Trials of residual insecticides against anophelines in African-type huts, *Bull. Entomol. Res.,* 48, 631, 1957.

339. **Hocking, K. S., Armstrong, J. A., and Downing, F. S.,** cited in Pampana E., *A Textbook of Malaria Eradication,* 2nd ed., Oxford University Press, 1969, 133.

340. **Cross, H. F. and Snyder, F. M.,** Impregnants against chiggers, *J. Econ. Entomol.,* 41, 936, 1948.

340a. **Ulmann, E., Ed.,** *Lindane, Monograph of an Insecticide,* Verlag K. Schillinger, Freiburg im Breisgau, 1972.

341. **Hassell, O.,** Stereochemistry of cyclohexane, *Q. Rev. Chem. Soc. (Lond.),* 7, 221, 1953.

342. **Orloff, H. D.,** The stereoisomerism of cyclohexane derivatives, *Chem. Rev.,* 54, 347, 1954.

343. **Riemschneider, R., Spat, M., Rausch, W., Bottger, E.,** Zur chemie von polyhalocyclohexanen XXIV. 1,2,3,4,5,6-hexachlorocyclohexan von schmp. 88-89°, *Monatsch. Chem.,* 84, 1068, 1953.

344. **Kauer, K., Du Vall, R., and Alquist, F.,** Epsilon isomer of 1,2,3,4,5,6-hexachlorocyclohexane, *Ind. Eng. Chem., Ind. Ed.,* 39, 1335, 1947.

345. **Visweswariah, K. and Majumder, S. K.,** A new insecticidal isomer of hexachlorocyclohexane (BHC), *Chem. Ind. (Lond.),* 379, 1969.

346. **Cristol, S. J.,** The structure of α-benzene hexachloride, *J. Am. Chem. Soc.,* 71, 1894, 1949.

347. **Bastiansen, O., Ellefson, O., and Hassel, O.,** Structure of α, β, γ, δ, and ε benzene hexachloride, *Research,* 2, 248, 1948.

348. **Bastiansen, O., Ellefson, O., and Hassell, O.,** Electron diffraction investigation of α, β, γ, δ and ε benzene hexachloride, *Acta Chem. Scand.,* 3, 918, 1949.

349. **Norman, N.,** The crystal structure of the epsilon isomer of 1,2,3,4,5,6-hexachlorocyclohexane, *Acta Chem. Scand.,* 4, 251, 1950.

350. **Hetland, E.,** Dipole moments of α, β, γ, and δ hexachlorocyclohexane and tetrachlorocyclohexane (mp 174°), *Acta Chem. Scand.,* 2, 678, 1948.

351. **Kolka, A., Orloff, H., and Griffing, M.,** Two new isomers of benzene hexachloride, *J. Am. Chem. Soc.,* 76, 3940, 1954.

352. **Whetstone, R. R., Davis, F., and Ballard, S.,** Interconversion of hexachlorocyclohexane isomers, *J. Am. Chem. Soc.,* 75, 1768, 1953.

353. **Guilhon, M. J.,** The insecticidal and toxic properties of some sulfur derivatives of hexachlorocyclohexane, *C. R. Acad. Agric. Fr.,* 23, 101, 1947; *C.A.,* 41, 6363, 1947.

354. **Cristol, S.,** The kinetics of the alkaline dehydrochlorination of the benzene hexachloride isomers. The mechanism of second order elimination reactions, *J. Am. Chem. Soc.,* 69, 338, 1947.

355. **Gunther, F. and Blinn, R. C.,** Alkaline degradation of benzene hexachloride, *J. Am. Chem. Soc.,* 69, 1215, 1947.

356. **La Clair, J. B.,** Determination of BHC by hydrolysable chlorine method, *Anal. Chem.,* 20, 241, 1948.

357. **Riemschneider, R.,** Konstitution und wirkung von insektiziden. Mitt XVI, *Z. Angew. Ent.,* 48, 423, 1961.

358. **Kurihara, N.,** BHC — its toxicity and its penetration, translocation and metabolism in insects and mammals, *Bochu Kagaku,* 35(II), 56, 1970.

359. **Riemschneider, R.,** Polyhalocyclohexanen und verwandten verbindungen. Mitt XXX: Relativ konfigurationsbestimmung. Konfigurations-bezeichnungen von polyhalocyclohexanen und cyclohexenen, *Osterreichische Chemiker-Zeitung,* 55, 102, 1954.

360. **Kurihara, N., Sanemitsu, Y., Tamura, Y., and Nakajima, M.,** Studies on BHC isomers and their related compounds. Part II. Isomerisation of 3,4,5,6-tetrachlorocyclohexene-1 (BTC) isomers, *Agric. Biol. Chem. (Tokyo),* 34, 790, 1970.

361. **Kurihara, N., Sanemitsu, Y., Nakajima, M., McCasland, G. E., and Johnson, L. F.,** Studies on BHC isomers and their related compounds. III. Proton magnetic resonance studies at 220 MHz of two new iodopentachlorocyclohexanes derived from tetrachlorocyclohexene (BTC), *Agric. Biol. Chem. (Tokyo),* 35, 71, 1971.

362. **Bridges, R. G.,** Retention of γ-benzene hexachloride by wheat and cheese, *J. Sci. Food Agric.,* 431, 1958.

363. **Schechter, M. S. and Hornstein, I.,** Colorimetric determination of benzene hexachloride, *Anal. Chem.,* 24, 544, 1952.

364. Physical and Chemical Properties of Hercules Toxaphene; Hercules Incorporated Agricultural Chemicals Technical Data Bulletin AP–103A; various Hercules technical bulletins.

365. **Klein, A. K. and Link, D. J.,** Field weathering of toxaphene and chlordan, *J. Assoc. Offic. Agric. Chem.,* 50, 586, 1967.

366. **Desalbres, L. and Rache, J.,** Les terpènes polychlorés et leurs propriétés insecticides, *Chim. Ind. (Paris),* 59, 236, 1948.

367. **Khanenia, F. S. and Zhuravlev, S. V.,** cited in Busvine, J. R., The insecticidal potency of γ-BHC and the chlorinated cyclodiene compounds and the significance of resistance to them, *Bull. Entomol. Res.,* 55, 271, 1964.

368. **Weinman, C. and Decker, G. C.,** Chlorinated hydrocarbon insecticides used alone and in combinations for grasshopper control, *J. Econ. Entomol.,* 42, 135, 1949.

369. **Schread, J. C.,** Persistence of DDT, parathion, etc., for *Popillia japonica* control, *J. Econ. Entomol.,* 42, 383, 1949.

370. The Use of Toxaphene Low Volume Formulations for Cotton Insect Control, Hercules Incorporated Bulletin, Wilmington, Delaware, 1969, 24 pp.

371. **Beckman, H. F., Ibert, E. R., Adams, B. B., Skoolin, D. O.,** Determination of total chlorine in pesticides by reduction with a liquid anhydrous ammonia-sodium mixture, *J. Agric. Food Chem.,* 6, 104, 1958.

372. **Graupner, A. J. and Dunn, C. L.,** Determination of toxaphene by a spectrophotometric diphenylamine procedure, *J. Agric. Food Chem.,* 8, 286, 1960.

373. **Brown, A. W. A.,** Insect resistance, *Farm Chemicals,* September 1969 et seq.

374. **Brown, A. W. A. and Pal, R.,** *Insect Resistance in Arthropods,* 2nd ed., W.H.O., Geneva, 1971, 446.

375. **Busvine, J. R.,** *A Critical Review of the Techniques Used for Testing Insecticides,* 2nd ed., Commonwealth Agricultural Bureaux, 1971.

376. **Bliss, C. I.,** Calculation of dose/mortality curve, *Ann. Appl. Biol.,* 22, 134, 1935.

377. **Dyte, C. E.,** Insecticide resistance in stored-product insects with special reference to *Tribolium castaneum, Trop. Stored Prod. Inf.,* (20), 13, 1970.

378. **Georghiou, G. P.,** Genetics of resistance to insecticides in houseflies and mosquitoes, *Exp. Parasitol.,* 26, 224, 1969.

379. **Davidson, G.,** Insecticide resistance in *Anopheles gambiae* Giles, a simple case of Mendelian inheritance, *Nature (Lond.),* 178, 861, 1956.

380. **Davidson, G.,** Studies on insecticide resistance in anopheline mosquitoes, *Bull. W.H.O.,* 18, 579, 1958.

381. **Guneidy, A. M. and Busvine, J. R.,** Genetical studies on dieldrin-resistance in *Musca domestica* L. and *Lucilia cuprina* (Wied.), *Bull. Entomol. Res.,* 55, 499, 1964.

382. **Georghiou, G. P., March, R. B., and Printy, G. E.,** A study of the genetics of dieldrin-resistance in the housefly (*Musca domestica* L.), *Bull. W.H.O.,* 29, 155, 1963.

383. Tadano, T. and Brown, A. W. A., Genetical linkage relationships of DDT resistance and dieldrin-resistance in *Culex pipiens fatigans* Wiedemann, *Bull. W.H.O.*, 36, 101, 1967.

384. Brown, A. W. A., Insecticide resistance-genetic implications and applications, *World. Rev. Pest Control*, 6, 104, 1967.

385. Busvine, J. R., Developments in pest control, *Pest Articles and News Summaries P.A.N.S.*, [A], 14, 310, 1968.

386. Tsukamoto, M., Biochemical genetics of insecticide resistance in the housefly, *Residue Rev.*, 25, 289, 1969.

387. Tsukamoto, M. and Suzuki, R., Genetic analysis of DDT-resistance in strains of the housefly *Musca domestica* L., *Botyu-Kagaku*, 29, 76, 1964.

388. Oppenoorth, F. J., Some Cases of Resistance Caused by the Alteration of Enzymes, Proc. 12th Int. Congr. Entomol., London, July 1964, 240.

389. Grigolo, A. and Oppenoorth, F. J., The importance of DDT-dehydrochlorinase for the effect of the resistance gene kdr in the housefly, *Genetica*, 37, 159, 1966.

390. Khan, M. A. Q. and Terriere, L. C., DDT-dehydrochlorinase activity in housefly strains resistant to various groups of insecticides, *J. Econ. Entomol.*, 61, 732, 1968.

391. Sawicki, R. M. and Farnham, A. W., Examination of the isolated autosomes of the SKA strain of house-flies (*Musca domestica* L.) for resistance to several insecticides with and without pretreatment with sesamex and TBTP, *Bull. Entomol. Res.*, 59, 409, 1968.

392. Hoyer, R. F. and Plapp, F. W., Jr., A gross genetic analysis of two DDT-resistant housefly strains, *J. Econ. Entomol.*, 59, 495, 1966.

393. Plapp, F. W., Jr. and Hoyer, R. F., Possible pleiotropism of a gene conferring resistance to DDT, DDT analogs, and pyrethrins in the House Fly and *Culex tarsalis*, *J. Econ. Entomol.*, 61, 761, 1968.

394. Plapp, F. W., Jr. and Hoyer, R. F., Insecticide resistance in the housefly: decreased rate of absorption as the mechanism of action of a gene that acts as an intensifier of resistance, *J. Econ. Entomol.*, 61, 1298, 1968.

395. Sawicki, R. M. and Farnham, A. W., Genetics of resistance to insecticides of the SKA strain of Musca domestica, III. Location and isolation of the factors of resistance to dieldrin, *Entomol. Exp. Appl.*, 11, 133, 1968.

396. Perry, A. S., Hennessy, D. J., and Miles, J. W., Comparative toxicity and metabolism of p, p'-DDT and various substituted DDT-derivatives by susceptible and resistant houseflies, *J. Econ. Entomol.*, 60, 568, 1967.

397. Oppenoorth, F. J., DDT-resistance in the housefly dependent on different mechanisms and the action of synergists, *Meded. Landbouwhogesch. Opzoekingsstn. Staat. Gent.*, 30, 1390, 1965.

398. El Basheir, S., Causes of resistance to DDT in a diazinon-selected and a DDT-selected strain of house flies, *Entomol. Exp. Appl.*, 10, 111, 1967.

399. Milani, R., Genetical aspects of insecticide resistance, *Bull. W.H.O.*, Supplement, 29, 77, 1963.

400. Oppenoorth, F. J. and Nasrat, G. E., Genetics of dieldrin and gamma-BHC (lindane) resistance in the housefly (*Musca domestica*), *Entomol. Exp. Appl.*, 9, 223, 1966.

401. Davidson, G., Resistance to Chlorinated Insecticides in Anopheline Mosquitoes, Proc. 12th Int. Congr. Entomol., London, July 1964, 236.

402. Oonnithan, E. S. and Miskus, R., Metabolism of C^{14}-dieldrin-resistant *Culex pipiens quinquefasciatus* Mosquitoes, *J. Econ. Entomol.*, 5, 425, 1964.

403. Tadano, T. and Brown, A. W. A., Development of resistance to various insecticides in *Culex pipiens fatigans* Wiedemann, *Bull. W.H.O.*, 35, 189, 1966.

404. Plapp, F. W., Jr., Chapman, G. A., and Morgan, J. W., *J. Econ. Entomol.*, 58, 1064, 1965.

405. Klassen, W. and Brown, A. W. A., Genetics of insecticide-resistance and several visible mutants in *Aedes aegypti*, *Can. J. Genet. Cytol.*, 6, 61, 1964.

406. Lockhart, W. L., Klassen, W., and Brown, A. W. A., Crossover values between dieldrin-resistance and DDT; resistance and linkage-group-2 genes in *Aedes aegypti*, *Can. J. Genet. Cytol.*, 12, 407, 1970.

407. Kimura, T. and Brown, A. W. A., DDT-dehydrochlorinase in *Aedes aegypti*, *J. Econ. Entomol.*, 57, 710, 1964.

408. Inwang, E. E., Khan, M. A. Q., and Brown, A. W. A., DDT-resistance in West African and Asian strains of *Aedes aegypti* (L), *Bull. W.H.O.*, 36, 409, 1967.

409. McDonald, I. C., Ross, M. H., and Cochran, D. G., Genetics and linkage of aldrin resistance in the German cockroach, *Blattella germanica* (L), *Bull. W.H.O.*, 40, 745, 1969.

410. Matsumura, F., Telford, J. N., and Hayashi, M., Effect of sesamex upon dieldrin resistance in the German cockroach, *J. Econ. Entomol.*, 60, 942, 1967.

411. Read, D. C. and Brown, A. W. A., Inheritance of dieldrin-resistance and adult longevity in the cabbage maggot, *Hylemya brassicae* (Bouche), *Can. J. Genet. Cytol.*, 8. 71, 1966.

412. Hooper, G. H. S. and Brown, A. W. A., Dieldrin-resistant and DDT-resistant strains of the spotted root maggot apparently restricted to heterozygotes for resistance, *J. Econ. Entomol.*, 58, 824, 1965.

413. Lineva, V. A. and Derbeneva-Uhova, V. P., cited in Zaghloul, T. M. A. and Brown, A. W. A., Effects of sublethal doses of DDT on the reproduction and susceptibility of *Culex pipiens* L., *Bull. W.H.O.*, 38, 459, 1968.

414. Zaghloul, T. M. A. and Brown, A. W. A., Effects of sublethal doses of DDT on the reproduction and susceptibility of *Culex pipiens* L., *Bull. W.H.O.*, 38, 459, 1968.

415. Moriarty, F., The sublethal effects of synthetic insecticides on insects, *Biol. Rev.*, 44, 321, 1969.

416. Hadaway, A. B., Cumulative effect of sublethal doses of insecticides on houseflies, *Nature (Lond.)*, 178, 149, 1956.

417. **Ahmad, N. and Brindley, W. A.,** Modification of parathion toxicity to wax moth larvae by chlorcyclizine, aminopyrine or phenobarbital, *Toxicol. Appl. Pharmacol.,* 15, 433, 1969.

418. **Agosin, M., Aravena, L., and Neghme, A.,** Enhanced protein synthesis in *Triatoma infestans* treated with DDT, *Exp. Parasitol.,* 16, 318, 1965.

419. **Plapp, F. W., Jr. and Casida, J. E.,** Induction by DDT and dieldrin of insecticide metabolism by house fly enzymes, *J. Econ. Entomol.,* 63, 1091, 1970.

420. **Walker, C. R. and Terriere, L. C.,** Induction of microsomal oxidases by dieldrin in *Musca domestica, Entomol. Exp. Appl.,* 13, 260, 1970.

421. **Yu, S. J. and Terriere, L. C.,** Induction of microsomal oxidases in the housefly and the action of inhibitors and stress factors, *Pestic. Biochem. Physiol.,* 1, 173, 1971.

422. **Melander, A. L.,** Strain of *Aspidiotus* resistant to lime-sulfur, *J. Econ. Entomol.,* 7, 167, 1914.

423. **Brown, A. W. A.,** Insecticide resistance comes of age, *Bull. Entomol. Soc. Am.,* 14, 3, 1968.

424. **Keiding, J.,** Persistence of resistant populations after the relaxation of the selection pressure, *World Rev. Pest Control,* 6, 115, 1967.

425. **Busvine, J. R.,** Mechanism of resistance to insecticides in house-flies, *Nature (Lond.),* 168, 193, 1951.

426. **Missiroli, A.,** Resistenza alli insecticidi di alcuni razze di *Musca domestica, Riv. Parassitol.,* 12, 5, 1951.

427. **Gilbert, I. H., Couch, M. D., and McDuffie, W. C.,** Development of resistance to insecticides in natural populations of house-flies, *J. Econ. Entomol.,* 46, 48, 1953.

428. **March, R. B. and Metcalf, R. L.,** Insecticide-resistant flies, *Soap Sanit. Chem.,* 26, 121, 1950.

429. **March, R. B.,** Summary of Research on Insects Resistant to Insecticides, in National Research Council, Division of Medical Sciences, *Conference on Resistance and Insect Physiology,* Washington, D.C., Dec. 8–9, 1951, 45 (National Research Council Publication 219).

430. **Metcalf, R. L.,** Physiological basis for insect resistance to insecticides, *Physiol. Rev.,* 35, 197, 1955.

431. **Georgopoulos, G. D.,** Extension to chlordane of the resistance to DDT observed in *Anopheles sacharovi, Bull. W.H.O.,* 11, 855, 1954.

432. **Davidson, G.,** Insecticide resistance in *Anopheles gambiae* Giles, *Nature (Lond.),* 178, 705, 1956.

433. **Wright, J. W.,** Present status of mosquito control, *Pestic. Sci.,* 3, 471, 1972.

434. **Busvine, J. R.,** Insecticide-resistance in mosquitoes, *Pestic. Sci.,* 3, 483, 1972.

435. **Hawkins, W. B.,** Tests on the Resistance of Anopheles Larvae in the Region of the Tennessee Valley Authority, in Seminar on the Susceptibility of Insects to Insecticides, Panama, 23–28 June 1958, Pan American Sanitary Bureau, W.H.O., Washington D.C., 1958.

436. **Cova-Garcia, P.,** Manifestations of behaviouristic resistance in Venezuela, in Seminar on the Susceptibility of Insects to Insecticides, Panama, 23, June 1958, Pan American Sanitary Bureau, W.H.O., Washington D.C., 1958.

437. **Gjullin, C. M. and Peters, R. F.,** Recent studies of mosquito resistance to insecticides in California, *Mosquito News,* 12, 1, 1952.

438. **Rachou, R. G.,** Some manifestations of behaviouristic resistance in Brazil, in Seminar on the Susceptibility of Insects to Insecticides, Panama, 23, 1958, Pan American Sanitary Bureau, W.H.O., Washington D.C., 1958.

439. **Wharton, R. H.,** The behaviour and mortality of *Anopheles maculatus* and *Culex fatigans* in experimental huts treated with DDT and BHC, *Bull. Entomol. Res.,* 42, 1, 1950.

440. **Mosna, E.,** *Culex pipiens* autogenicus DDT-resistenti e loro controllo con Octaklor e esachlorocicloesano, *Riv. Parassitol.,* 9, 19, 1948.

441. **Kerr, J. A., de Camargo, S., and Abedi, Z. H.,** cited in Busvine, J. R. and Pal, R., The impact of insecticide-resistance on control of vectors and vector-borne diseases, *Bull. W.H.O.,* 40, 731, 1969.

442. **Fay, R. W.,** Insecticide resistance in *Aedes aegypti, Am. J. Trop. Med. Hyg.,* 5, 378, 1956.

443. **Wright, J. W. and Brown, A. W. A.,** Survey of possible insecticide resistance in body lice, *Bull. W.H.O.,* 16, 9, 1957.

444. **Wright, J. W. and Pal, R.,** Second survey of insecticide resistance in body-lice, 1958–63, *Bull. W.H.O.,* 33, 485, 1965.

445. **Grayson, J. M.,** Effects on the German cockroach of twelve generations of selection for survival to treatments with DDT and benzene hexachloride, *J. Econ. Entomol.,* 46, 124, 1953.

446. **Grayson, J. M.,** Differences between a resistant and a non-resistant strain of the German cockroach, *J. Econ. Entomol.,* 47, 253, 1954.

447. **Whitnall, A. B. M. and Bradford, B.,** An arsenic resistant tick and its control with gammexane dips, *Bull. Entomol. Res.,* 40, 207, 1949.

448. **Whitnall, A. B. M., Thorburn, J. A., Mellardy, W. M., Whitehead, G. B., and Meerholz, F.,** A BHC-resistant tick, *Bull. Entomol. Res.,* 43, 51, 1952.

449. **Stone, B. F. and Meyers, R. A. J.,** Dieldrin-resistant cattleticks, *Boophilus microplus* (Canestrini) in Queenland, *Aust. J. Agric. Res.,* 8, 312, 1957.

450. **Wharton, R. H. and Roulston, W. J.,** Resistance of ticks to chemicals, *Ann. Rev. Entomol.,* 15, 381, 1970.

451. **Adkisson, P. L.,** Development of resistance by the tobacco budworm to endrin and carbaryl, *J. Econ. Entomol.,* 61, 37, 1968.

452. **Plapp, F. W., Jr.,** Insecticide resistance in *Heliothis*: tolerance in larvae of *H. viriscens* as compared with *H. zea* to organophosphate insecticides, *J. Econ. Entomol.,* 64, 999, 1971.

453. **Ishikura, H.,** High yield rice cultivation and the use of pesticides, *Japan Pesticide Information,* No. 5, Japan Plant Protection Association, October 1970, 5.

454. **Fukaya, F.,** Insecticide resistance, detection methods and counter-measures, *Japan Pesticide Information,* No. 6, Japan Plant Protection Association, January 1971, 25.

455. **Vernon, A. J.,** Control of cocoa capsids in West Africa, *Chem. Ind. (Lond.),* 1219, 1961.

456. **Rushton, M. P.,** Cocoa capsid control in Sierra Leone, *Span,* 6, 23, 1963.

457. **Dunn, J. A.,** Insecticide resistance in the cocoa capsid, *Distantiella theobroma, Nature (Lond.),* 199, 1207, 1963.

458. **Harris, C. R., Mazurek, J. H., and Svec, H. J.,** Cross-resistance shown by aldrin-resistant seed maggot flies, *Hylemya* spp., to other cyclodiene insecticides and related materials, *J. Econ. Entomol.,* 57, 702, 1964.

459. **McClanahan, R. J., Harris, C. R., and Miller, L. A.,** Resistance to aldrin, dieldrin, and heptachlor in the onion maggot, *Hylemya antiqua* (Meig.) in Ontario, *Rep. Entomol. Soc. Ont.,* 89, 55, 1958.

460. **Harris, C. R., Manson, G. F., and Mazurek, J. H.,** Development of insecticidal resistance by soil insects in Canada, *J. Econ. Entomol.,* 55, 777, 1962.

461. **Coaker, T. H., Mowat, D. J., and Wheatley, G. A.,** Insecticide resistance in the cabbage root fly in Britain, *Nature (Lond.),* 200, 664, 1963.

462. **Gostick, K. G. and Coaker, T. H.,** Monitoring for Insecticide Resistance in Root Flies, *Proc. 5th Br. Insectic. Fungic. Conf.,* 1969, 89.

463. **Kilpatrick, J. W. and Schoof, H. F.,** Interrelationship of water and *Hermetia illucens* breeding to *Musca domestica* production in human excrement, *Am. J. Trop. Med. Hyg.,* 8, 103, 1959.

464. **Knutson, H.,** Changes in reproductive potential in houseflies in response to dieldrin, *Misc. Publ. Entomol. Soc. Am.,* 1, 27, 1959.

465. **Parkin, E. A.,** The onset of insecticide resistance among field populations of stored-product insects, *J. Stored Prod. Res.,* 1, 3, 1965.

466. **Lindgren, D. L. and Vincent, L. E.,** The susceptibility of laboratory reared and field-collected cultures of *Tribolium confusum* and *T. castaneum,* to ethylene dibromide, hydrocyanic acid and methyl bromide, *J. Econ. Entomol.,* 58, 551, 1965.

467. **Dyte, C. E. and Blackman, D. G.,** The spread of insecticide resistance in *Tribolium castaneum* (Herbst), *J. Stored Prod. Res.,* 6, 255, 1970.

468. **Lloyd, C. J.,** Studies on the cross-tolerance to DDT-related compounds of a pyrethrin-resistant strain of *Sitophilus granarius* (L.), *J. Stored Prod. Res.,* 5, 337, 1969.

469. **Abedi, Z. H. and Brown, A. W. A.,** Development and reversion of DDT-resistance in *Aedes aegypti, Can. J. Genet. Cytol.,* 2, 252, 1960.

470. **Georghiou, G. P., March, R. B., and Printy, G. E.,** Induced regression of dieldrin-resistance in the housefly (*Musca domestica* L.), *Bull. W.H.O.,* 29, 167, 1963.

471. **Georghiou, G. P. and Metcalf, R. L.,** Dieldrin susceptibility: partial restoration in anopheles selected with a carbamate, *Science,* 140, 301, 1963.

472. **Georghiou, G. P. and Bowen, W. R.,** An analysis of housefly resistance to insecticides in California, *J. Econ. Entomol.,* 59, 204, 1966.

473. **Stone, B. F.,** The inheritance of DDT-resistance in the cattle tick, Boophilus microplus, *Aust. J. Agric. Res.,* 13, 984, 1962.

474. **Schnitzerling, H. J., Roulston, W. J., and Schuntner, C. A.,** The absorption and metabolism of [^{14}C] DDT in DDT-resistant and susceptible strains of the cattle tick *Boophilus microplus, Aust. J. Biol. Sci.,* 23, 219, 1970.

475. **Winteringham, F. P. W.,** Mechanisms of selective insecticidal action, *Ann. Rev. Entomol.,* 14, 409, 1969.

476. **Gerolt, P.,** Mode of entry of contact insecticides, *J. Insect Physiol.,* 15, 563, 1969.

477. **Moriarty, F. and French, M. C.,** The uptake of dieldrin from the cuticular surface of *Periplaneta Americana, Pestic. Biochem. Physiol.,* 1, 286, 1971.

478. **Narahashi, T.,** Effects of insecticides on excitable tissues, in *Advances in Insect Physiology,* Vol. 8, Beament, J. W. L., Treherne, J. E., and Wigglesworth, V. B., Eds., Academic Press, London and New York, 1971.

479. **Winteringham, F. P. W. and Lewis, S. E.,** On the mode of action of insecticides, *Ann. Rev. Entomol.,* 4, 303, 1959.

480. **Soto, A. R. and Deichmann, W. B.,** Major metabolism and acute toxicity of aldrin, dieldrin and endrin, *Environ. Res.,* 1, 307, 1967.

481. **Robinson, J.,** Persistent pesticides, *Ann. Rev. Pharmacol.,* 10, 353, 1970.

482. **Brooks, G. T.,** The metabolism of diene-organochlorine (cyclodiene) insecticides, *Residue Rev.,* 27, 81, 1969.

483. **Brooks, G. T.,** The fate of chlorinated hydrocarbons in living organisms, in *Pesticide Terminal Residues,* Invited papers from the IUPAC International Symposium on Terminal Residues, Tel Aviv, 1971, Tahori, A. S., Ed., Butterworths, London, 1971, 111.

484. **Brooks, G. T.,** Pathways of enzymatic degradation of pesticides, in *Environmental Quality and Safety,* Vol. I, Coulston, F. and Korte, F., Eds., Georg Thieme, Stuttgart, Academic Press, New York, 1972, 106.

485. **Matsumura, F.,** Metabolism of Pesticides in Higher Plants, in *Environmental Quality and Safety,* Vol. I, Coulston, F. and Korte, F., Eds., Georg Thieme, Stuttgart, Academic Press, New York, 1972, 106.

486. **Klein, W.,** Metabolism of Insecticides in Microorganisms and Insects, in *Environmental Quality and Safety,* Vol. I, Coulston, F., Korte, F., Eds., George Thieme, Stuttgart, Academic Press, 1972, 1964.

487. **Schaefer, C. H. and Sun, Y. P.,** A study of dieldrin in the housefly central nervous system in relation to dieldrin resistance, *J. Econ. Entomol.,* 60, 1580, 1967.

488. **Sellers, L. G. and Guthrie, F. E.,** Localisation of dieldrin in the housefly thoracic ganglion by electron microscopic autoradiography, *J. Econ. Entomol.,* 64, 352, 1971.

489. **Telford, J. N. and Matsumura, F.,** Dieldrin binding in subcellular nerve components of cockroaches. An electron microscopic and autoradiographic study, *J. Econ. Entomol.,* 63, 795, 1970.

490. **Kurihara, N., Nakajima, E., and Shindo, H.,** Whole body autoradiographic studies on the distribution of BHC and nicotine in the American cockraoch, in *Biochemical Toxicology of Insecticides,* O'Brien, R. D. and Yamamoto, I., Eds., Academic Press, New York, 1970, 41.

491. **Sun, Y. P.,** Dynamics of insect toxicology – a mathematical and graphical evaluation of the relationship between insect toxicity and rates of penetration and detoxication of insecticides, *J. Econ. Entomol.,* 61, 949, 1968.

492. **Glasstone, S.,** *A Textbook of Physical Chemistry,* 2nd ed., MacMillan, London, 1948, 1075.

493. **Hewlett, P. S.,** Interpretation of dosage-mortality data for DDT-resistant houseflies, *Ann. Appl. Biol.,* 46, 37, 1958.

494. **Menzie, C. M.,** Fate of pesticides in the environment, *Ann. Rev. Entomol.,* 17, 199, 1972.

495. **Hamaker, J. W.,** Mathematical prediction of cumulative levels of pesticides in soil, in Organic Pesticides in the Environment, *Am. Chem. Soc., Adv. Chem. Ser.,* 60, 122, 1966.

496. **Decker, G. C., Bruce, N. W., and Bigger, J. H.,** The accumulation and dissipation of residues resulting from the use of aldrin in soils, *J. Econ. Entomol.,* 58, 266, 1965.

497. **Lichtenstein, E. P.,** Persistence and degradation of pesticides in the environment, in *Scientific Aspects of Pest Control,* Publication 1402, National Academy of Sciences, National Research Council, Washington, D.C., 1966.

498. **Harris, C. R.,** Factors influencing the effectiveness of soil insecticides, *Ann. Rev. Entomol.,* 17, 177, 1972.

499. **Harris, C. R. and Sans, W. W.,** Behaviour of dieldrin in soil: microplot field studies on the influence of soil type on biological activity and absorption by carrots, *J. Econ. Entomol.,* 65, 333, 1972.

500. **Harris, C. R.,** Factors influencing the biological activity of technical chlordane and some related compounds in soil, *J. Econ. Entomol.,* 65, 341, 1972.

501. **Strickland, A. H.,** Some estimates of insecticide and fungicide usage in agriculture and horticulture in England and Wales 1960–64. Pesticides in the environment and their effects on wildlife, *J. Appl. Ecol.,* 3[suppl.], 3, 1966.

502. *Third Report of the Research Committee on Toxic Chemicals,* Agricultural Research Council, HMSO London, 1970, 69.

503. **Plapp, F. W., Jr.,** On the molecular biology of insecticide resistance, in *Biochemical Toxicology of Insecticides,* O'Brien, R. D. and Yamamoto, I., Eds., Academic Press, New York, 1970, 179.

504. **Lichtenstein, E. P. and Corbett, J. R.,** Enzymatic conversion of aldrin to dieldrin with subcellular components of pea plants, *J. Agric. Food Chem.,* 17, 589, 1969.

505. **Yu, S. J., Kiigemagi, U., and Terriere, L. C.,** Oxidative metabolism of aldrin and isodrin by bean root fractions, *J. Agric. Food Chem.,* 19, 5, 1971.

506. **Lyr, H. and Ritter, G.,** Zum wirkungsmechanismus von Hexachlorocyclohexan – Isomeren in Hefezellen, *Z. Allge. Mikrobiol.,* 9, 545, 1969.

507. **Tu, C. M., Miles, J. R. W., and Harris, C. R.,** Soil microbial degradation of aldrin, *Life Sci.,* [Oxford] 7, 311, 1968.

508. **Korte, F.,** Metabolism of ^{14}C-labelled insecticides in microorganisms, insects and mammals, *Botyu-kagaku,* 32, 46, 1967.

508a. **Rice, C. P. and Sikka, H. C.,** Uptake and metabolism of DDT by six species of marine algae, *J. Agric. Food Chem.,* 21, 148, 1973.

509. **Robinson, J.,** The burden of chlorinated hydrocarbon pesticides in man, *Can. Med. Assoc. J.,* 100, 180, 1969.

510. **Quaife, M. L., Winbush, J. S., and Fitzhugh, O. G.** Survey of quantitative relationships between ingestion and storage of aldrin and dieldrin in animals and man, *Food Cosmet. Toxicol.,* 5, 39, 1967.

511. Department of Health, Education, and Welfare, Report of the Secretary's Commission on Pesticides and their Relationship to Environmental Health, Parts I and II, U.S. Govt. Print. Off., Washington, D.C., 1969, 263.

512. **Bitman, J., Cecil, H. C., Harris, S. J., and Fries, G. F.,** Comparison of DDT effect on pentobarbital metabolism in rats and quail, *J. Agric. Food Chem.,* 19, 333, 1971.

513. **Robinson, J., Roberts, M., Baldwin, M., and Walker, A. I. T.,** The pharamacokinetics of HEOD [dieldrin] in the rat, *Food Cosmet. Toxicol.,* 7, 317, 1969.

514. **Brown, V. K. H., Robinson, J., and Richardson, A.,** Preliminary studies on the acute and subacute toxicities of a photoisomerisation product of HEOD, *Food Cosmet. Toxicol.,* 5, 771, 1967.

515. **Robinson, J. and Richardson, A. R.,** personal communication.

516. **Hunter, C. G. and Robinson, J.,** Pharmacodynamics of dieldrin [HEOD]. I. Ingestion by human subjects for 18 months, *Arch. Environ. Health,* 15, 614, 1967.

517. **Klein, W., Muller, W., and Korte, F.,** Ausscheidung, verteilung und stoffwechsel von endrin-^{14}C in ratten, *Justus Liebigs Ann. Chem.,* 713, 180, 1968.

518. **Kaul, R., Klein, W., and Korte, F.,** Metabolismus und kinetic der verteilung von β-dihydroheptachlor-^{14}C in mannlichen ratten, *Tetrahedron,* 26, 99, 1970.

519. **Robinson, J. and Roberts, M.,** Accumulation, distribution and elimination of organochlorine insecticides by vertebrates, *Soc. Chem. Ind. (Lond.) Monogr.,* No. 29, 106, 1968.

520. Abbott, D. C., Collins, G. B., and Goulding, R., Organochlorine pesticide residues in human fat in the United Kingdom 1969–71, *Br. Med. J.,* 553, 1972.

521. Coulson, J. C., Deans, I. R., Potts, G. R., Robinson, J., and Crabtree, A. N., Changes in organochlorine contamination of the marine environment of Eastern Britain monitored by Shag eggs, *Nature (Lond.),* 236, 454, 1972.

522. Harrison, H. L., Loucks, O. L., Mitchell, J. W., Parkhurst, D. F., Tracy, C. R., Watts, D. G., and Yannacone, V. J., Jr., Systems studies of DDT transport, *Science,* 170, 503, 1970.

523. Woodwell, M., Craig, P. P., and Johnson, H. A., DDT in the biosphere: where does it go? *Science,* 174, 1101, 1971.

524. Kapoor, I. P., Metcalf, R. L., Hirwe, A. S., Po-Yung Lu, Coats, J. R., and Nystrom, R. F., Comparative metabolism of DDT, methylchlor, and ethoxychlor in mouse, insects, and in a model ecosystem, *J. Agric. Food Chem.,* 20, 1, 1972.

525. Ivie, G. W. and Casida, J. E., Enhancement of photoalteration of cyclodiene insecticide chemical residues by rotenone, *Science,* 167, 1620, 1970.

526. Bailey, S., Bunyan, P. J., Rennison, B. D., and Taylor, A., The metabolism of 1,1-di[p-chlorophenyl]-2,2-dichloro-ethylene and 1,1-di[p-chlorophenyl]-2-chloroethylene in the pigeon, *Toxicol. Appl. Pharmacol.,* 14, 23, 1969.

527. Bailey, S., Bunyan, P. J., and Taylor, A., The metabolism of p,p'-DDT in some avian species, in *Environmental Quality and Safety,* Vol. I, Coulston, F. and Korte, F., Eds., Georg Thieme, Stuttgart, Academic Press, New York, 1972, 244.

528. Datta, P. R., In vivo detoxication of p,p'-DDT via p,p'-DDE to p,p'-DDA in rats, *Ind. Med.,* 39, 49, 1970.

529. Peterson, J. E. and Robinson, W. H., Metabolic products of p,p'-DDT in the rat, *Toxicol. Appl. Pharmacol.,* 6, 321, 1964.

530. Morgan, D. P. and Roan, C. C., Absorption, storage and metabolic conversion of ingested DDT and DDT metabolites in man, *Arch. Environ. Health,* 22, 301, 1971.

531. Agosin, M., Michaeli, D., Miskus, R., Nagasawa, S., and Hoskins, W. M., A new DDT metabolising enzyme in the German cockroach, *J. Econ. Entomol.,* 54, 340, 1961.

532. Rowlands, D. G. and Lloyd, C. J., DDT metabolism in susceptible and pyrethrin resistant *Sitophilus granarius* [L.], *J. Stored Prod. Res.,* 5, 413, 1969.

533. Leeling, N. C., personal communication, 1972.

534. Oppenoorth, F. J. and Houx, N. W. H., DDT resistance in the housefly caused by microsomal degradation, *Entomol. Exp. Appl.,* 11, 81, 1968.

535. McKinney, J. D. and Fishbein, L., DDE formation; dehydrochlorination or dehypochlorination, *Chemosphere,* No. 2, 67, 1972.

536. Hassall, K. A., Reductive dechlorination of DDT; the effect of some physical and chemical agents on DDD production by pigeon liver preparations, *Pestic. Biochem. Physiol.,* 1, 259, 1971.

537. Hathway, D. E., Biotransformations, in *Foreign Compound Metabolism in Mammals,* Hathway, D. E., Brown, S. S., Chasseaud, L. F., and Hutson, D. H., Reporters, *Specialist Periodical Report of the Chemical Society,* Vol. 1, London, 1970, 295.

538. Wedemeyer, G., Dechlorination of 1,1,1-trichloro-2,2-bis[p-chlorophenyl] ethane by *Aerobacter aerogenes, Appl. Microbiol.,* 15, 569, 1967.

539. Alexander, M., Microbial degradation of pesticides, in *Environmental Toxicology of Pesticides,* Matsumura, F., Boush, G. M., and Misato, T., Eds., Academic Press, New York and London, 1972, 365.

539a. Pfaender, F. K. and Alexander, M., Extensive microbial degradation of DDT *in vitro* and DDT metabolism by natural communities, *J. Agric. Food Chem.,* 20, 842, 1972.

539b. Albone, E. S., Eglinton, G., Evans, N. C., and Rhead, M. M., Formation of bis(p-chlorophenyl) acetonitrile (p,p'-DDCN) from p,p'-DDT in anaerobic sewage sludge, *Nature (Lond.),* 240, 420, 1972.

539c. Jensen, S., Gothe, R., and Kindstedt, M. O., Bis-(p-chlorophenyl)-acetonitrile (DDCN), a new DDT derivative formed in anaerobic digested sewage sludge and lake sediment, *Nature (Lond.),* 240, 421, 1972.

540. Oppenoorth, F. J., Resistance to gamma-hexachlorocyclohexane in *Musca domestica* L., *Arch. Neerl. Zool.,* 12, 1, 1956.

541. Reed, W. T. and Forgash, A. J., Lindane: metabolism to a new isomer of pentachlorocyclohexene, *Science,* 160, 1232, 1968.

542. Reed, W. T. and Forgash, A. J., Metabolism of lindane to organic-soluble products by houseflies, *J. Agric. Food Chem.,* 18, 475, 1970.

543. Grover, P. L. and Sims, P., The metabolism of γ-2,3,4,5,6-pentachlorocyclohex-1-ene and γ-hexachlorocyclohexane in rats, *Biochem. J.,* 96, 521, 1965.

544. Saha, J., cited in Hurtig, H., Significance of conversion products and metabolites of pesticides in the environment, in *Environmental Quality and Safety,* Coulston, F. and Korte, F., Eds., Georg Thieme, Stuttgart and Academic Press, New York, 1972, 67.

545. Duxbury, J. M., Tiedje, J. M., Alexander, M., and Dawson, J. E., 2,4-D metabolism; enzymatic conversion of chloromaleylacetic acid to succinic acid, *J. Agric. Food Chem.,* 18, 199, 1970.

546. Bowman, M. C., Acree, F., Jr., Lofgren, C. S., and Beroza, M., Chlorinated insecticides; Fate in aqueous suspensions containing mosquito larvae, *Science,* 146, 1480, 1964.

547. **Brooks, G. T., Harrison, A., and Lewis, S. E.,** Cyclodiene epoxide ring hydration by microsomes from mammalian liver and houseflies, *Biochem. Pharmacol.,* 19, 255, 1970.

548. **Kaul, R., Klein, W., and Korte, F.,** Verteilung, ausscheidung und metabolismus von Telodrin und heptachlor in ratten und mannlichen kaninchen, endprodukt des warmblutermetabolismus von heptachlor, *Tetrahedron,* 26, 331, 1970.

549. **Matsumura, F. and Nelson, J. O.,** Identification of the major metabolic product of heptachlor epoxide in rat faeces, *Bull. Environ. Contam. Toxicol.,* 5, 489, 1971.

550. **Schwemmer, B., Cochrane, W. P., and Polen, P. B.,** Oxychlordane, animal metabolite of chlordane: isolation and synthesis, *Science,* 169, 1087, 1970.

551. **Lawrence, J. H., Barron, R. P., Chen, J. Y. T., Lombardo, P., and Benson, W. R.,** Note on identification of a chlordane metabolite found in milk and cheese, *J. Assoc. Offic. Agric. Chem.,* 53, 261, 1970.

552. **Street, J. C. and Blau, S. E.,** Oxychlordane: accumulation in rat adipose tissue on feeding chlordane isomers or technical chlordane, *J. Agric. Food Chem.,* 20, 395, 1972.

553. **Davidow, B., Hagan, E., and Radomski, J. L.,** A metabolite of chlordane in tissues of animals, *Fed. Proc. Abstr.,* 10, 291, 1951.

554. **Georgacakis, E. and Khan, M. A. Q.,** Toxicity of the photoisomers of cyclodiene insecticides to freshwater animals, *Nature (Lond.),* 233, 120, 1971.

555. **Gorbach, S., Haaring, R., Knauf, W., and Werner, H. J.,** Residue analysis and biotests in rice fields of East Java treated with Thiodan, *Bull. Environ. Contam. Toxicol.,* 6, 193, 1971.

556. **Brooks, G. T.,** Perspectives of cyclodiene metabolism, in Proceedings of the Symposium on the Science and Technology of Residual Insecticides in Food Production with Special Reference to Aldrin and Dieldrin, Shell Chemical Co. New York, Library of Congress Cat. No. 68-27527, 1968, 89.

557. **Wong, D. T. and Terriere, L. C.,** Epoxidation of aldrin, isodrin and heptachlor by rat liver microsomes, *Biochem. Pharmacol.,* 14, 375, 1965.

558. **Nakatsugawa, T., Ishida, M., and Dahm, P. A.,** Microsomal oxidation of cyclodiene insecticides, *Biochem. Pharmacol.,* 14, 1853, 1965.

559. **Heath, D. F.,** [36] Cl-dieldrin in mice, in *Radioisotopes and Radiation in Entomology,* Proc. Bombay Symposium 1960, International Atomic Energy Agency, Vienna, 1962, 83.

560. **Heath, D. F. and Vandekar, M.,** Toxicity and metabolism of dieldrin in rats, *Br. J. Ind. Med.,* 21, 269, 1964.

561. **Korte, F. and Arent, H.,** Isolation and identification of dieldrin metabolites from urine of rabbits after oral administration of dieldrin-^{14}C, *Life Sci.,* [Oxford], 4, 2017, 1965.

562. **Richardson, A., Baldwin, M. K., and Robinson, J.,** Metabolites of dieldrin [HEOD] in the urine and faeces of rats, *Chem. Ind. (Lond.),* 588, 1968.

563. **Klein, A. K., Link, J. D., and Ives, N. F.,** Isolation and purification of metabolites found in the urine of male rats fed aldrin and dieldrin, *J. Assoc. Offic. Agric. Chem.,* 51, 895, 1968.

564. **Feil, V. J., Hedde, R. D., Zaylskie, R. G., and Zachrison, C. H.,** Identification of *trans*-6,7-dihydroxydihydroaldrin and 9-[*syn*-epoxy]hydroxy-1,2,3,4,10,10-hexachloro-6,7-epoxy-1,4,4a, 5,6,7,8,8a-octahydro-1,4-endo-5,8-exo-dimethanonaphthalene, *J. Agric. Food Chem.,* 18, 120, 1970.

564a. **Bedford, C. T. and Harrod, R. K.,** Synthesis of 9-hydroxy-HEOD, a major mammalian metabolite of HEOD (dieldrin), *Chem. Commun.,* 735, 1972.

565. **Baldwin, M. K.,** The metabolism of the chlorinated insecticides aldrin, dieldrin, endrin and isodrin, Ph.D. Thesis, University of Surrey, 1971.

565a. **Richardson, A. and Robinson, J.,** Identification of a major metabolite of HEOD (dieldrin) in human faeces, *Xenobiotica,* 1, 213, 1971.

566. **Oda, J. and Muller, W.,** Identification of a mammalian breakdown product of dieldrin, in *Environmental Quality and Safety,* Vol. 1, Coulston, F. and Korte, F., Eds., George Thieme, Stuttgart, Academic Press, New York, 1972, 248.

567. **McKinney, J. D.,** personal communication, June, 1972.

568. **McKinney, J. D., Matthews, H. B., and Fishbein, L.,** Major faecal metabolite of dieldrin in rat, structure and chemistry, *J. Agric. Food Chem.,* 20, 597, 1972.

569. **Rosen, J. D.,** Conversion of pesticides under environmental conditions, in *Environmental Quality and Safety,* Vol. I, Coulston, F. and Korte, F., Eds., George Thieme, Stuttgart, Academic Press, New York, 1972, 85.

570. **Matsumura, F., Patil, K. C., and Boush, G. M.,** Formation of photodieldrin by microorganisms, *Science,* 170, 1206, 1970.

571. **Baldwin, M. K. and Robinson, J.,** Metabolism in the rat of the photoisomerisation product of dieldrin, *Nature (Lond.),* 224, 283, 1969.

572. **Khan, M. A. Q., Sutherland, D. J., Rosen, J. D., and Carey, W. F.,** Effect of sesamex on the toxicity and metabolism of cyclodienes and their photoisomers in the housefly, *J. Econ. Entomol.,* 63, 470, 1970.

573. **Matthews, H. B. and Matsumura, F.,** Metabolic fate of dieldrin in the rat, *J. Agric. Food Chem.,* 17, 845, 1969.

574. **Klein, A. K., Dailey, R. E., Walton, M. S., Beck, V., and Link, J. D.,** Metabolites isolated from urine of rats fed ^{14}C-photodieldrin, *J. Agric. Food Chem.,* 18, 705, 1970.

575. **Dailey, R. E., Klein, A. K., Brouwer, E., Link, J. D., and Braunberg, R. C.,** Effect of testosterone on metabolism of ^{14}C-photodieldrin in normal, castrated and oophorectomised rats, *J. Agric. Food Chem.,* 20, 371, 1972.

576. **El Zorgani, G. A., Walker, C. H., and Hassall, K. A.,** Species differences in the *in vitro* metabolism of HEOM, a chlorinated cyclodiene epoxide, *Life Sci.,* [Oxford], 9, (Part II), 415, 1970.

577. **Korte, F., Ludwig, G., and Vogel, J.,** Umwandlung von Aldrin-^{14}C and Dieldrin-^{14}C durch mikroorganismen, leberhomogenate und moskito-larven, *Justus Liebigs Ann. Chem.,* 656, 135, 1962.

578. **Matsumura, F., Boush, G. M., and Tai, A.,** Breakdown of dieldrin by a soil microorganism, *Nature (Lond.),* 219, 965, 1968.

579. **Mehendale, H. M., Skrentny, R. F., and Dorough, H. W.,** Oxidative metabolism of aldrin by subcellular root fractions of several plant species, *J. Agric. Food Chem.,* 20, 398, 1972.

580. **Giannotti, O.,** Estudos sobre o mecanismo de acao do aldrin, dieldrin e endrin em *Periplaneta americana* [L.]. *Arq. Inst. Biol. Sao Paulo,* 25, 253, 1958.

581. **Brooks, G. T.,** Mechanisms of resistance of the adult housefly [*Musca domestica*] to 'cyclodiene' insecticides, *Nature (Lond.),* 186, 96, 1960.

582. **Richardson, A., Robinson, J., and Baldwin, M. K.,** Metabolism of endrin in the rat, *Chem. Ind. (Lond.),* 502, 1970.

583. **Pimentel, D.,** *Ecological Effects of Pesticides on Non-target Species,* Office of Science and Technology, U.S. Govt. Print. Off., Washington, D.C., June, 1972.

584. **Brooks, G. T. and Harrison, A.,** Relations between structure, metabolism and toxicity of the cyclodiene insecticides, *Nature (Lond.),* 198, 1169, 1963.

585. **Schupfan, I., Sajko, B., and Ballschmiter, K.,** The chemical and photochemical degradation of the cyclodiene insecticides aldrin, dieldrin, endosulfan and other hexachloronorbornene derivatives, *Z. Naturforsch. Teil B,* 27b, 147, 1972.

586. **Schupfan, I. and Ballschmiter, K.,** Metabolism of polychlorinated norbornenes by *Clostridium butyricum, Nature (Lond.),* 237, 100, 1972.

587. **Parke, D. V.,** *The Biochemistry of Foreign Compounds,* Pergamon Press, Oxford, 1968.

588. **Estabrook, R. W., Baron, J., and Hildebrandt, A.,** A new spectral species associated with cytochrome P_{450} in liver microsomes, *Chem.-Biol. Interactions,* 3, 260, 1971.

589. **Tsukamoto, M.,** Metabolic fate of DDT in *Drosophila melanogaster;* identification of a non-DDE metabolite, *Botyu-kagaku,* 24, 141, 1959.

590. **Sun, Y. P. and Johnson, E. R.,** Synergistic and antagonistic actions of insecticide-synergist conbinations and their mode of action, *J. Agric. Food Chem.,* 8, 261, 1960.

591. **Ray, J. W.,** Insect microsomal cytochromes, in *Pest Infestation Research 1965,* Agricultural Research Council, H.M.S.O., London, 1966, 59.

592. **Lewis, S. E.,** Effect of carbon monoxide on metabolism of insecticides *in vivo, Nature (Lond.),* 215, 1408, 1967.

593. **Hodgson, E. and Plapp, F. W., Jr.,** Biochemical characteristics of insect microsomes, *J. Agric. Food Chem.,* 18, 1048, 1970.

594. **Casida, J. E.,** Mixed-function oxidase involvement in the biochemistry of insecticide synergists, *J. Agric. Food Chem.,* 18, 753, 1970.

595. **Krieger, R. I. and Wilkinson, C. F.,** Localisation and properties of an enzyme system effecting aldrin epoxidation in larvae of the southern armyworm [*Prodenia eridania*], *Biochem. Pharmacol.,* 18, 1403, 1969.

596. **Krieger, R. I., Feeny, P. P., and Wilkinson, C. F.,** Detoxication enzymes in the guts of caterpillars: an evolutionary answer to plant defenses? *Science,* 172, 579, 1971.

597. **Wilkinson, C. F. and Brattsten, L. B.,** Microsomal drug metabolizing enzymes in insects, *Drug Metab. Rev.,* 1, 153, 172. 1972.

598. **Evans, W. C., Smith, B. S. W., Moss, P., and Fernley, H. N.,** Bacterial metabolism of 4-chlorophenoxyacetate, *Biochem. J.,* 122, 509, 1971, et. seq.

599. **Rowlands, D. G.,** The metabolism of contact insecticides in stored grain, *Residue Rev.,* 17, 105, 1967.

600. **Glass, B. L.,** Relation between the degradation of DDT and the iron redox system in soils, *J. Agric. Food Chem.,* 20, 324, 1972.

601. **Brooks, G. T.,** Progress in metabolic studies of the cyclodiene insecticides and its relevance to structure-activity correlations, *World Rev. Pest Control,* 5, 62, 1966.

602. **Oesch, F., Jerina, D. M., and Daly, J. W.,** Substrate specificity of hepatic epoxide hydrase in microsomes and in a purified preparation: evidence for homologous enzymes, *Arch. Biochem. Biophys.,* 144, 253, 1971.

603. **Boyland, E.,** in *Special Publication No. 5,* Williams, R. T., Ed., The Biochemical Society, London, 1950, 40.

604. **Oesch, F., Kaubisch, N., Jerina, D. M., and Daly, J. W.,** Hepatic epoxide hydrase. Structure-activity relationships for substrates and inhibitors, *Biochemistry,* 10, 4858, 1971.

605. **Boyland, E. and Chasseaud, L. F.,** Glutathione S-aralkyltransferase, *Biochem. J.,* 115, 985, 1969.

606. **Schonbrod, R. D., Khan, M. A. Q., Terriere, L. C., and Plapp, F. W., Jr.,** Microsomal oxidases in the housefly: a survey of fourteen strains, *Life Sci.* (Oxford), 7 (Part 1), 681, 1968.

607. **Khan, M. A. Q.,** Some biochemical characteristics of the microsomal cyclodiene epoxidase system and its inheritance in the housefly, *J. Econ. Entomol.,* 62, 388, 1969.

608. **Matthews, H. B. and Casida, J. E.,** Properties of housefly microsomal cytochromes in relation to sex, strain, substrate specificity, and apparent inhibition and induction by synergist and insecticide chemicals, *Life Sci.* (Oxford), 9 (Part 1), 989, 1970.

609. **Ishida, M.,** Comparative studies on BHC metabolising enzymes, DDT dehydrochlorinase and glutathione S-transferases, *Agric. Biol. Chem. (Tokyo),* 32, 947, 1968.

610. **Goodchild, B. and Smith, J. N.,** The separation of multiple forms of housefly 1,1,1-trichloro-2,2-bis-(p-chloro-phenyl)ethane (DDT) dehydrochlorinase from glutathione S-aryltransferase by electrofocusing and electrophoresis, *Biochem. J.,* 117, 1005, 1970.

611. **Dinamarca, M. L., Levenbook, L., and Valdes, E.,** DDT-dehydrochlorinase II. Subunits, sulfhydryl groups, and chemical composition, *Arch. Biochem. Biophys.,* 147, 374, 1971.

612. **Sternburg, J., Kearns, C. W., and Moorefield, H.,** DDT-dehydrochlorinase, an enzyme found in DDT-resistant flies, *J. Agric. Food Chem.,* 2, 1125, 1954.

613. **Lipke, H. and Kearns, C. W.,** DDT-dehydrochlorinase, in *Advances in Pest Control Research,* Vol. 3, Metcalf, R. L., Ed., Interscience, New York, 1960, 253.

614. **Mullins, L. J.,** The structure of nerve cell membranes, in *Molecular Structure and Functional Activity of Nerve Cells,* Grenell, R. G. and Mullins, L. J., Eds., Publication No. 1, American Institute of Biological Sciences, Washington, D.C., 1956.

615. **Hennessy, D. J., Fratantoni, J., Hartigan, J., Moorefield, H. H., and Weiden, M. H. J.,** Toxicity of 2-(2-halogen-4-chlorophenyl)-2-(4-chlorophenyl)-1,1,1-trichloroethanes to normal and to DDT-resistant houseflies, *Nature (Lond.),* 190, 341, 1961.

616. **Kimura, T., Duffy, J. R., and Brown, A. W. A.,** Dehydrochlorination and DDT-resistance in Culex mosquitos, *Bull. W.H.O.,* 32, 557, 1965.

617. **Pillai, M. K. K., Hennessy, D. J., and Brown, A. W. A.,** Deuterated analogues as remedial insecticides against DDT-resistant *Aedes aegypti, Mosquito News,* 23, 118, 1963.

618. **Pillai, M. K. K. and Brown, A. W. A.,** Physiological and genetical studies on resistance to DDT substitutes in *Aedes aegypti, J. Econ. Entomol.,* 58, 255, 1965.

619. **Busvine, J. R. and Townsend, M. G.,** The significance of BHC degradation in resistant houseflies, *Bull. Entomol. Res.,* 53, 763, 1963.

620. **Decker, G. C. and Bruce, W. N.,** Housefly resistance to chemicals, *Am. J. Trop. Med. Hyg.,* 1, 395, 1952.

621. **Bridges, R. G. and Cox, J. T.,** Resistance of houseflies to γ-benzene hexachloride and dieldrin, *Nature (Lond.),* 184, 1740, 1959.

622. **Balabaskeran, S. and Smith, J. N.,** The inhibition of 1,1,1-trichloro-2,2-bis(p-chlorophenyl)ethane(DDT)dehydro-chlorinase and glutathione S-aryltransferase in grass-grub and housefly preparations, *Biochem. J.,* 117, 989, 1970.

623. **Brooks, G. T. and Harrison, A.,** The oxidative metabolism of aldrin and dihydroaldrin by houseflies, housefly microsomes and pig liver microsomes and the effect of inhibitors, *Biochem. Pharmacol.,* 18, 557, 1969.

624. **Gillette, J. R., Conney, A. H., Cosmides, G. J., Estabrook, R. W., Fouts, J., and Mannering, G. J., Eds.,** *Microsomes and Drug Oxidations,* Academic Press, New York, London, 1969.

625. **Gillett, J. W. and Chan, T. M.,** Cyclodiene insecticides as inducers, substrates and inhibitors of microsomal epoxidation, *J. Agric. Food Chem.,* 16, 590, 1968.

626. **Kinoshita, F. K., Frawley, J. P., and DuBois, K. P.,** Quantitative measurement of induction of hepatic microsomal enzymes by various dietary levels of DDT and toxaphene in rats, *Toxicol. Appl. Pharmacol.,* 9, 505, 1966.

627. **Poland, A., Smith, D., Kuntzman, R., Jacobson, M., and Conney, A. H.,** Effect of intensive occupational exposure to DDT on phenylbutazone and cortisol metabolism in human subjects, *Clin. Pharmacol. Ther.,* 11, 724, 1970.

628. **Cram, R. L. and Fouts, J. R.,** The influence of DDT and γ-chlordane on the metabolism of hexobarbital and zoxazolamine in two mouse strains, *Biochem. Pharmacol.,* 16, 1001, 1967.

629. **Stephen, B. J., Gerlich, J. D., and Guthrie, F. E.,** Effect of DDT on induction of microsomal enzymes and deposition of calcium in the domestic hen, s,Bull. Environ. Contam. Toxicol., *5, 569, 1971.*

630. deposition of calcium in the domestic hen, *Bull. Environ. Contam. Toxicol.,* 5, 569, 1971.

630. **Alary, J. G. and Brodeur, J.,** Studies on the mechanism of phenobarbital-induced protection against parathion in adult female rats, *J. Pharmacol. Exp. Ther.,* 169, 159, 1969.

631. **Triolo, A. J., Mata, E., and Coon, J. M.,** Effects of organochlorine insecticides on the toxicity and *in vitro* plasma detoxication of paraoxon, *Toxicol. Appl. Pharmacol.,* 17, 174, 1970.

632. **Chapman, S. K. and Leibman, K. C.,** The effect of chlordane, DDT and 3-methylcholanthrene upon the metabolism and toxicity of diethyl-4-nitrophenyl phosphorothionate [parathion], *Toxicol. Appl. Pharmacol.,* 18, 977, 1971.

633. **Wright, A. S., Potter, D., Wooder, M. F., Donninger, C., and Greenland, R. D.,** The effects of dieldrin on the subcellular structure and function of mammalian liver cells, *Food Cosmet. Toxicol.,* 10, 311, 1972.

634. **Gillett, J. W., Chan, T. M., and Terriere, L. C.,** Interactions between DDT analogues and microsomal epoxidase systems, *J. Agric. Food Chem.,* 14, 540, 1966.

635. **Street, J. C.,** Modification of animal responses to toxicants, in *Enzymatic Oxidations of Toxicants,* Hodgson, E., Ed., North Carolina State University, 1968, 197.

636. **Street, J. C., Urry, F. M., Wagstaff, D. J., and Blau, S. E.,** Induction by different inducers: structure-activity relationships among DDT analogues, in *Proc. 2nd Int. IUPAC Congress of Pesticide Chemistry,* Tahori, A. S., Ed., Gordon Breach, London, 1972.

637. **Abernathy, C. O., Hodgson, E., and Guthrie, F. E.,** Structure-activity relations on the induction of hepatic microsomal enzymes in the mouse by 1,1,1-trichloro-2,2-bis(p-chlorophenyl)ethane(DDT) analogues, *Biochem. Pharmacol.,* 20, 2385, 1971.

638. **Chadwick, R. W., Cranmer, M. F., and Peoples, A. J.,** Comparative stimulation of γ-HCH metabolism by pretreatment of rats with γ-HCH, DDT, and DDT plus γ-HCH, *Toxicol. Appl. Pharmacol.,* 18, 685, 1971.

639. **Mayer, F. L., Jr., Street, J. C., and Neuhold, J. M.,** Organochlorine insecticide interactions affecting residue storage in rainbow trout, *Bull. Environ. Contam. Toxicol.,* 5, 300, 1970.

640. **Kuntzman, R.,** Drugs and enzyme induction, *Ann. Rev. Pharmacol.,* 9, 21, 1969.

641. **Welch, R. M., Levin, W., and Conney, A. H.,** Insecticide inhibition and stimulation of steroid hydroxylases in rat liver, *J. Pharmacol. Exp. Ther.,* 155, 167, 1967.

642. **Bitman, J., Cecil, H. C., Harris, S. J., and Fries, G. F.,** Estrogenic activity of o,p'-DDT in the mammalian uterus and avian oviduct, *Science,* 162, 371, 1968.

643. **Bitman, J. and Cecil, H. C.,** Estrogenic activity of DDT analogs and polychlorinated biphenyls, *J. Agric. Food Chem.,* 18, 1108, 1970.

644. **Bitman, J., Cecil, H. C., Harris, S. J., and Fries, E. F.,** DDT induces a decrease in eggshell calcium, *Nature (Lond.),* 224, 44, 1969.

645. **Heath, R. G., Spann, J. W., and Kreitzer, J. F.,** Marked DDE impairment of Mallard reproduction in controlled studies, *Nature (Lond.),* 224, 47, 1969.

646. **Porter, R. D. and Wiemeyer, S. N.,** Dieldrin and DDT: effects on Sparrow Hawk eggshells and reproduction, *Science,* 165, 199, 1969.

647. **Robinson, J.,** Birds and pest control chemicals, *Bird Study,* 17, 195, 1970.

648. **Sun, Y. P.,** Correlation between laboratory and field data on testing insecticides, *J. Econ. Entomol.,* 59, 1131, 1966.

649. **Sternburg, J. and Kearns, C. W.,** Metabolic fate of DDT when applied to certain naturally tolerant insects, *J. Econ. Entomol.,* 45, 497, 1952.

650. **Brooks, G. T., Harrison, A., and Power, S. V.,** in Pest Infestation Research 1964, Annual Report of the Pest Infestation Laboratory, Agricultural Research Council, HMSO London, 1965, 58.

651. **Holan, G.,** Rational design of degradable insecticides, *Nature (Lond.),* 232, 644, 1971.

652. **Holan, G.,** Rational design of insecticides, *Bull. W.H.O.,* 44, 355, 1971.

652a. **Hirwe, A. S., Metcalf, R. L., and Kapoor, I. P.,** α-Trichloromethylbenzylanilines and α-trichloromethylbenzyl phenyl ethers with DDT-like insecticidal action, *J. Agric. Food Chem.,* 20, 818, 1972.

653. **Metcalf, R. L. and Georghiou, G. P.,** Cross tolerances of dieldrin-resistant flies and mosquitos to various cyclodiene insecticides, *Bull. W.H.O.,* 27, 251, 1962.

654. **Brooks, G. T., Harrison, A., and Cox, J. T.,** Significance of the epoxidation of the isomeric insecticides aldrin and isodrin by the adult housefly *in vivo, Nature (Lond.),* 197, 311, 1963.

655. **Brooks, G. T.,** The design of insecticidal chlorohydrocarbon derivatives, in *Drug Design,* Vol. IV, Ariens, E. J., Ed., Academic Press, New York, 1973, 379.

656. **Wang, C. M., Narahashi, T., and Yamada, M.,** The neurotoxic action of dieldrin and its derivatives in the cockroach, *Pestic. Biochem. Physiol.,* 1, 84, 1971.

657. **Kurihara, N.,** personal communication, 1973.

658. **Spector, W. S., Ed.,** *Handbook of Toxicology,* Vol. 1, NAS-NRS Wright Air Development Center Tech. Rep. 55-16, 1955, 408.

659. **Rogers, A. J.,** Eagles, affluence and pesticides, *Mosquito News,* 32, 151, 1972.

660. **Robinson, J.,** Organochlorine insecticides and bird populations in Britain, in *Chemical Fallout,* Miller, M. W., and Berg, G. C., Eds., Charles C Thomas, Springfield, Illinois, 1969, 113.

661. **Gaines, T. B.,** Acute toxicity of pesticides, *Toxicol. Appl. Pharmacol.,* 14, 515, 1969.

662. *Summary of Toxicology of Aldrin and Dieldrin,* Shell International Chemical Company, Toxicology Division, 1971.

663. **Deichmann, W. B., MacDonald, W. E., Blum, E., Bevilacqua, M., Radomski, J., Keplinger, M., and Balkus, M.,** Tumorigenicity of aldrin, dieldrin and endrin in the albino rat, *Ind. Med.,* 39, 37, 1970.

664. **Alabaster, J. S.,** Evaluating Risks of Pesticides to Fish, Proc. 5th Br. Insectic. Fungic. Conf. Vol. 2, 1969, 370.

665. Hercules Incorporated, (Agricultural Chemicals Department), *Hercules Toxaphene, Summary of Toxicological Investigations,* Bulletin T-105, 1962.

666. **Winteringham, F. P. W. and Barnes, J. M.,** Comparative response of insects and mammals to certain halogenated hydrocarbons used as insecticides, *Physiol. Rev.,* 35, 701, 1955.

667. **Dale, W. E., Gaines, T. B., Hayes, W. J., Jr., and Pearce, G. W.,** Poisoning by DDT: relation between clinical signs and concentration in rat brain, *Science,* 142, 1474, 1963.

668. **Yeager, J. F. and Munson, S. C.,** Site of action of DDT in Periplaneta, *Science,* 102, 305, 1945.

669. **Hoffman, R. A. and Lindquist, A. W.,** Temperature coefficients for chlorinated hydrocarbons, *J. Econ. Entomol.,* 42, 891, 1949.

670. **Guthrie, F. E.,** Holding temperature and insecticidal mortality; Blattella, *J. Econ. Entomol.,* 43, 559, 1950.

671. **Holan, G.,** New halocyclopropane insecticides and the mode of action of DDT, *Nature (Lond.),* 221, 1025, 1969.

672. **Eaton, J. L. and Sternburg, J. G.,** Temperature effects on nerve activity in DDT-treated American cockroaches, *J. Econ. Entomol.,* 60, 1358, 1967.

673. **Eaton, J. L. and Sternburg, J. G.,** Uptake of DDT by the American cockroach central nervous system, *J. Econ. Entomol.,* 60, 1699, 1967.

674. **Sternburg, J. and Kearns, C. W.,** The presence of toxins other than DDT in the blood of DDT-poisoned roaches, *Science,* 116, 144, 1952.

675. **Casida, J. E. and Maddrell, S. H. P.,** Diuretic hormone release on poisoning Rhodnius with insecticide chemicals, *Pestic. Biochem. Physiol.,* 1, 71, 1971.

676. Sternburg, J. G. and Hewitt, P., In vivo protection of cholinesterase against inhibition by TEPP and its methyl homologue by prior treatment with DDT, *J. Insect Physiol.*, 8, 643, 1962.

677. Winteringham, F. P. W., Hellyer, G. C., and McKay, M. A., Effects of the insecticides DDT and dieldrin on phosphorus metabolism of the adult housefly *Musca domestica* L., *Biochem. J.*, 76, 543, 1960.

678. Matsumura, F. and O'Brien, R. D., Interactions of DDT with components of American cockroach nerve, *J. Agric. Food Chem.*, 14, 39, 1966.

679. Matsumura, F. and Patil, K. C., Adenosine triphosphatase sensitive to DDT in synapses of rat brain, *Science*, 166, 121, 1969.

680. Koch, R. B., Cutkomp, L. K., and Do, F. M., Chlorinated hydrocarbon insecticide inhibition of cockroach and honeybee ATPase, *Life Sci.* (Oxford), 8, 289, 1969.

681. Weiss, D. E., A molecular mechanism for the permeability changes in nerve during the passage of an action potential, *Aust. J. Biol. Sci.*, 22, 1355, 1969.

682. Martin, H. and Wain, R. L., Properties versus toxicity of DDT analogues, *Nature (Lond.)*, 154, 512, 1944.

683. Riemschneider, R., Chemical structure and activity of DDT analogs, with special consideration of their spatial structures, in *Advances in Pest Control Research*, Vol. 2, Metcalf, R. L., Ed., Interscience, New York, 1958, 307.

684. Gunther, F. A., Blinn, R. C., Carman, G. E., and Metcalf, R. L., Mechanisms of insecticidal action. The structural topography theory and DDT-type compounds, *Arch. Biochem. Biophys.*, 50, 504, 1954.

685. Mullins, L. J., Structure-toxicity in hexachlorocyclohexane isomers, *Science*, 122, 118, 1955.

685a. Fahmy, M. A. H., Fukuto, T. R., Metcalf, R. L., and Holmstead, R. L., Structure-activity correlations in DDT analogs, *J. Agric. Food Chem.*, 21, 585, 1973.

686. Wilson, W. E., Fishbein, L., and Clements, S. T., DDT; participation in ultraviolet-detectable, charge-transfer complexation, *Science*, 171, 180, 1971.

687. O'Brien, R. D. and Matsumura, F., DDT; A new hypothesis of its mode of action, *Science*, 146, 657, 1964.

688. Baranyovits, F. L. C., cited in Winteringham, F. P. W. and Barnes, J. M., Comparative response of insects and mammals to certain halogenated hydrocarbons used as insecticides, *Physiol. Rev.*, 35, 701, 1955.

689. McNamara, B. P. and Krop, S., Pharmacology of BHC isomers, *J. Pharmacol. Exp. Ther.*, 92, 140, 1948.

690. Nakajima, E., Shindo, H., and Kurihara, N., Whole body autoradiographic studies in the distribution of α-, β- and γ-BHC in mice, *Radioisotopes*, 19, 532, 1970.

691. Telford, J. N. and Matsumura, F., Electron microscopic and autoradiographic studies on distribution of dieldrin in the intact nerve tissues of German cockroaches, *J. Econ. Entomol.*, 64, 230, 1971.

692. Cherkin, A., Mechanisms of general anaesthesia by non-hydrogen-bonding molecules, *Ann. Rev. Pharmacol.*, 9, 259, 1969.

693. Wang, C. M., Narahashi, T., and Yamada, M., The neurotoxic action of dieldrin and its derivatives in the cockroach, *Pestic. Biochem. Physiol.*, 1, 84, 1971.

694. Ryan, W. H. and Shankland, D. L., Synergistic action of cyclodiene insecticides with DDT on the membrane of giant axons of the American cockroach, *Periplaneta americana, Life Sci. (Oxford)*, 10, 193, 1971.

695. Matsumura, F., Studies on the Biochemical Mechanisms of Resistance in the German Cockroach Strains, paper presented at the 2nd International Congress of Pesticide Chemistry, Tel Aviv, 1971.

696. Narahashi, T., personal communication, 1971.

697. Brooks, G. T. and Harrison, A., The metabolism of some cyclodiene insecticides in relation to dieldrin resistance in the adult housefly, *Musca domestica* L., *J. Insect Physiol.*, 10, 633, 1964.

698. Sellers, L. G. and Guthrie, F. E., Distribution and metabolism of ¹⁴C-dieldrin in the resistant and susceptible housefly, *J. Econ. Entomol.*, 65, 378, 1972.

699. Colhoun, E. H., Approaches to mechanisms of insecticidal action, *J. Agric. Food Chem.*, 8, 252, 1960.

700. Yamasaki, T. and Narahashi, T., Resistance of houseflies to insecticides and the susceptibility of nerve to insecticides. Studies on the mechanism of action of insecticides XVII, *Botyu-kagaku*, 23, 146, 1958.

701. Sun, Yun-Pei, Schaefer, C. H., and Johnson, E. R., Effects of application methods on the toxicity and distribution of dieldrin in houseflies, *J. Econ. Entomol.*, 60, 1033, 1967.

702. Ray, J. W., Insecticide absorbed by the central nervous system of susceptible and resistant cockroaches exposed to dieldrin, *Nature (Lond.)*, 197, 1226, 1963.

703. Matsumura, F. and Hayashi, M., Dieldrin resistance; biochemical mechanisms in the German cockroach, *J. Agric. Food Chem.*, 17, 231, 1969.

704. Hayashi, M. and Matsumura, F., Insecticide mode of action: effect of dieldrin on ion movement in the nervous system of *Periplaneta americana* and *Blattella germanica* cockroaches, *J. Agric. Food Chem.*, 16, 622, 1967.

705. Welsh, J. H. and Gordon, H. T., The mode of action of certain insecticides on the arthropod nerve axon, *J. Cell. Comp. Physiol.*, 30, 147, 1947.

706. Hodge, H. C., Boyce, A. M., Deichmann, W. B., and Kraybill, H. F., Toxicology and no-effect levels of aldrin and dieldrin, *Toxicol. Appl. Pharmacol.*, 10, 613, 1967.

707. Hathway, D. E. and Mallinson, A., Effect of Telodrin on the liberation and utilisation of ammonia in rat brain, *Biochem. J.*, 90, 51, 1964.

708. Hathway, D. E., Mallinson, A., and Akintonwa, D. A. A., Effects of dieldrin, picrotoxin and Telodrin on the metabolism of ammonia in brain, *Biochem. J.*, 94, 676, 1965.

709. Hathway, D. E., The biochemistry of Telodrin and dieldrin, *Arch. Environ. Health*, 11, 380, 1965.

710. Pocker, Y., Beug, W. M., and Ainardi, V. R., Carbonic anhydrase interaction with DDT, DDE and Dieldrin, *Science*, 174, 1336, 1971.

711. Gair, R., Agricultural Development and Advisory Service, Cambridge, U.K., cited in Hessayon, D. G., Tennant Memorial Lecture — *Homo Sapiens, the species the conservationist forgot, Chem. Ind. (Lond.)*, 407, 1972.

712. Sherma, J. and Zweig, G., *Paper Chromatography: Paper Chromatography and Electrophoresis*, Vol. 2, Zweig, G. and Whitaker, J. R., Eds., Academic Press, New York, 1971.

713. DeLacy, T. P. and Kennard, C. H. L., Crystal structure of 1,1-bis(p-chlorophenyl)-2,2-dichloropropane and 1,1-bis(p-ethoxyphenyl)-2,2-dimethylpropane, *J. Chem. Soc. (Lond.)*, Perkin II, 2141, 1972.

713a. DeLacy, T. P. and Kennard, C. H. L., Crystal structures of 1,1-bis(p-chlorophenyl-2,2,2-trichloroethane (p,p'-DDT) and 1-(o-chlorophenyl)-1-(p-chlorophenyl)-2,2,2-trichloroethane (o,p'-DDT), *J. Chem. Soc. (Lond.)*, Perkin II, 2148, 1972.

714. Hoskins, W. M. and Gordon, H. T., Arthropod resistance to chemicals, *Ann. Rev. Entomol.*, 1, 89, 1956.

715. Hill, R. L. and Teipel, J. W., Fumarase and Crotonase, in *The Enzymes*, 3rd ed., Boyer, P. D., Ed., Academic Press, New York, London, 1971, 539.

716. Moriarty, F., The sublethal effects of synthetic insecticides on insects, *Biol. Rev.*, 44, 321, 1969.

717. Kagan, Yu. S., Fudel-Ossipova, S. I., Khaikina, B. J., Kuzminskaya, U. A., and Kouton, S. D., On the problem of the harmful effect of DDT and its mechanism of action, *Residue Rev.*, 27, 43, 1969.

718. Davies, J. E., Edmundson, W. F., Eds., Community Studies on Pesticides, Dade County, Florida: *Epidemiology of DDT*, Futura Publishing, Mount Kisco, New York, 1972, Chap. 10, 11, 13.

719. Lykken, L., Relation of U.S. Food Production to Pesticides, paper presented at the Symposium on Nutrition and Public Policy in the U.S., 163rd American Chemical Society National Meeting, Boston, April 1972.

720. Van Tiel, N., Pesticides in environment and food, in *Environmental Quality and Safety*, Vol. I, Coulston, F. and Korte, F., Eds., Georg Thieme, Stuttgart, Academic Press, New York, 1972, 180.

721. Hurtig, H., A commentary on pesticides and perspective, *Chem. Ind. (Lond.)*, 888, 1969.

722. Polen, P. B., Fate of insecticidal chlorinated hydrocarbons in storage and processing of foods, in *Pesticide Terminal Residues*, Invited papers from the IUPAC International Symposium on Terminal Residues, Tel Aviv, 1971, Tahori, A. S., Ed., Butterworth's London, 1971, 137.

723. Kenaga, E. E., Factors related to the bioconcentration of pesticides, in *Environmental Toxicology of Pesticides*, Matsumura, F., Boush, G. M., and Misato, T., Eds., Academic Press, New York and London, 1972, 193.

724. Robinson, J., Richardson, A., Crabtree, A. N., Coulson, J. C., and Potts, G. R., Organochlorine residues in marine organisms, *Nature (Lond.)*, 214, 1307, 1967.

725. Hurtig, H., Significance of conversion products and metabolites of pesticides in the environment, in *Environmental Quality and Safety*, Vol. I, Coulston, F. and Korte, F., Eds., Georg Thieme, Stuttgart, Academic Press, New York, 1972, 58.

726. Walker, C. H., *Environmental Pollution by Chemicals*, Hutchinson Educational, London, 1971, Chap. V.

727. Robinson, J., Residues of organochlorine insecticides in dead birds in the United Kingdom, *Chem. Ind. (Lond.)*, 1974, 1967.

728. Rudd, R. L. and Herman, S. G., Ecosystemic transferal of pesticides residues in an aquatic environment, in *Environmental Toxicology of Pesticides*, Matsumura, F., Boush, G. M., and Misato, T., Eds., Academic Press, New York and London, 1972, 471.

729. Rudd, R. L., *Pesticides and the Living Landscape*, Faber and Faber, London, 1965, 41.

SYSTEMATIC NAME INDEX

A

Acrocercops sp., 173
Aedes aegypti, 24, 152, 175
Aedes sp., 38, 152
Agriotes sp., 32, 172, 193
Agromyza phaseoli, 32
Agrotis orthogonia, 31
Agrotis sp., 172
Agrotis ypsilon, 31
Alabama argillacea, 29
Allisonotum impressicolle, 173
Amblyomma americanum, 36
Amblyomma maculatum, 36
Amphimallon sp., 172
Amsacta sp., 174
Anasa tristis, 152
Anastrephe fraterculus, 173
Anomala sp., 172
Anopheles albimanus, 16, 152
Anopheles gambiae, 195
Anopheles quadrimaculatus, 14, 37, 152
Anopheles sp., 38, 195
Anoxia sp., 172
Anthonomus grandis, 154, 173
Anthrenus vorax, 26
Aphaenogaster sp., 173
Aphis gossypii, 160
Aphis spiraecola, 152
Apis mellifera, 28
Argyrotaenia citrana, 29, 31
Argyrotaenia velutinana, 31
Astylus sp., 172
Attagenus piceus, 171, 196

B

Bacillus thuringiensis, 208
Blaniulus guttulatus, 173
Blatta orientalis, 23, 24, 175
Blattella germanica, 23, 24, 25, 36, 170, 175
Blissus leucopterus, 171
Boophilus annulatus, 35
Boophilus decoloratus, 195
Boophilus microplus, 35, 195
Boophilus sp., 35, 195
Bothynoderes sp., 173
Bovicola (Damalinia) sp., 36
Buphonella sp., 172

C

Calandra granaria, 25
Calandra oryzae, 26
Carpocapsa pomonella, 31, 171
Cephus cinctus, 32
Caratatis capitata, 173
Chaetopsis debilis, 154

Chalcodermus seneus, 208
Chilo suppressalis, 174
Cimex lectularius, 37, 152
Cimex sp., 152, 175, 195
Cirphis unipuncta, 30
Cleonus sp., 173
Coccidae, 173
Coccus viridis, 173
Colias eurytheme, 30
Conorrynchus sp., 173
Cosmopolites sordidus, 173
Crioceris duodecimpunctata, 29
Culex fatigans, 15, 16, 152
Culex sp., 38, 195
Culicoides sp., 39, 175, 194
Cyclocephala, 172

D

Dermanyssus gallinae, 35
Dermestes lardarius, 26
Dermestes maculatus, 26
Dermolepida sp., 173
Diabrotica sp., 33, 173
Diatraea saccharalis, 29, 32
Dorysthenis sp., 173
Drosophila melanogaster, 24, 83
Dyscinetus sp., 173
Dysdercus sp., 160

E

Earias insulana, 173
Elasmopalpus lignosellus, 173
Eleodes sp., 172
Ephestia kuehniella, 35
Epilachna varivestis, 29, 83
Erioischia brassicae, 172
Estigmene acraea, 29, 30
Eulepida sp., 172
Euxoa messoria, 154
Euxoa sp., 172

F

Folsomia candida, 28
Folsomia fimetaria, 83
Formicidae, 172

G

Glossina morsitans, 39
Glossina sp., 39, 175
Gonocephalum sp., 172
Gryllotalpida sp., 172
Gryllus pennsylvanicus, 154

H

Haematobia irritans, 39
Heliothis armigera, 31
Heliothis sp., 33, 193
Heliothis uirescens, 154
Heliothis zea, 154
Heliothrips haemorrhoidalis, 30
Hemitarsonemus latus, 174
Heteronychus sp., 173
Holotrichia sp., 172
Hylemyia antiqua, 174
Hylemia brassicae, 174
Hylemia sp., 172
Hypera postica, 154

I

Idocerus sp., 174
Ixodes persulcatus, 195
Ixodes ricinus, 39

L

Lachnosterna sp., 172
Laphygma frugiperda, 30, 154
Lepisma saccharina, 26
Leptinotarsa decemlineata, 152
Leptotrombidium sp., 176
Leucopholis sp., 172, 173
Limonius sp., 32, 154
Locusta migratoria, 154
Locusta pardaline, 154
Lucilia cuprina, 205
Lucilia sericata, 39
Ludius sp., 32
Lygidea mendax, 30
Lygus pabulinus, 30
Lymantria dispar, 30

M

Macrosiphum pisi, 152
Mallophaga sp., 36
Melanoplus differentialis, 152, 170
Melanoplus sp., 172, 193
Melolontha sp., 172
Melophagus ovinus, 39
Musca domestica, 15, 18, 24, 38, 152, 205
Myzus persicae, 29

O

Oncopeltus fasciatus, 170
Ornithodoros moubata, 195, 205
Ornithodoros sp., 36, 175, 205
Ornithodoros tholozani, 195
Otiorrynchus sp., 173

P

Pantomorus leucoloma, 33
Pectinophora gossypiella, 32, 160, 173
Pediculus humanus capitis, 37, 152
Pediculus humanus corporis, 36, 152, 205
Periplaneta americana, 24, 36, 175
Phlebotomus sp., 39, 175, 194
Phormia regina, 15, 16
Phyllophaga sp., 172
Phyllotetra sp., 193
Pieris brassicae, 30
Pieris rapae, 30
Plesiocoris nigricollis, 30
Plodia interpunctella, 35
Popillia japonica, 28, 32
Premnotrypes solani, 173
Prodenia eridania, 30
Protaparce quinquemaculata, 30
Protoparce sexta, 29, 30, 33
Psila rosae, 172
Ptinus tectus, 26
Pyrausta nubilalis, 31

R

Rhipicephalus sanquineus, 36
Rhodnius prolixus, 196
Ryania speciosa, 4

S

Sarcoptes scabei, 195
Scapteriscus sp., 173
Schistocerca gregaria, 154
Serica sp., 172
Simulium sp., 28, 38, 175, 194
Siphunculata (Anoplura) sp., 36
Sitophilus granarius, 35
Spodoptera eridania, 154
Steneotarsonemus pallidus, 172
Stomoxys calcitrans, 76

T

Temnorrynchus sp., 173
Tenebrio molitor, 35
Tetranychus sp., 208
Thermobia domestica, 26
Tineola biselliella, 25, 171
Tipula paludosa, 154
Tipulida sp., 172
Triatoma infestans, 37, 196
Triatoma sp., 175, 196
Tribolium castaneum, 35

Tribolium confusum, 26, 35
Trombicula sp., 195

Xenopsylla sp., 175
Xylomyges eridania, 173

X

Xenopsylla cheopis, 37

Z

Zygogramma exclamationis, 154

INDEX

Beer's Law, 73, 74, 80
Bees, 13, 28, 34, 159, 187–188
Beetles, 153, 159, 173–174, 193, 198
  ambrosia, 29
  asparagus, 29
  bark, 33
  blossom, 192–193
  cabbage flea, 10
  carpet, 17, 26, 153, 171, 196, 209
  Colorado potato, 10, 14, 33–34, 104, 152–153
  confused flour, 26, 35
  cucumber, 33
  elm bark, 28, 33
  flea, 14, 33–34
  grain, 26
  Japanese, 14, 27, 32, 34, 208
  lady bird, 159
  larder, 26
  leaf, 33, 42, 285
  May, 200–201
  Mexican bean, 29, 32, 34, 154, 166, 172–173
  mustard, 192–193
  parasitic, 159
  powder post, 29
  raspberry, 10
  rust red flour, 35
  spider, 26, 35
  tobacco flea, 31
  turnip flea, 185–187, 192
  white-fringed, 28, 33
Benzene, chlorinated derivatives of, 188–191, 198–201
Bioassay of insecticides, 83
Bird poisoning episodes, 155, 179–180
  seed dressings in, 155, 179–180
Blowflies, 15–16, 39, 180, 194–195, 205–206
Boots Pure Drug Company, 17
Borers, 14, 29, 31–32, 172–174, 207–208
  European corn, 14, 31, 172, 207–208
  rice stem, 174
  sugar cane, 29, 32, 207–208
Bornyl chloride, chlorination of, 205–206
British Standards Institution (BSI), 163–164, 205
Bugs, 14, 29–30, 144–145, 152–154, 159, 170–176,
    188, 196, 208–209
  apple capsid, 30
  apple red, 30
  chinch, 170–171
  lygus, 14, 30
  meadow spittle, 34
  mealy, 29
  milkweed, 144–145, 170–171, 176
  pentatomid, 29, 154, 208
  plant, 29–30, 173–174, 188, 208–209
  reduviid, 37, 175, 196
  squash, 152–153
Bulan®
  analysis of, 72
  as a constituent of Dilan®, 72
  chemical decomposition of, 67
  synthesis of, 50
  toxicity to insects, 16
  toxicity to mammals, 16

C

Cabbage, 10, 12, 14, 29–30, 32, 153, 171–172,
    180–182, 194
Camphene, chlorination of, 205–206
Cane sugar, 159–160, 173–174, 180–181, 208
Carbamate insecticides, 37, 43–45, 140–144, 176
Carbaryl (Sevin®), uses in pest control, 37, 43
Carbon disulfide, 4
Caterpillars, 153–154, 173–174, 208–209
  alfalfa, 30, 34
  cabbage, 14, 30
  salt-marsh, 29, 30, 34
Cela Landwirtschaftliche Chemikalien Gesellschaft, 188
Cereals, protection by chlorinated insecticides, 179–180
Chafers, 153
Chiggers, 153, 175–176, 195, 210
Chloral, preparation of, 52, 53
Chlorbenside
  analysis of, 75
  synthesis of, 52
Chlordane
  analysis of, 143–145, 147, 155–158
    by colorimetric methods, 155–158
    by gas-liquid chromatography, 143–145, 147,
      156–157
    by hydrolyzable chlorine in, 155–156
    by infrared spectrometry, 143–144, 155–158
    by total chlorine in, 155–156
    in crops (residues), 156–157
  formulations of, 147, 149–151
    aerosols, 153
    deactivation of carriers for, 150–151
    emulsions, 150–154
    granules, 150–151
    solutions, 150–151, 153
    wettable powders, 150–151
    with DDT, 30
    with methyl parathion, 151
    with Phosvel®, 151
    with pyrethrum, 150
  insect resistance to, 176, 196
  isomers of, 90, 110–112, 124–127, 130–131, 141,
      144, 156–157
    dechlorination of, 129–131
    dehydrochlorination of, 130, 145, 157
    environmental stability of, 129–130
    metabolites of, 130–131
    nomenclature of, 144–145
    oxidation of, 131 (see "oxychlordane")
    photochemistry of, 124, 129–130
    physical properties of, 143–145
    separation of, 144–147, 156–157
    stereochemistry of, 90–91, 129, 144–145
    structure of, 111–112, 144–145
    synthesis of, 110–112, 126, 141
    synthesis of, ^{14}C-labeled, 120–121
    toxicity of, 144, 154–155
  technical product
    composition of, 112, 125–126, 129–130, 143–149,
      154–155
    definition of, 145, 148–149

Dieldrin, formulations of, 162–163, 166–167
  aerosols, 167–168
  as baits, 171
  as seed dressings, 167–168, 171–172
  cost for mosquito control, 175
  deactivation of carriers for, 168
  granules, 167–168, 172
  solutions, 167–168
  wettable powders, 167–169, 171–172, 175–176
  with DDT, 168
  with fertilizers, 167–168
  with fungicide, 171–173
  with sulfur, 168
Dieldrin (*see* HEOD)
  composition of, 163–165
  conversion into aldrin, 134–135
  definition of, 93, 163–164, 169
  insect resistance to, 175–176, 180–182
  manufacture of, 162–163
  molecular rearrangements of, 94, 124, 177–178
  physical properties of, 162–165
  problems with labels for, 169–170
  residues in bird tissues, 155
  residue tolerances on crops, 181
  restrictions on the use of, 41, 155, 179–180, 182
  toxicity to insects, 162, 169–171, 195
  toxicity to mammals, 174–175
  toxicological investigations on, 181–182
Dieldrin (*see also* aldrin), uses in pest control, 32, 34, 37, 39, 42, 162–163, 169–171, 174, 179–181
  against agricultural pests, 32, 34, 162, 170–174
  against public health pests, 37, 39, 42, 163, 169–170, 174–176, 194–195
  against stored product pests, 171, 196
Diels-Alder reaction (Diene synthesis), 85–87, 91, 102, 104–105, 140
  dienophiles in, 85, 105
  stereochemistry of, 85–87
Difluorotetrachlorocyclopentadiene, 103–105, 117, 124, 141
Dihydroaldrin
  synthesis of, 114, 162
  synthesis of derivatives of, 116–117, 134, 136, 162, 176–177
Dihydrochlordene
  chlorination of, 110–111
  synthesis of, 107, 123, 125
Dihydroheptachlor (DHC) isomers
  constitution of, 124
  dechlorination of, 127
  nomenclature of, 90–91
  nuclear magnetic resonance spectra of, 110
  photochemistry of, 127
  synthesis of, 110–111
Dilan®
  Bulan® and Prolan® in, 58
  formulation of, 58–59
  physical and chemical properties of, 58, 67
  synthesis of, 50
  toxicity to insects, 16, 34
  toxicity to mammals, 16
  uses against agricultural pests, 31–32, 34

Dinitrophenols as insecticides, 4, 154, 170–171
Dioxathion, 207–208
Dow Chemical Company, 17
Dutch elm disease, 32
Dysentery, 12
DMC (Dimite®)
  as acaricide and DDT synergist, 34, 51
  formulation of, 58–59
  physical and chemical properties of, 58–59
  synthesis of, 51, 58–59

E

Earwigs, 153
Ecology of soil insects, influence of DDT on, 27–29
Economic Entomology, The American Association of, 13, 26–27
Ectoparasites, 35
E.I. du Pont de Nemours Company, 16
Electron-capture detector, 79–80, 82, 125–126, 143–144, 160–161, 203
Endosulfan (Thiodan®)
  analysis of, 132–134, 160–161
    individual isomers, 160–161
    in residues, 132–134, 160–161
  biodegradability of, 158–160
  chemical reactions of, 132–134
  constitution of, 124, 131–132
  effect on non-target organisms, 159–160
  formulations of, 158–160
    with DDT, 159–160
    with methyl parathion, 159–160
  insecticidal activity of, 109
  isomers of, 98–99, 109, 131–134, 158–159
  photochemistry of, 132–134
  physical properties of, 158–160
  synthesis of, 109, 131, 158
  systematic name of, 98–99
  toxicity of, 158–160
  uses in pest control, 159
    against agricultural pests, 159–160, 180
    against public health insects, 159–160
  as fish control agent, 159–160
Endrin, analysis of
  by gas-liquid chromatography, 80–82, 123–124, 177–178
    confirmatory derivatization in, 178
  chemical reactions of, 136–138
  colorimetric, 139–140, 176–177
  composition of technical, 166
  constitution of, 136–138
  dechlorinated derivatives of, 139–140
  definition of, 93, 166
  formulation of (*see* aldrin, dieldrin), 168–169, 174
    compatability with other pesticides, 166
    deactivation of carriers for, 166, 168–169
    stability of, 166, 168–169
    with DDT, 174–175
    with methyl parathion, 168–169
  insect resistance to, 182
  manufacture of, 162–163, 166

identification of residues of, 125–126, 157–158
insect resistance to, 155
in technical chlordane, 144–147, 154–155
manufacture of, 149–150
molecular rearrangements of, 95–96, 127–128
nomenclature of, 87–91, 95–98
photochemistry of, 124, 127–128
physical properties of, 149–150
polycyclo-name for, 95–96
restrictions on the use of, 41, 155, 179–180
stereochemistry of, 87–91, 127
synthesis of, 109, 142, 149–150
synthesis of ^{14}C-labeled, 120–121
toxicity to insects, 154–155, 169–171
uses in pest control, 155, 169–171, 175–176
  against agricultural pests, 155, 170–171
  against public health pests, 155, 169–171, 175–176
Heptachlor epoxide
analysis of, 157–158
bacterial deoxygenation of, 155
biological degradation of, 155
biological formation of, 126, 155
chemical reactions of, 126–127, 157
nomenclature of, 90–91
residues in bird tissues, 155
synthesis of, 109, 110, 126
Heptachlorocyclohexanes, 198–203
Heptachloronorbornene, 106
Hercules Powder Company, 206
Hexabromocyclopentadiene, 99, 112–113, 142, 165–166
Hexachlorocyclohexane (HCH)
analysis of (*see also* lindane)
  by column chromatography, 77
  by polarography, 76
biological degradation of, 201
chemical reactions of, 198–201, 203
  conversion into chlorinated benzenes, 200–201
chlorinated derivatives of, 186–187, 189–192
colorimetric analysis of, 203
comparative insecticidal efficiency of, 83, 151–153, 169–171, 192–194
composition of technical, 189–191, 198
conformations of, 197–201
definitions of, 185, 191–192
effects on non-target organisms, 194
formulations of (*see also* uses of HCH and lindane against pests)
  aerosols, 30, 33
  dusts, 33, 36, 175–176, 192, 194–196
  emulsions, 33, 192, 194–196
  granules, 195
  seed dressings, 193–194
  solutions, 33, 192, 194–196
  stability of, 192
  wettable powders, 192–196
fumigant properties, 38, 185, 192–193
isolation of lindane from technical, 187–191
isomers of, 185–192, 194, 198–203
  electron affinity of, 203
  physical properties of, 189–191
  separation of, 189–192

stability of, 191–192
  structural determination of, 82, 198–200
isomerization of, 200
isosteres of, 201–202
isotope-labeled, 202
Japanese manufacturers of, 188
manufacture of, 185–191
manufacturers of, 188
mixed halogen analogues of, 201–202
odor of, 187–188, 191–192, 196
physical properties of, 189
phytotoxicity of, 32–33, 194
residues in human tissue, 187–188
restrictions on the use of, 42
tainting of crops by, 192–194, 196, 208
tainting of food commodities by, 35, 192–193
toxicity to insects, 83, 151–154, 169–171, 192
stereochemistry of, 197–201
U.S. manufacturers of, 188
uses in pest control
  against agricultural pests, 29–31, 34–35, 153–154, 170–171, 185–188, 192, 208–209
  against orchard pests, 193–194
  against public health pests, 35, 37–39, 42–43, 45, 151–153, 169–171, 175–176
  against soil pests, 29, 32–33, 192, 196, 208–209
  against stored product pests, 35, 171, 185–186, 192, 196
  against veterinary pests, 36, 195–196
W.H.O. specification for, 191–192
Hexachlorocyclopentadiene
chemical reactions of, 101–105, 123–124, 140–141, 149, 159, 161
pesticidal properties of the products (*see* individual cyclodiene insecticides), 102, 104–105, 107, 109, 111–116
in technical chlordane, 147–149, 156
physical properties of, 99, 101–102
synthesis of, 99–102
synthesis of ^{14}C-labeled, 101, 118
Hexachloronorbornadiene, 87–90, 120–121
as intermediate in insecticide synthesis, 106, 116, 164–165
dipole moment of, 124
synthesis of, 106
Hexachloronorbornene, photolysis of, 139–140
Hexafluorocyclopentadiene, 117–118
HHDN
analogues, 112, 114–117, 161–162
  synthesis of, 114–117, 161–162
  toxicity of, 116, 161–162
  with mixed halogen content, 112, 117
analysis of (*see* aldrin)
chemical reactions of, 114–117, 132–135
constitution of, 132–134
dechlorinated derivatives of, 116–117, 134–136, 138–139
definition of, 93, 113–115, 163, 169
epoxide formation from, 114, 132–134, 161–162
from HEOD, 134–135
molecular rearrangements of, 135–137
photochemistry of, 116, 135

Triphenylmethane dyes, 7
Typhus, 10–12, 43, 195

U

United States Department of Agriculture, 12–14, 27–29,
    141, 152–153, 206, 209
  Beltsville laboratory of, 12
  Bureau of Entomology and Plant Quarantine of, 13–14,
    27, 152–153
  Orlando laboratory of, 12
United States Department of Health, Education, and
    Welfare, 41
United States Food and Drug Administration, 147–149

V

"Velsicol 1068", 142, 147–148, 151–152

W

Wagner-Meerwein rearrangement, 135, 138
Wasps, 153, 159
Weevils, 10, 14, 29, 153–154, 187, 208–209
  alfalfa, 14, 33–34, 154–155, 208–209
  apple blossom, 10, 192–194
  bean, 192–193
  boll, 28–29, 33–34, 153–154, 173–174, 192–193,
    207–209
  clover leaf, 33–34

grain, 10, 25–26, 35–36, 192–193, 196
  pea, 192–193
  rice, 26, 35–36
Wisconsin Alumni Research Foundation, 9
Woodward-Hoffman orbital symmetry rules, 85
World Food Supply
  components of, 4–5, 179
  cost of crop losses, 179
  inadequacy of, 178–179
  role of pesticides in maintaining, 4–5, 179, 182
World Health Organization, 42–43, 53–54
Worms
  army, 30–31, 153–154, 192–193, 207
  bag, 32, 174, 208–209
  corn ear, 31, 34
  cotton boll, 14, 28–29, 31, 154–155, 173–174, 193
  cotton leaf, 29
  cut, 29–31, 42, 153–155, 171–172
  earth, 153
  golden meal, 35–36
  pink boll, 32, 159–160, 173–174
  root, 32–33, 155
  tobacco bud, 31, 154–155
  tobacco horn, 29, 32–33
  wire, 29, 32, 153–155, 171–172, 193–194

Z

Zeidler synthesis, 7, 10, 47, 52